George John Romanes

Animal Intelligence

George John Romanes

Animal Intelligence

ISBN/EAN: 9783743333581

Manufactured in Europe, USA, Canada, Australia, Japa

Cover: Foto ©berggeist007 / pixelio.de

Manufactured and distributed by brebook publishing software (www.brebook.com)

George John Romanes

Animal Intelligence

THE INTERNATIONAL SCIENTIFIC SERIES.

ANIMAL INTELLIGENCE.

BY
GEORGE J. ROMANES, M. A., LL. D., F. R. S.,
ZOOLOGICAL SECRETARY OF THE LINNEAN SOCIETY.

PREFACE.

When I first began to collect materials for this work it was my intention to divide the book into two parts. Of these I intended the first to be concerned only with the facts of animal intelligence, while the second was to have treated of these facts in their relation to the theory of Descent. Finding, however, as I proceeded, that the material was too considerable in amount to admit of being comprised within the limits of a single volume, I have made arrangements with the publishers of the 'International Scientific Series' to bring out the second division of the work as a separate treatise, under the title 'Mental Evolution.' This treatise I hope to get ready for press within a year or two.

My object in the work as a whole is twofold. First, I have thought it desirable that there should be something resembling a text-book of the facts of Comparative Psychology, to which men of science, and also metaphysicians, may turn whenever they may have occasion to acquaint themselves with the particular level of intelligence to which this or that species of animal attains. Hitherto the endeavour of assigning these levels has been almost exclusively in the hands of popular writers; and as these have, for the most part, merely strung together, with discrimination more or less inadequate, innumerable anecdotes of the display of animal intelligence, their books are valueless as works of reference. So much, indeed, is this the case, that Comparative Psychology has been virtually excluded from the hierarchy of the sciences. If we except the methodical researches of a few distinguished naturalists, it would appear that the phenomena of mind in animals, having constituted so much and so long the theme of unscientific authors, are now considered well-nigh unworthy of serious treatment by scientific methods. But it is surely needless to point out that the phenomena which constitute the subject-matter of Comparative Psychology, even if we regard them merely as facts in Nature, have at least as great a claim to accurate classification as those phenomena of structure which constitute the subject-matter of Comparative Anatomy. Leaving aside, therefore, the reflection that within the last twenty years the facts of animal intelligence have suddenly acquired a new and profound importance, from the proved probability of their genetic continuity with those of human intelligence, it would remain true that their systematic arrangement is a worthy object of scientific endeavour. This, then, has been my first object, which, otherwise stated, amounts merely to passing the animal kingdom in review in order to give a trustworthy account of the grade of psychological development which is presented by each group. Such is the scope of the present treatise.

My second, and much more important object, is that of considering the facts of animal intelligence in their relation to the theory of Descent. With the exception of Mr. Darwin's admirable chapters on the mental powers and moral sense, and Mr. Spencer's great work on the Principles of Psychology, there has hitherto been no earnest attempt at tracing the principles which have been probably concerned in the genesis of Mind. Yet there is not a doubt that, for the present generation at all events, no subject of scientific inquiry can present a higher degree of interest; and therefore it is mainly with the view of furthering this inquiry that I have undertaken this work. It will thus be apparent that the present volume, while complete in itself as a statement of the facts of Comparative Psychology, has for its more ultimate purpose the laying of a firm foundation for my future treatise on Mental Evolution. But although, from what I have just said, it will be apparent that the present treatise is preliminary to a more important one, I desire to emphasise this statement, lest the critics, in being now presented only with a groundwork on which the picture is eventually to be painted, should deem that the art displayed is of somewhat too commonplace a kind. If the present work is read without reference to its ultimate object of supplying facts for the subsequent deduction of principles, it may well seem but a small improvement upon the works of the anecdote-mongers. But if it is remembered that my object in these pages is the mapping out of animal psychology for the purposes of a subsequent synthesis, I may fairly claim to receive credit for a sound scientific intention, even where the only methods at my disposal may incidentally seem to minister to a mere love of anecdote.

It remains to add a few words on the principles which I have laid down for my own guidance in the selection and arrangement of facts. Considering it desirable to cast as wide a net as possible, I have fished the seas of popular literature as well as the rivers of scientific writing. The endless multitude of alleged facts which I have thus been obliged to read, I have found, as may well be imagined, excessively tedious; and as they are for the most part recorded by wholly unknown observers, the labour of reading them would have been useless without some trustworthy principles of selection. The first and most obvious principle that occurred to me was to regard only those facts which stood upon the authority of observers well known as competent; but I soon found that this principle constituted much too close a mesh. Where one of my objects was to determine the upper limit of intelligence reached by this and that class, order, or species of animals, I usually found that the most remarkable instances of the display of intelligence were recorded by persons bearing names more or less unknown to fame. This, of course, is what we might antecedently expect, as it is obvious that the chances must always be greatly against the more intelligent individuals among animals happening to fall under the observation of the more intelligent individuals among men. Therefore I soon found that I had to choose between neglecting all the more important part of the evidence—and consequently in most cases feeling sure that I had fixed the upper limit of intelligence too low—or supplementing the principle of looking to authority alone with some other principles of selection, which, while embracing the enormous class of alleged facts recorded by unknown observers, might be felt to meet the requirements of a reasonably critical method. I therefore adopted the following principles as a filter to this class of facts. First, never to accept an alleged fact without the authority of some name. Second, in the case of the name being unknown, and the alleged fact of sufficient importance to be entertained, carefully to consider whether, from all the circumstances of the case as recorded, there was any considerable opportunity for malobservation; this principle generally demanded that the alleged fact, or action on the part of the animal, should be of a particularly marked and unmistakable kind, looking to the end which the action is said to have accomplished.

Third, to tabulate all important observations recorded by unknown observers, with the view of ascertaining whether they have ever been corroborated by similar or analogous observations made by other and independent observers. This principle I have found to be of great use in guiding my selection of instances, for where statements of fact which present nothing intrinsically improbable are found to be unconsciously confirmed by different observers, they have as good a right to be deemed trustworthy as statements which stand on the single authority of a known observer, and I have found the former to be at least as abundant as the latter. Moreover, by getting into the habit of always seeking for corroborative cases, I have frequently been able to substantiate the assertions of known observers by those of other observers as well or better known.

So much, then, for the principles by which I have been guided in the selection of facts. As to the arrangement of the facts, I have taken the animal kingdom in ascending order, and endeavoured to give as full a sketch as the selected evidence at my disposal permitted of the psychology which is distinctive of each class, or order, and, in some cases, family, genus, or even species. The reason of my entering into greater detail with some natural groups than with others scarcely requires explanation. For it is almost needless to say that if the animal kingdom were classified with reference to Psychology instead of with reference to Anatomy, we should have a very different kind of zoological tree from that which is now given in our diagrams. There is, indeed, a general and, philosophically considered, most important parallelism running through the whole animal kingdom between structural affinity and mental development; but this parallelism is exceedingly rough, and to be traced only in broad outlines, so that although it is convenient for the purpose of definite arrangement to take the animal kingdom in the order presented by zoological classification, it would be absurd to restrict an inquiry into Animal Psychology by any considerations of the apparently disproportionate length and minute subdivision with which it is necessary to treat some of the groups. Anatomically, an ant or a bee does not require more consideration than a beetle or a fly; but psychologically there is need for as great a difference of treatment as there is in the not very dissimilar case of a monkey and a man.

Throughout the work my aim has been to arrive at definite principles rather than to chronicle mere incidents—an aim which will become more apparent when the work as a whole shall have been completed. Therefore it is that in the present volume I have endeavoured, as far as the nature and circumstances of the inquiry would permit, to suppress anecdote. Nevertheless, although I have nowhere introduced anecdotes for their own sake, I have found it unavoidable not to devote much the largest part of the present essay to their narration. Hence, with the double purpose of limiting the introduction of anecdotes as much as possible, and of not repeating more than I could help anecdotes already published, I have in all cases, where I could do so without detriment to my main object, given the preference to facts which have been communicated to me by friends and correspondents. And here I may fitly take the opportunity of expressing my thanks and obligations to the latter, who in astonishing numbers have poured in their communications during several years from all quarters of the globe. I make this statement because I desire to explain to all my correspondents who may read this book, that I am not the less sensible of their kindness because its bounty has rendered it impossible for me to send acknowledgments in individual cases. However, I should like to add in this connection that it does not follow, because I have only quoted a small percentage of the letters which I have received, that all of the remainder have been useless. On the contrary, many of these have served to convey information and suggestions which, even if not reserved for express quotation in my forthcoming work, have been of use in guiding my judgment on particular points. Therefore I hope that the publication of these remarks may serve to swell the stream of communications into a yet larger flow.[1]

In all cases where I have occasion to quote statements of fact, which in the present treatise are necessarily numerous, I have made a point of trying to quote *verbatim*. Only where I have found that the account given by an author or a correspondent might profitably admit of a considerable degree of condensation have I presented it in my own words.

And here I have to express my very special obligations to Mr. Darwin, who not only assisted me in the most generous manner with his immense stores of information, as well as with his valuable judgment on sundry points of difficulty, but has also been kind enough to place at my disposal all the notes and clippings on animal intelligence which he has been collecting for the last forty years, together with the original MS. of his wonderful chapter on 'Instinct.' This chapter, on being re-cast for the 'Origin of Species,' underwent so merciless an amount of compression that the original draft constitutes a rich store of hitherto unpublished material. In my second work I shall have occasion to draw upon this store more largely than in the present one, and it is needless to add

[1] Letters may be addressed to me directly at 18 Cornwall Terrace, Regent's Park, London, N W.

that in all cases where I do draw upon it I shall be careful to state the source to which I am indebted.

[The above was written when I sent this work to the publishers several months ago, and I have thought it best to leave the concluding paragraph as it originally stood. But in making this explanation, I cannot allude to the calamity which has since occurred without paying my tribute, not alone to the memory of the greatest genius of our age, but still more, and much more, to the memory of a friend so inexpressibly noble, kind, and generous, that even my immense admiration of the naturalist was surpassed by my loving veneration for the man.]

CONTENTS.

	PAGE
INTRODUCTION	1

CHAPTER I.
APPLICATION OF THE FOREGOING PRINCIPLES TO THE LOWEST ANIMALS 18

CHAPTER II.
MOLLUSCA 25

CHAPTER III.
ANTS 31

CHAPTER IV.
BEES AND WASPS 143

CHAPTER V.
TERMITES 198

CHAPTER VI.
SPIDERS AND SCORPIONS 204

CHAPTER VII.
REMAINING ARTICULATA 226

CHAPTER VIII.
FISH 241

CHAPTER IX.
BATRACHIANS AND REPTILES 254

CONTENTS.

CHAPTER X.
BIRDS 266

CHAPTER XI.
MAMMALS 326

CHAPTER XII.
RODENTS 353

CHAPTER XIII.
ELEPHANT 386

CHAPTER XIV.
THE CAT 411

CHAPTER XV.
FOXES, WOLVES, JACKALS, &C. . . . 426

CHAPTER XVI.
THE DOG 437

CHAPTER XVII.
MONKEYS, APES, AND BABOONS . . . 471

INDEX 499

INTRODUCTION.

BEFORE we begin to consider the phenomena of mind throughout the animal kingdom it is desirable that we should understand, as far as possible, what it is that we exactly mean by mind. Now, by mind we may mean two very different things, according as we contemplate it in our own individual selves, or in other organisms. For if we contemplate our own mind, we have an immediate cognizance of a certain flow of thoughts or feelings, which are the most ultimate things, and indeed the only things, of which we are cognisant. But if we contemplate mind in other persons or organisms, we have no such immediate cognizance of thoughts or feelings. In such cases we can only *infer* the existence and the nature of thoughts and feelings from the activities of the organisms which appear to exhibit them. Thus it is that we may have a subjective analysis of mind and an objective analysis of mind - the difference between the two consisting in this, that in our subjective analysis we are restricted to the limits of a single isolated mind which we call our own, and within the territory of which we have immediate cognizance of all the processes that are going on, or at any rate of all the processes that fall within the scope of our introspection. But in our objective analysis of other or foreign minds we have no such immediate cognizance; all our knowledge of their operations is derived, as it were, through the medium of ambassadors—these ambassadors being the activities of the organism. Hence it is evident that in our study of animal intelligence we are wholly restricted to the objective method. Starting from what I know subjectively of the operations of my own individual mind, and the activities which in my own organism they prompt, I proceed by analogy to infer from the observable activities of other organisms what are the mental operations that underlie them.

Now, in this mode of procedure what is the kind of activities which may be regarded as indicative of mind? I certainly do not so regard the flowing of a river or the blowing of the wind. Why? First, because the objects are too remote in kind from my own organism to admit of my drawing any reasonable analogy between them and it; and, secondly, because the activities which they present are of invariably the same kind under the same circumstances; they afford no evidence of feeling or purpose. In other words, two conditions require to be satisfied before we even begin to imagine that observable activities are indicative of mind: first, the activities must be displayed by a living organism; and secondly, they must be of a kind to suggest the presence of two elements which we recognise as the distinctive characteristics of mind as such—consciousness and choice.

So far, then, the case seems simple enough. Wherever we see a living organism apparently exerting intentional choice, we might infer that it is conscious choice, and therefore that the organism has a mind. But further reflection shows us that this is just what we cannot do; for although it is true that there is no mind without the power of conscious choice, it is not true that all apparent choice is due to mind. In our own organisms, for instance, we find a great many adaptive movements performed without choice or even consciousness coming into play at all—such, for instance, as in the beating of our hearts. And not only so, but physiological experiments and pathological lesions prove that in our own and in other organisms the mechanism of the nervous system is sufficient, without the intervention of consciousness, to produce muscular movements of a highly co-ordinate and apparently intentional character. Thus, for instance, if a man has his back broken in such a way as to sever the nervous connection between his brain and lower extremities, on pinching or tickling his feet they are drawn suddenly away from the irritation, although the man is quite unconscious of the adaptive movement of his muscles; the lower nerve-centres of the spinal cord are competent to bring about this movement of adaptive response without requiring to be directed by the brain. This non-mental operation of the lower nerve-centres in the production of apparently intentional movements is called Reflex Action, and the cases of its occurrence, even within the limits of our own organism, are literally numberless. Therefore, in view of such non-mental nervous adjustment, leading to movements which are only in appearance intentional, it clearly becomes a matter of great difficulty to say in the case of the lower animals whether any action which appears to indicate intelligent choice is not really action of the reflex kind.

On this whole subject of mind-like and yet not truly mental action I shall have much to say in my subsequent treatise, where I shall be concerned among other things with tracing the probable genesis of mind from non-mental antecedents. But here it is sufficient merely to make this general statement of the fact, that even within the experience supplied by our own organisms adaptive movements of a highly complex and therefore apparently purposive character may be performed without any real purpose, or even consciousness of their performance. It thus becomes evident that before we can predicate the bare existence of mind in the lower animals, we need some yet more definite criterion of mind than that which is supplied by the adaptive actions of a living organism, howsoever apparently intentional such actions may be. Such a criterion I have now to lay down, and I think it is one that is as practically adequate as it is theoretically legitimate.

Objectively considered, the only distinction between adaptive movements due to reflex action and adaptive movements due to mental perception, consists in the former depending on inherited mechanisms within the nervous system being so constructed as to effect *particular* adaptive movements in response to *particular* stimulations, while the latter are independent of any such inherited adjustment of special mechanisms to the exigencies of special circumstances. Reflex actions under the influence of their appropriate stimuli may be compared to the actions of a machine under the manipulations of an operator; when certain springs of action are touched by certain stimuli, the whole machine is thrown into appropriate movement; there is no room for choice, there is no room for uncertainty; but as surely as any of these inherited mechanisms are affected by the stimulus with reference to which it has been constructed to act, so surely will it act in precisely the same way as it always has acted. But the case with conscious mental adjustment is quite different. For, without at present going into the question concerning the relation of body and mind, or waiting to ask whether cases of mental adjustment are not really quite as *mechanical* in the sense of being the necessary result or correlative of a chain of physical sequences due to a physical stimulation, it is enough to point to the variable and incalculable character of mental adjustments as distinguished from the constant and foreseeable character of reflex adjustments. All, in fact, that in an objective sense we can mean by a mental adjustment is an adjustment of a kind that has not been definitely fixed by heredity as the only adjustment possible in the given circumstances of stimulation. For were there no alternative of adjustment, the case, in an animal at least, would be indistinguishable from one of reflex action.

It is, then, adaptive action by a living organism in cases where the inherited machinery of the nervous system does not furnish data for our prevision of what the adaptive action must necessarily be—it is only here that we recognise the objective evidence of mind. The criterion of mind, therefore, which I propose, and to which I shall adhere throughout the present volume, is as follows:— Does the organism learn to make new adjustments, or to modify old ones, in accordance with the results of its own individual experience? If it does so, the fact cannot be due merely to reflex action in the sense above described, for it is impossible that heredity can have provided in advance for innovations upon, or alterations of, its machinery during the lifetime of a particular individual.

In my next work I shall have occasion to consider this criterion of mind more carefully, and then it will be shown that as here stated the criterion is not rigidly exclusive, either, on the one hand, of a possibly mental element in apparently non-mental adjustments, or, conversely, of a possibly non-mental element in apparently mental adjustments. But, nevertheless, the criterion is the best that is available, and, as it will be found sufficient for all the purposes of the present work, its more minute analysis had better be deferred till I shall have to treat of the probable evolution of mind from non-mental antecedents. I may, however, here explain that in my use of this criterion I shall always regard it as fixing only the upper limit of non-mental action; I shall never regard it as fixing the lower limit of mental action. For it is clear that long before mind has advanced sufficiently far in the scale of development to become amenable to the test in question, it has probably begun to dawn as nascent subjectivity. In other words, because a lowly organised animal does *not* learn by its own individual experience, we may not therefore conclude that in performing its natural or ancestral adaptations to appropriate stimuli consciousness, or the mind-element, is wholly absent; we can only say that this element, if present, reveals no evidence of the fact. But, on the other hand, if a lowly organised animal *does* learn by its own individual experience, we are in possession of the best available evidence of conscious memory leading to intentional adaptation. Therefore our criterion applies to the upper limit of non-mental action, not to the lower limit of mental.

Of course to the sceptic this criterion may appear unsatisfactory, since it depends, not on direct knowledge, but on inference. Here, however, it seems enough to point out, as already observed, that it is the best criterion available; and further, that scepticism of this kind is logically bound to deny evidence of mind, not only in the case of the lower animals, but also in that of the higher, and even in that of men other than the sceptic himself. For all objections which could apply to the use of this criterion of mind in the animal kingdom would apply with equal force to the evidence of any mind other than that of the individual objector. This is obvious, because, as I have already observed, the only evidence we can have of objective mind is that which is furnished by objective activities; and as the subjective mind can never become assimilated with the objective so as to learn by direct feeling the mental processes which there accompany the objective activities, it is clearly impossible to satisfy any one who may choose to doubt the validity of inference, that in any case other than his own mental processes ever do accompany objective activities. Thus it is that philosophy can supply no demonstrative refutation of idealism, even of the most extravagant form. Common sense, however, universally feels that analogy is here a safer guide to truth than the sceptical demand for impossible evidence; so that if the objective existence of other organisms and their activities is granted—without which postulate comparative psychology, like all the other sciences, would be an unsubstantial dream—common sense will always and without question conclude that the activities of organisms other than our own, when analogous to those activities of our own which we know to be accompanied by certain mental states, are in them accompanied by analogous mental states.

The theory of animal automatism, therefore, which is usually attributed to Descartes (although it is not quite clear how far this great philosopher really entertained the theory), can never be accepted by common sense; and even as a philosophical speculation it will be seen, from what has just been said, that by no feat of logic is it possible to make the theory apply to animals to the exclusion of man. The expression of fear or affection by a dog involves quite as distinctive and complex a series of neuro-muscular actions as does the expression of similar emotions by a human being; and therefore, if the evidence of corresponding mental states is held to be inadequate in the one case, it must in consistency be held similarly inadequate in the other. And likewise, of course, with all other exhibitions of mental life.

It is quite true, however, that since the days of Descartes—or rather, we might say, since the days of Joule—the question of animal automatism has assumed a new or more defined aspect, seeing that it now runs straight into the most profound and insoluble problem that has ever been presented to human thought—viz. the relation of body to mind in view of the doctrine of the conservation of energy. I shall subsequently have occasion to consider this problem with the close attention that it demands; but in the present volume, which has to deal only with the phenomena of mind as such, I expressly pass the problem aside as one reserved for separate treatment. Here I desire only to make it plain that the mind of animals must be placed in the same category, with reference to this problem, as the mind of man; and that we cannot without gross inconsistency ignore or question the evidence of mind in the former, while we accept precisely the same kind of evidence as sufficient proof of mind in the latter.

And this proof, as I have endeavoured to show, is in all cases and in its last analysis the fact of a living organism showing itself able to learn by its own individual experience. Wherever we find an animal able to do this, we have the same right to predicate mind as existing in such an animal that we have to predicate it as existing in any human being other than ourselves. For instance, a dog has always been accustomed to eat a piece of meat when his organism requires nourishment, and when his olfactory nerves respond to the particular stimulus occasioned by the proximity of the food. So far, it may be said, there is no evidence of mind; the whole series of events comprised in the stimulations and muscular movements may be due to reflex action alone. But now suppose that by a number of lessons the dog has been taught not to eat the meat when he is hungry until he receives a certain verbal signal: then we have exactly the same kind of evidence that the dog's actions are prompted by mind as we have that the actions of a man are so prompted.[1] Now we find

[1] Of course it may be said that we have no evidence of *prompting*

that the lower down we go in the animal kingdom, the more we observe reflex action, or non-mental adjustment, to predominate over volitional action, or mental adjustment. That is to say, the lower down we go in the animal kingdom, the less capacity do we find for changing adjustive movements in correspondence with changed conditions; it becomes more and more hopeless to *teach* animals—that is, to establish associations of ideas; and the reason of this, of course, is that ideas or mental units become fewer and less definite the lower we descend through the structure of mind.

It is not my object in the present work to enter upon any analysis of the operations of mind, as this will require to be done as fully as possible in my next work. Nevertheless, a few words must here be said with regard to the main divisions of mental operation, in order to define closely the meanings which I shall attach to certain terms relating to these divisions, and the use of which I cannot avoid.

The terms sensation, perception, emotion, and volition need not here be considered. I shall use them in their ordinary psychological significations; and although I shall subsequently have to analyse each of the organic or mental states which they respectively denote, there will be no occasion in the present volume to enter upon this subject. I may, however, point out one general consideration to which I shall throughout adhere. Taking it for granted that the external indications of mental processes which we observe in animals are trustworthy, so that we are justified in inferring particular mental states from particular bodily actions, it follows that in consistency we must everywhere apply the same criteria.

For instance, if we find a dog or a monkey exhibiting marked expressions of affection, sympathy, jealousy, rage, &c., few persons are sceptical enough to doubt that the complete analogy which these expressions afford with

in either case; but this is the side issue which concerns the general relation of body and mind, and has nothing to do with the guarantee of inferring the presence of mind in particular cases.

those which are manifested by man, sufficiently prove the existence of mental states analogous to those in man of which these expressions are the outward and visible signs. But when we find an ant or a bee apparently exhibiting by its actions these same emotions, few persons are sufficiently non-sceptical not to doubt whether the outward and visible signs are here trustworthy as evidence of analogous or corresponding inward and mental states. The whole organisation of such a creature is so different from that of a man that it becomes questionable how far analogy drawn from the activities of the insect is a safe guide to the inferring of mental states—particularly in view of the fact that in many respects, such as in the great preponderance of 'instinct' over 'reason,' the psychology of an insect is demonstrably a widely different thing from that of a man. Now it is, of course, perfectly true that the less the resemblance the less is the value of any analogy built upon the resemblance, and therefore that the inference of an ant or a bee feeling sympathy or rage is not so valid as is the similar inference in the case of a dog or a monkey. Still it *is* an inference, and, so far as it goes, a valid one—being, in fact, the only inference available. That is to say, if we observe an ant or a bee apparently exhibiting sympathy or rage, we must either conclude that some psychological state resembling that of sympathy or rage is present, or else refuse to think about the subject at all; from the observable facts there is no other inference open. Therefore, having full regard to the progressive weakening of the analogy from human to brute psychology as we recede through the animal kingdom downwards from man, still, as it is the only analogy available, I shall follow it throughout the animal series.

It may not, however, be superfluous to point out that if we have full regard to this progressive weakening of the analogy, we must feel less and less certain of the real similarity of the mental states compared ; so that when we get down as low as the insects, I think the most we can confidently assert is that the known facts of human psychology furnish the best available pattern of the probable facts of insect psychology. Just as the theologians tell us—and logically enough—that if there is a Divine Mind, the best, and indeed only, conception we can form of it is that which is formed on the analogy, however imperfect, supplied by the human mind; so with 'inverted anthropomorphism' we must apply a similar consideration with a similar conclusion to the animal mind. The mental states of an insect may be widely different from those of a man, and yet most probably the nearest conception that we can form of their true nature is that which we form by assimilating them to the pattern of the only mental states with which we are actually acquainted. And this consideration, it is needless to point out, has a special validity to the evolutionist, inasmuch as upon his theory there must be a psychological, no less than a physiological, continuity extending throughout the length and breadth of the animal kingdom.

In these preliminary remarks only one other point requires brief consideration, and this has reference to the distinction between what in popular phraseology is called 'Instinct' and 'Reason.' I shall not here enter upon any elaborate analysis of a distinction which is undoubtedly valid, but shall confine my remarks to explaining the sense in which I shall everywhere use these terms.

Few words in our language have been subject to a greater variety of meanings than the word instinct. In popular phraseology, descended from the Middle Ages, all the mental faculties of the animal are termed instinctive, in contradistinction to those of man, which are termed rational. But unless we commit ourselves to an obvious reasoning in a circle, we must avoid assuming that all actions of animals are instinctive, and then arguing that because they are instinctive, therefore they differ from the rational actions of man. The question really lies in what is here assumed, and we can only answer it by examining in what essential respect instinct differs from reason.

INTRODUCTION.

Again, Addison says:—

I look upon instinct as upon the principle of gravitation in bodies, which is not to be explained by any known qualities inherent in the bodies themselves, nor from any laws of mechanism, but as an immediate impression from the first Mover, and the Divine energy acting in the creatures.

This mode of 'looking upon instinct' is merely to exclude the subject from the sphere of inquiry, and so to abstain from any attempt at definition.

Innumerable other opinions might be quoted from well-known writers, 'looking upon instinct' in widely different ways; but as this is not an historical-work, I shall pass on at once to the manner in which science looks upon it, or, at least, the manner in which it will always be looked upon throughout the present work.

Without concerning ourselves with the origin of instincts, and so without reference to the theory of evolution, we have to consider the most conspicuous and distinctive features of instinct as it now exists. The most important point to observe in the first instance is that instinct involves *mental* operations; for this is the only point that serves to distinguish instinctive action from reflex. Reflex action, as already explained, is non-mental neuromuscular adaptation to appropriate stimuli; but instinctive action is this and something more; there is in it the element of mind. Such, at least, is instinctive action in the sense that I shall always allude to it. I am, of course, aware that the limitation which I thus impose is one which is ignored, or not recognised, by many writers even among psychologists; but I am persuaded that if we are to have any approach to definiteness in the terms which we employ—not to say of clearness in our ideas concerning the things of which we speak —it is most desirable to restrict the word instinct to mental as distinguished from non-mental activity. No doubt it is often difficult, or even impossible, to decide whether or not a given action implies the presence of the mind-element—*i.e.*, conscious as distinguished from unconscious adaptation; but this is altogether a separate matter, and has nothing to do with the question of defining instinct in a manner which shall be formally exclusive, on the one hand of reflex action, and on the other of reason. As Virchow truly observes, 'it is difficult or impossible to draw the line between instinctive and reflex action;' but at least the difficulty may be narrowed down to deciding in particular cases whether or not an action falls into this or that category of definition; there is no reason why the difficulty should arise on account of any ambiguity of the definitions themselves. Therefore I endeavour to draw as sharply as possible the line which *in theory* should be taken to separate instinctive from reflex action; and this line, as I have already said, is constituted by the boundary of non-mental or unconscious adjustment, with adjustment in which there is concerned consciousness or mind.

Having thus, I hope, made it clear that the difficulty of drawing a distinction between reflex and instinctive actions as a class is one thing, and that the difficulty of assigning particular actions to one or the other of our categories is another thing, we may next perceive that the former difficulty is obviated by the distinction which I have imposed, and that the latter only arises from the fact that on the objective side there is no distinction imposable. The former difficulty is obviated by the distinction which I have drawn, simply because the distinction is itself a definite one. In particular cases of adjustive action we may not always be able to affirm whether consciousness of their performance is present or absent; but, as I have already said, this does not affect the validity of our definition; all we can say of such cases is that if the performance in question is attended with consciousness it is instinctive, and if not it is reflex.

And the difficulty of assigning particular actions to one or other of these two categories arises, as I have said, merely because on the objective side, or the side of the nervous system, there is no distinction to be drawn. Whether or not a neural process is accompanied by a mental process, it is in itself the same. The advent and development of consciousness, although progressively converting reflex action into instinctive, and instinctive into rational, does this exclusively in the sphere of subjectivity; the nervous processes engaged are throughout the same in kind, and differ only in the relative degrees of their complexity. Therefore, as the dawn of consciousness or the rise of the mind-element is gradual and undefined, both in the animal kingdom and in the growing child, it is but necessary that in the early morning, as it were, of consciousness any distinction between the mental and the non-mental should be obscure, and generally impossible to determine. Thus, for instance, a child at birth does not close its eyes upon the near approach of a threatening body, and it only learns to do so by degrees as the result of experience ; at first, therefore, the action of closing the eyelids in order to protect the eyes may be said to be instinctive, in that it involves the mind-element:[1] yet it afterwards becomes a reflex which asserts itself even in opposition to the will. And, conversely, sucking in a new-born child, or a child *in utero*, is, in accordance with my definition, a reflex action; yet in later life, when consciousness becomes more developed and the child *seeks* the breast, sucking may properly be called an instinctive action. Therefore it is that, as in the ascending scale of objective complexity the mind-element arises and advances gradually, many particular cases which occupy the undefined boundary between reflex action and instinct cannot be assigned with confidence either to the one region or to the other.

We see then the point, and the only point, wherein instinct can be consistently separated from reflex action; viz., in presenting a mental constituent. Next we must consider wherein instinct may be separated from reason. And for this purpose we may best begin by considering what we mean by reason.

The term 'reason' is used in significations almost as various as those which are applied to 'instinct.' Some-

---

[1] *I.e.*, ancestral as well as individual. If the race had not always had occasion to close the eyelids to protect the eyes, it is certain that the young child would not so quickly learn to do so in virtue of its own individual experience alone; and as the action cannot be attributed to any process of conscious inference, it is not rational; but we have seen that it is not originally reflex; therefore it is instinctive.

times it stands for all the distinctively human faculties taken collectively, and in antithesis to the mental faculties of the brute; while at other times it is taken to mean the distinctively human faculties of intellect.

Dr. Johnson defines it as 'the power by which man deduces one proposition from another, and proceeds from premises to consequences.' This definition presupposes language, and therefore ignores all cases of inference not thrown into the formal shape of predication. Yet even in man the majority of inferences drawn by the mind never emerge as articulate propositions; so that although, as we shall have occasion fully to observe in my subsequent work, there is much profound philosophy in identifying reason with speech as they were identified in the term Logos, yet for purposes of careful definition so to identify intellect with language is clearly a mistake.

More correctly, the word reason is used to signify the power of perceiving analogies or ratios, and is in this sense equivalent to the term 'ratiocination,' or the faculty of deducing inferences from a perceived equivalency of relations. Such is the only use of the word that is strictly legitimate, and it is thus that I shall use it throughout the present treatise. This faculty, however, of balancing relations, drawing inferences, and so of forecasting probabilities, admits of numberless degrees; and as in the designation of its lower manifestations it sounds somewhat unusual to employ the word reason, I shall in these cases frequently substitute the word intelligence. Where we find, for instance, that an oyster profits by individual experience, or is able to perceive new relations and suitably to act upon the result of its perceptions, I think it sounds less unusual to speak of the oyster as displaying intelligence than as displaying reason. On this account I shall use the former term to signify the lower degrees of the ratiocinative faculty; and thus in my usage it will be opposed to such terms as instinct, reflex action, &c., in the same manner as the term reason is so opposed. This is a point which, for the sake of clearness, I desire the reader to retain in his memory. I shall always speak of intelligence and intellect in antithesis to instinct, emo-

INTRODUCTION. 15

tion, and the rest, as implying mental faculties the same in kind as those which in ourselves we call rational.

Now it is notorious that no distinct line can be drawn between instinct and reason. Whether we look to the growing child or to the ascending scale of animal life, we find that instinct shades into reason by imperceptible degrees, or, as Pope expresses it, that these principles are 'for ever separate, yet for ever near.' Nor is this other than the principles of evolution would lead us to expect, as I shall afterwards have abundant occasion to show. Here, however, we are only concerned with drawing what distinction we can between instinct and reason as these faculties are actually presented to our observation. And this in a general way it is not difficult to do.

We have seen that instinct involves 'mental operations,' and that by this feature it is distinguished from reflex action; we have now to consider the features by which it is distinguished from reason. These are accurately, though not completely, conveyed by Sir Benjamin Brodie, who defines instinct as 'a principle by which animals are induced, independently of experience and reasoning, to the performances of certain voluntary acts, which are necessary to their preservation as individuals, or to the continuance of the species, or in some other way convenient to them.'[1] This definition, as I have said, is accurate as far as it goes, but it does not state with sufficient generality and terseness that all instinctive action is adaptive; nor does it clearly bring out the distinction between instinct and reason which is thus well conveyed by the definition of Hartmann, who says in his 'Philosophy of the Unconscious,' that 'instinct is action taken in pursuance of an end, but without conscious perception of what the end is.' This definition, however, is likewise defective in that it omits another of the important differentiæ of instinct—namely, the uniformity of instinctive action as performed by different individuals of the same species. Including this feature, therefore, we may more accurately and completely define instinct as mental action (whether in animals or human beings),

[1] *Psychological Researches*, p. 187.

directed towards the accomplishing of adaptive movement, antecedent to individual experience, without necessary knowledge of the relation between the means employed and the ends attained, but similarly performed under the same appropriate circumstances by all the individuals of the same species. Now in every one of these respects, with the exception of containing a mental constituent and in being concerned in adaptive action, instinct differs from reason. For reason, besides involving a mental constituent, and besides being concerned in adaptive action, is always subsequent to individual experience, never acts but upon a definite and often laboriously acquired knowledge of the relation between means and ends, and is very far from being always similarly performed under the same appropriate circumstances by all the individuals of the same species.

Thus the distinction between instinct and reason is both more definite and more manifold than is that between instinct and reflex action. Nevertheless, in particular cases there is as much difficulty in classifying certain actions as instinctive or rational, as there is in cases where the question lies between instinct and reflex action. And the explanation of this is, as already observed, that instinct passes into reason by imperceptible degrees; so that actions in the main instinctive are very commonly tempered with what Pierre Huber calls 'a little dose of judgment or reason,' and *vice versâ*. But here, again, the difficulty which attaches to the classification of particular actions has no reference to the validity of the distinctions between the two classes of actions; these are definite and precise, whatever difficulty there may be in applying them to particular cases.

Another point of difference between instinct and reason may be noticed which, although not of invariable, is of very general applicability. It will have been observed, from what has already been said, that the essential respect in which instinct differs from reason consists in the amount of conscious deliberation which the two processes respectively involve. Instinctive actions are actions which, owing to their frequent repetition, become so habitual in the course of generations that all the individuals of the same species automatically perform the same actions under the stimulus supplied by the same appropriate circumstances. Rational actions, on the other hand, are actions which are required to meet circumstances of comparatively rare occurrence in the life-history of the species, and which therefore can only be performed by an intentional effort of adaptation. Consequently there arises the subordinate distinction to which I allude, viz., that instinctive actions are only performed under particular circumstances which have been frequently experienced during the life-history of the species; whereas rational actions are performed under varied circumstances, and serve to meet novel exigencies which may never before have occurred even in the life-history of the individual.

Thus, then, upon the whole, we may lay down our several definitions in their most complete form.

Reflex action is non-mental neuro-muscular adjustment, due to the inherited mechanism of the nervous system, which is formed to respond to particular and often recurring stimuli, by giving rise to particular movements of an adaptive though not of an intentional kind.

Instinct is reflex action into which there is imported the element of consciousness. The term is therefore a generic one, comprising all those faculties of mind which are concerned in conscious and adaptive action, antecedent to individual experience, without necessary knowledge of the relation between means employed and ends attained, but similarly performed under similar and frequently recurring circumstances by all the individuals of the same species.

Reason or intelligence is the faculty which is concerned in the intentional adaptation of means to ends. It therefore implies the conscious knowledge of the relation between means employed and ends attained, and may be exercised in adaptation to circumstances novel alike to the experience of the individual and to that of the species.

CHAPTER I.

APPLICATION OF THE FOREGOING PRINCIPLES TO THE LOWEST ANIMALS.

Protozoa.

No one can have watched the movements of certain Infusoria without feeling it difficult to believe that these little animals are not actuated by some amount of intelligence. Even if the manner in which they avoid collisions be attributed entirely to repulsions set up in the currents which by their movements they create, any such mechanical explanation certainly cannot apply to the small creatures seeking one another for the purposes of prey, reproduction, or, as it sometimes seems, of mere sport. There is a common and well-known rotifer whose body is of a cup shape, provided with a very active tail, which is armed at its extremity with strong forceps. I have seen a small specimen of this rotifer seize a much larger one with its forceps, and attach itself by this means to the side of the cup. The large rotifer at once became very active, and swinging about with its burden until it came to a piece of weed, it took firm hold of the weed with its own forceps, and began the most extraordinary series of movements, which were obviously directed towards ridding itself of the encumbrance. It dashed from side to side in all directions with a vigour and suddenness which were highly astonishing, so that it seemed as if the animalcule would either break its forceps or wrench its tail from its body. No movements could possibly be better suited to jerk off the offending object, for the energy with which the jerks were given, now in one direction and now in another, were, as I have said, most surprising. But not less surprising was the tenacity with which the smaller rotifer retained its hold; for although one might think that it was being almost jerked to pieces, after each bout of jerking it was seen to be still attached. This trial of strength, which must have involved an immense expenditure of energy in proportion to the size of the animals, lasted for several minutes, till eventually the small rotifer was thrown violently away. It then returned to the conflict, but did not succeed a second time in establishing its hold. The entire scene was as like intelligent action on the part of both animals as could well be imagined, so that if we were to depend upon appearances alone, this one observation would be sufficient to induce me to attribute conscious determination to these microscopical organisms.

But, without denying that conscious determination may here be present, or involving ourselves in the impossible task of proving such a negative, we may properly affirm that until an animalcule shows itself to be teachable by individual experience, we have no sufficient evidence derived or derivable from any number of such apparently intelligent movements, that conscious determination is present. Therefore, I need not wait to quote the observations of the sundry microscopists who detail facts more or less similar to the above, with expressions of their belief that microscopical organisms display a certain degree of instinct or intelligence as distinguished from mechanical, or wholly non-mental adjustment. But there are some observations relating to the lowest of all animals, and made by a competent person, which are so remarkable that I shall have to quote them in full. These observations are recorded by Mr. H. J. Carter, F.R.S., in the 'Annals of Natural History,' and in his opinion prove that the beginnings of instinct are to be found so low down in the scale as the Rhizopoda. He says:—'Even *Athealium* will confine itself to the water of the watch-glass in which it may be placed when away from sawdust and chips of wood among which it has been living; but if the watch-glass be placed upon the sawdust, it will very soon make its way over the side of the watch-glass and get to it.'

This is certainly a remarkable observation: for it seems

to show that the rhizopod distinguishes the presence of the sawdust outside the watch-glass, and crawls over the brim of the latter in order to get into more congenial quarters, while it is contented with the water in the watch-glass so long as there is no sawdust outside. But to proceed:

On one occasion, while investigating the nature of some large, transparent, spore-like elliptical cells (fungal?) whose protoplasm was rotating, while it was at the same time charged with triangular grains of starch, I observed some actinophorous rhizopods creeping about them, which had similarly shaped grains of starch in their interior; and having determined the nature of these grains in both by the addition of iodine, I cleansed the glasses, and placed under the microscope a new portion of the sediment from the basin containing these cells and actinophryans for further examination, when I observed one of the spore-like cells had become ruptured, and that a portion of its protoplasm, charged with the triangular starch-grains, was slightly protruding through the crevice. It then struck me that the actinophryans had obtained their starch-grains from this source; and while looking at the ruptured cell, an *actinophrys* made its appearance, and creeping round the cell, at last arrived at the crevice, from which it extricated one of the grains of starch mentioned, and then crept off to a good distance. Presently, however, it returned to the same cell; and although there were now no more starch-grains protruding, the *actinophrys* managed again to extract one from the interior through the crevice. All this was repeated several times, showing that the *actinophrys* instinctively knew that those were nutritious grains, that they were contained in this cell, and that, although each time after incepting a grain it went away to some distance, it knew how to find its way back to the cell again which furnished this nutriment.

On another occasion I saw an *actinophrys* station itself close to a ripe spore-cell of *pythium*, which was situated upon a filament of *Spirogyra crassa*; and as the young ciliated monadic germs issued forth, one after another, from the dehiscent spore-cell, the *actinophrys* remained by it and caught every one of them, even to the last, when it retired to another part of the field, as if instinctively conscious that there was nothing more to be got at the old place.

But by far the greatest feat of this kind that ever presented itself to me was the catching of a young *acineta* by an old sluggish *amœba*, as the former left its parent; and this took place as follows:—

In the evening of the 2nd of June, 1858, in Bombay, while looking through a microscope at some *Euglenæ*, &c., which had been placed aside for examination in a watch-glass, my eye fell upon a stalked and triangular *acineta* (*A. mystacina*?), around which an *amœba* was creeping and lingering, as they do when they are in quest of food. But knowing the antipathy that the *amœba*, like almost every other infusorian, has to the tentacles of the *acineta*, I concluded that the *amœba* was not encouraging an appetite for its whiskered companion, when I was surprised to find that it crept up the stem of the *acineta*, and wound itself round its body. This mark of affection, too much like that frequently evinced at the other end of the scale, even where there is a mind for its control, did not long remain without interpretation. There was a young *acineta*, tender, and without poisonous tentacles (for they are not developed at birth), just ready to make its exit from the parent, an exit which takes place so quickly, and is followed by such rapid bounding movements of the non-ciliated *acineta*, that who would venture to say, *à priori*, that a dull, heavy, sluggish *amœba* could catch such an agile little thing? But the *amœba* are as unerring and unrelaxing in their grasp as they are unrelenting in their cruel inceptions of the living and the dead, when they serve them for nutrition; and thus the *amœba*, placing itself round the ovarian aperture of the *acineta*, received the young one, nurse-like, in its fatal lap, incepted it, descended from the parent, and crept off. Being unable to conceive at the time that this was such an act of atrocity on the part of the *amœba* as the sequel disclosed, and thinking that the young *acineta* might yet escape, or pass into some other form in the body of its host, I watched the *amœba* for some time afterwards, until the tale ended by the young *acineta* becoming divided into two parts, and thus in their respective digestive spaces ultimately becoming broken down and digested.[1]

With regard to these remarkable observations it can only, I think, be said that although certainly very suggestive of something more than mechanical response to stimulation, they are not sufficiently so to justify us in ascribing to these lowest members of the zoological scale any rudiment of truly mental action. The subject, how-

[1] H. J. Carter, F.R.S., *Annals of Natural History*, 3rd Series, 1863, pp. 45–6.

ever, is here full of difficulty, and not the least so on account of the *amœba* not only having no nervous system, but no observable organs of any kind; so that, although we may suppose that the adaptive movements described by Mr. Carter were non-mental, it still remains wonderful that these movements should be exhibited by such apparently unorganised creatures, seeing that as to the remoteness of the end attained, no less than the complex refinement of the stimulus to which their adaptive response was due, the movements in question rival the most elaborate of non-mental adjustments elsewhere performed by the most highly organised of nervous systems.

Cœlenterata.

Dr. Eimer attributes ' voluntary action ' to the Medusæ, and indeed draws a sharp distinction between what he considers their ' involuntary ' and ' voluntary ' movements. In this distinction, however, I do not at all concur; for although I am well acquainted with the difference between the active and slow rhythm upon which the distinction is founded, I see no evidence whatever for supposing that the difference involves any psychological element. The active swimming is produced by stimulation, and is no doubt calculated to lead to the escape of the organism; but this fact certainly does not carry us beyond the ordinary possibilities of reflex action. And even when, as in some species is constantly the case, bouts of active swimming appear to arise spontaneously or without observable stimulation, the fact is to be attributed to a liberation of overplus ganglionic energy, or to some unobservable stimulation; it does not justify the supposition of any psychical element being concerned.[1]

M'Crady gives an interesting account of a medusa which carries its larvæ on the inner sides of its bell-shaped body. The manubrium, or mobile digestive cavity

---

[1] For an account of the natural movements of the Medusæ and the effects of stimulation upon them, see Croonian Lecture in *Phil. Trans.* 1875, and also *Phil. Trans.* 1877 and 1879.

of the animal, depends, as in the other Medusæ, from the summit of the concave surface of the bell, like a clapper or tongue. Now M'Crady observed this depending organ to be moved first to one side and then to the other side of the bell, in order to give suck to the larvæ on the sides of the bell—the larvæ dipping their long noses into the nutrient fluids which that organ of the parent's body contained. I cite this case, because if it occurred in one of the higher animals it would probably be called a case of instinct; but as it occurs in so low an animal as a jelly-fish, it is unreasonable to suppose that intelligence can ever have played any part in originating the action. Therefore we may set it down as the uncompounded result of natural selection.

Some species of medusæ—notably *Sarsia*—seek the light, crowding into the path of a beam, and following it actively if moved. They derive advantage from so doing, because certain small crustacea on which they feed likewise crowd into the light. The seeking of light by these medusæ is therefore doubtless of the nature of a reflex action which has been developed by natural selection in order to bring the animals into contact with their prey. Paul Bert has found that *Daphnia pulex* seeks the light (especially the yellow ray), and Engelmann has observed the same fact with regard to certain protoplasmic organisms. But in none of these or other such cases is there any evidence of a psychical element being concerned in the process.

Echinodermata.

Some of the natural movements of these animals, as also some of their movements under stimulation, are very suggestive of purpose; but I have satisfied myself that there is no adequate evidence of the animals being able to profit by individual experience, and therefore, in accordance with our canon, that there is no adequate evidence of their exhibiting truly mental phenomena. On the other hand, the study of reflex action in these organisms is full of interest—so much so that in my next work I shall take them as typical organisms in this connection.[1]

[1] See Croonian Lecture, 1881, in forthcoming issue of *Phil. Trans.*

Annelida.

Mr. Darwin has now in the press a highly interesting work on the habits of earth-worms. It appears from his observations that the manner in which these animals draw down leaves, &c., into their burrows is strongly indicative of instinctive action, if not of intelligent purpose—seeing that they always lay hold of the part of the leaf (even though an exotic one) by the traction of which the leaf will offer least resistance to being drawn down. But as this work will so shortly be published, I shall not forestall any of the facts which it has to state, nor should I yet like to venture an opinion as to how far these facts, when considered altogether, would justify any inference to a truly mental element as existing in these animals.

Of the land leeches in Ceylon, Sir E. Tennent gives an account which likewise seems to bespeak intelligence as occurring in annelids. He says:—

In moving, the land leeches have the power of planting one extremity on the earth and raising the other perpendicularly to watch for their victim. Such is their vigilance and instinct, that on the approach of a passer-by to a spot which they infest, they may be seen amongst the grass and fallen leaves on the edge of a native path, poised erect, and preparing for their attack on man and horse. On descrying their prey they advance rapidly by semicircular strides, fixing one end firmly and arching the other forwards, till by successive advances they can lay hold of the traveller's foot, when they disengage themselves from the ground and ascend his dress in search of an aperture to enter. In these encounters the individuals in the rear of a party of travellers in the jungle invariably fare worst, as the leeches, once warned of their approach, congregate with singular celerity.[1]

[1] *Natural History of Ceylon*, p. 481.

CHAPTER II.

MOLLUSCA.

I SHALL treat of the Mollusca before the Articulata, because as a group their intelligence is not so high. Indeed, it is not to be expected that the class of animals wherein the 'vegetative' functions of nutrition and reproduction predominate so largely over the animal functions of sensation, locomotion, &c., should present any considerable degree of intelligence. Nevertheless, in the only division of the group which has sense organs and powers of locomotion highly developed —viz., the Cephalopoda—we meet with large cephalic ganglia, and, it would appear, with no small development of intelligence: Taking, however, the sub-kingdom in ascending order, I shall first present all the trustworthy evidence that I have been able to collect, pointing to the highest level of intelligence that is attained by the lower members.

The following is quoted from Mr. Darwin's MS. :—

Even the headless oyster seems to profit from experience, for Dicquemase ('Journal de Physique,' vol. xxviii. p. 244) asserts that oysters taken from a depth never uncovered by the sea, open their shells, lose the water within, and perish; but oysters taken from the same place and depth, if kept in reservoirs, where they are occasionally left uncovered for a short time, and are otherwise incommoded, learn to keep their shells shut, and then live for a much longer time when taken out of the water.[1]

---

[1] This fact is also stated by Bingley, *Animal Biography*, vol. iii. p. 454, and is now turned to practical account in the so-called 'Oyster-schools' of France. The distance from the coast to Paris being too great for the newly dredged oysters to travel without opening their shells, they are first taught in the schools to bear a longer and longer exposure to the air without gaping, and when their education in this respect is completed they are sent on their journey to the metropolis, where they arrive with closed shells, and in a healthy condition.

Some evidence of intelligence seems to be displayed by the razor-fish. For the animals dislike salt, so that when this is sprinkled above their burrows in the sand, they come to the surface and quit their habitations. But if the animal is once seized when it comes to the surface and afterwards allowed to retire into its burrow, no amount of salt will force it again to come to the surface.[1]

With regard to snails, L. Agassiz writes: 'Quiconque a eu l'occasion d'observer les amours des limaçons, ne saurait mettre en doute la séduction déployée dans les mouvements et les allures qui préparent et accomplissent le double embrassement de ces hermaphrodites.'[2]

Again, Mr. Darwin's MS. quotes from Mr. W. White[3] a curious exhibition of intelligence in a snail, which does not seem to have admitted of mal-observation. This gentleman 'fixed a land-shell mouth uppermost in a chink of rock; in a short time the snail protruded itself to its utmost length, and, attaching its foot vertically above, tried to pull the shell out in a straight line. Not succeeding, it rested for a few minutes and then stretched out its body on the right side and pulled its utmost, but failed. Resting again, it protruded its foot on the left side, pulled with its full force, and freed the shell. This exertion of force in three directions, which seems so geometrically suitable, must have been intentional.'

If it is objected that snail shells must frequently be liable to be impeded by obstacles, and therefore that this display of manœuvring on the part of their occupants is to be regarded as a reflex, I may remark that here again we have one of those incessantly recurring cases where it is difficult to draw the line between intelligence and non-intelligence. For, granting that the action is to a certain extent mechanical, we must still recognise that the animal while executing it must have remembered each of the two directions in which it had pulled ineffectually before it began to pull in the third direction; and it is improbable that snail shells are so frequently caught in positions from which a pull in only one direction will

---

[1] Bingley, *loc. cit.*, vol. iii. p. 449.
[2] *De l'Espèce et de la Classe*, &c., 1869, p. 106.
[3] *A Londoner's Walk to Edinburgh*, p. 155 (1856).

release them, that natural selection would have developed a special instinct to try pulling successively in three directions at right angles to one another.

The only other instance that I have met with of the apparent display of intelligence in snails is the remarkable one which Mr. Darwin gives in his 'Descent of Man,' on the authority of Mr. Lonsdale. Although the interpretation which is assigned to the fact seems to me to go beyond anything that we should have reason to expect of snail intelligence, I cannot ignore a fact which stands upon the observation of so good an authority, and shall therefore quote it in Mr. Darwin's words:—

These animals appear also susceptible of some degree of permanent attachment: an accurate observer, Mr. Lonsdale, informs me that he placed a pair of land-snails (*Helix pomatia*), one of which was weakly, into a small and ill-provided garden. After a short time the strong and healthy individual disappeared, and was traced by its track of slime over a wall into an adjoining well-stocked garden. Mr. Lonsdale concluded that it had deserted its sickly mate; but after an absence of twenty-four hours it returned, and apparently communicated the result of its successful exploration, for both then started along the same track, and disappeared over the wall.[1]

In this case the fact must be accepted, seeing that it stands on the authority of an accurate observer, and is of so definite a kind as not to admit of mistake. Consequently we are shut up to the alternative of supposing the return of the healthy snail to its mate a mere accident, and their both going over the wall into the well-stocked garden another mere accident, or acquiescing in the interpretation which Mr. Darwin assigns. Now, if we look closely into the matter, the chances against the double accident in question are certainly so considerable as to render the former supposition almost impossible. On the other hand, there is evidence to prove, as I shall immediately show, that a not distantly allied animal is unquestionably able to remember a particular locality as its home, and habitually to return to this locality after feeding. Therefore, in view of this analogous and cor-

---

[1] *Descent of Man*, pp. 262–3.

roborative case, the improbability of the snail remembering for twenty-four hours the position of its mate is very much reduced; while the subsequent communication, if it took place, would only require to have been of the nature of 'follow me,' which, as we shall repeatedly find, is a degree of communicative ability which many invertebrated animals possess. Therefore, in view of these considerations, I incline to Mr. Darwin's opinion that the facts can only be explained by supposing them due to intelligence on the part of the snails. Thus considered, these facts are no doubt very remarkable; for they would appear to indicate not merely accurate memory of direction and locality for twenty-four hours, but also no small degree of something akin to 'permanent attachment,' and sympathetic desire that another should share in the good things which one has found.[1]

The case to which I have just alluded as proving beyond all doubt that some Gasteropoda are able to retain a very precise and accurate memory of locality, is that of the common limpet.

Mr. J. Clarke Hawkshaw publishes in the Journal of the Linnæan Society the following account of the habits in question:—

The holes in the chalk in which the limpets are often to be found are, I believe, excavated in a great measure by rasping from the lingual teeth, though I doubt whether the object is to form a cavity to shelter in, though the cavities, when formed, may be of use for that purpose. It must be of the greatest importance to a limpet that, in order that it may insure a firm adherence to the rock, its shell should fit the rock accurately; when the shell does fit the rock accurately, a small amount of muscular contraction of the animal would cause the shell to adhere so firmly to a smooth surface as to be practically immoveable without fracture. As the shells cannot be adapted daily to different forms of surface, the limpets generally return to the same place of attachment. I am sure this is the case with many; for I found shells perfectly adjusted to the uneven surfaces of flints, the growth of the shells being in some parts

[1] The facts, however, in order to sustain such conclusions, of course require corroboration, and it is therefore to be regretted that Mr. Lonsdale did not experimentally repeat the conditions.

distorted and indented to suit inequalities in the surface of the flints. . . .

I noticed signs that limpets prefer a hard, smooth surface to a pit in the chalk. On one surface of a large block, over all sides of which limpets were regularly and plentifully distributed, there were two flat fragments of a fossil shell about 3 inches by 4 inches, each embedded in the chalk. The chalk all round these fragments was free from limpets; but on the smooth surface of the pieces of shell they were packed as closely as they could be. I noticed another case, which almost amounts, to my mind, to a proof that they prefer a smooth surface to a hole. A limpet had formed a clearing on one of the sea-weed-covered blocks before referred to. In the midst of this clearing was a pedestal of flint rather more than one inch in diameter, standing up above the surface of the chalk; it projected so much that a tap from my hammer broke it off. On the top of the smooth fractured surface of this flint the occupant of the clearing had taken up its abode. The shell was closely adapted to the uneven surface, which it would only fit in one position. The cleared surface was in a hollow with several small natural cavities, where the limpet could have found a pit ready made to shelter in; yet it preferred, after each excursion, to climb up to the top of the flint, the most exposed point in all its domain.[1]

It appears certain from these observations, which to some extent were anticipated by those of Mr. F. C. Lukis,[2] that limpets, after every browsing excursion, return to one particular spot or home; and the precise memory of direction and locality implied by this fact seems to justify us in regarding these actions of the animal as of a nature unquestionably intelligent.

Coming now to the cephalopoda, there is no doubt that if a larger sphere of opportunity permitted, adequate observation of these animals would prove them to be much the most intelligent members of the sub-kingdom. Unfortunately, however, this sphere of opportunity has hitherto been very limited. The following meagre account is all that I have been able to gather concerning the psychology of these interesting animals.

According to Schneider,[3] the Cephalopoda show un-

---

[1] *Journal Linn. Soc.* vol. xiv. p. 406 *et seq.*
[2] *Mag. Nat. Hist.* 1831, vol. iv. p. 346.
[3] *Thiersche Wille*, § 78.

mistakable evidence of consciousness and intelligence. This observer had an opportunity of watching them for a long time in the zoological station at Naples; and he says that they appeared to recognise their keeper after they had for some time received their food from him. Hollmann narrates that an octopus, which had had a struggle with a lobster, followed the latter into an adjacent tank, to which it had been removed for safety, and there destroyed it. In order to do this the octopus had to climb up a vertical partition above the surface of the water and descend the other side.[1] According to Schneider, the Cephalopoda have an abstract idea of water, seeking to return to it when removed, even though they do not see it. But this probably arises from the sense of discomfort due to exposure of their skin to the air; and if we can call it an 'idea,' it is doubtless shared by all other aquatic Mollusca when exposed to air.

[1] *Leben der Cephalopoden*, s. 21.

CHAPTER III.

ANTS.

WITHIN the last ten or twelve years our information on the habits and intelligence of these insects has been so considerably extended, that in here rendering a condensed epitome of our knowledge in this most interesting branch of comparative psychology, it will be found that the chapter is constituted principally of a statement of observations and experiments which have been conducted during the short period named. The observers to whom we are mainly indebted for this large increase of our knowledge are Messrs. Bates, Belt, Müller, Moggridge, Lincecum, MacCook, and Sir John Lubbock. From the fact that these naturalists conducted their observations in different parts of the world and on widely different species of ants, it is not surprising that their results should present many points of difference; for this only shows, as we might have expected, that different species of ants differ considerably in habits and intelligence. Therefore, in now drawing all these numerous observations to a focus, I shall endeavour to show clearly their points of difference as well as their points of agreement; and in order that the facts to be considered may be arranged in some kind of order, I shall deal with them under the following heads:—Powers of special sense; Sense of direction; Powers of memory; Emotions; Powers of communication; Habits general in sundry species; Habits peculiar to certain species; General intelligence of various species.

Powers of Special Sense.

Taking first the sense of sight, Sir John Lubbock made a number of experiments on the influence of light coloured by passing through various tints of stained glass, with the following results. The ants which he observed greatly dislike the presence of light within their nests, hurrying about in search of the darkest corners when light is admitted. The experiments showed that the dislike is much greater in the case of some colours than in that of others. Thus under a slip of red glass there were congregated on one occasion 890 ants, under green 544, under yellow 495, and under violet only 5. To our eyes the violet is as opaque as the red, more so than the green, and much more so than the yellow. Yet, as the numbers show, the ants had scarcely any tendency to congregate under it: there were nearly as many under the same area of the uncovered portion of the nest as under that shaded by the violet glass. It is curious that the coloured glasses appear to act on the ants in a graduated series, which corresponds with the order of their influence on a photographic plate. Experiments were therefore made to test whether it might not be the actinic rays that were so particularly distasteful to the ants; but with negative results. Placing violet glass above red produces the same effect as red glass alone. Obviously, therefore, the ants avoid the violet glass because they dislike the rays which it transmits, and do not prefer the other colours because they like the rays which they transmit. Sodium, barium, strontium, and lithium flames were also tried, but not with so much effect as the coloured glass.

It has just been observed that the relative dislike which Sir John Lubbock's ants showed to lights of different colours seems to be determined by the position of the colour in the spectrum—there being a regular gradation of intolerance shown from the red to the violet end. As these ants dislike light, the question suggests itself that the reason of their graduated intolerance to light of different colours may be due to their eyes not being so much affected by the rays of low as by those of high refrangibility. In this connection it would be interesting to ascertain whether ants of the genus *Atta* show a similarly graduated intolerance to the light in different parts of the spectrum; for both Moggridge and MacCook record of this genus that it not only does not shun the light, but seeks it—coming to the glass sides of their artificial nests to enjoy the light of a lamp. Possibly, therefore, the scale of preference to lights of different colours would be found in this genus to be the reverse of that which Sir John Lubbock has found in the case of the British species.

As regards hearing, Sir John Lubbock found that sounds of various kinds do not produce any effect upon the insects. Tuning-forks and violin notes, shouting, whistling, &c., were all equally inefficient in producing the slightest influence upon the animals; and experiments with sensitive flames, microphone, telephone, &c., failed to yield any evidence of ants emitting sounds inaudible to human ears.

Lastly, as regards the sense of smell, Sir John Lubbock found that on bringing a camel's-hair brush steeped in various strong scents near where ants were passing, "some went on without taking any notice, but others stopped, and evidently perceiving the smell, turned back. Soon, however, they returned, and passed the scented pencil. After doing this two or three times, they generally took no further notice of the scent. This experiment left no doubt on my mind." In other cases the ants were observed to wave about and throw back their antennæ when the scented pencil was brought near.

That ants track one another by scent was long ago mentioned by Huber, and also that they depend on this sense for their power of finding supplies which have been previously found by other ants. Huber proved their power of tracking a path previously pursued by their friends, by drawing his finger across the trail, so obliterating the scent at that point, and observing that when the ants arrived at that point they became confused and ran about in various directions till they again came upon the trail on the other side of the interrupted space, when they proceeded on their way as before. The more numerous and systematic experiments of Sir John Lubbock have fully corroborated Huber's observations, so far as these points are concerned. Thus, to give only one or two of these experiments; in the accompanying woodcut (Fig. 1) A is the nest, B a board, $n\,f\,g$ slips of paper, h and m

similar slides of glass, on one of which, *h*, there was placed pupæ, while the other, *m*, was left empty. Sir John Lubbock watched two particular (marked) ants proceeding from A to *h* and back again, carrying the pupæ on *h* to the nest A. Whenever an ant came out of A upon B he transposed the slips *f* and *g*. Therefore at the angle below *n* there was a choice presented to the ant of taking the unscented pathway leading to the full glass *h*, or the scented pathway leading to the empty glass *m*. The two marked ants, knowing their way, always took the right turn at the angle; but the stranger ants, being guided only by scent, for the most part took the wrong turn at the angle, so going to the empty glass *m*. For out of 150 stranger ants only 21 went to *h*, while the remaining 129 went to *m*. Still the fact that all the stranger ants did not follow the erroneous scent-trail to *m*, may be taken to indicate that they are also assisted in finding treasure by the sense of sight, though in a lesser degree. Therefore Sir John Lubbock concludes that in finding treasure 'they are guided in some cases by sight, while in others they track one another by scent.'

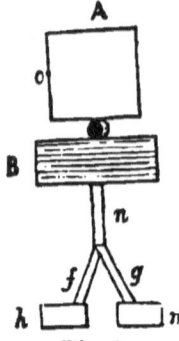

Fig. 1.

As further evidence showing how much more ants depend upon scent than upon sight in finding their way, the following experiment may be quoted. In the accompanying woodcut (Fig. 2) the line marked 1, 2, 3 represents the edge of a paper bridge leading to the nest; A the top of a pencil which is standing perpendicularly upon a board, represented by the general black surface; B the top of the same pencil when moved a distance of a few inches from its first position A. On the top of this pencil were placed some pupæ. Sir John Lubbock, after contriving this arrangement, marked an ant and put it upon the pupæ on the top of the pencil. After she had made two journeys carrying pupæ from the pencil to the nest (the tracks she pursued being represented by the two thick white lines), while she was in the nest he moved the pencil to its position at B. The thin white line represents the course then pursued by the ant in its endeavours to find the pencil, which was shifted only a few inches from A to B. That is, 'the ants on their journey to the shifted object travelled very often back-

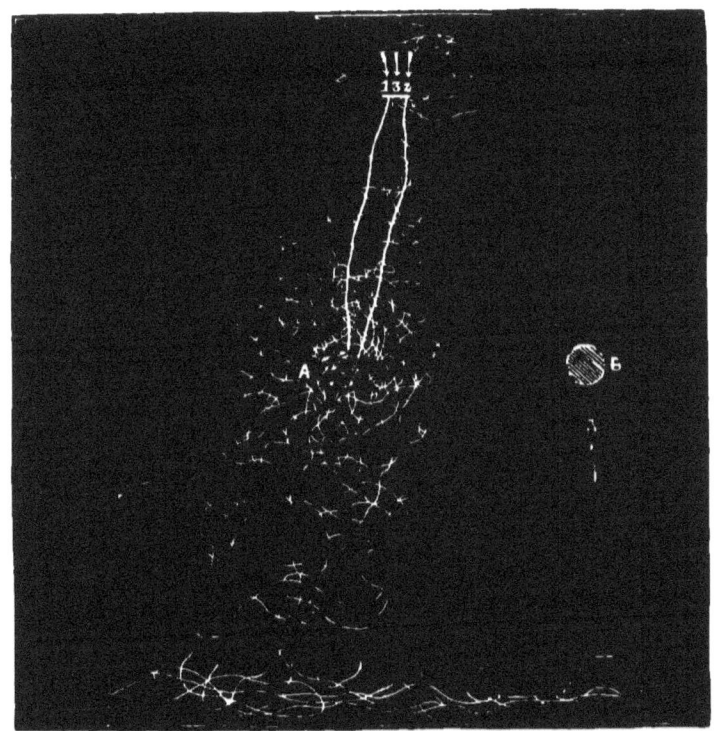

Fig. 2.

wards and forwards and round the spot where the coveted object first stood. Then they would retrace their steps towards the nest, wander hither and thither from side to side between the nest and the point A, and only after very repeated efforts around the original site of the larvæ reach, as it were, accidentally the object desired at B.' Therefore the ants were clearly not guided by the *sight* of the pencil.

The same thing is well shown by another form of experiment. 'Some food was placed at the point *a* (Figs. 3 and 4) on a board measuring 20 inches by 12 inches, and so arranged that the ants in going straight from it to the nest would reach the board at the point *b*, and after passing under the paper tunnel *c*, would proceed between five pairs of wooden bricks, each 3 inches in length and 1¾ inches in height. When they got to know their way they went quite straight along the line *d e* to *a*. The board was then twisted as shown in Fig. 4. 'The bricks and tunnel being arranged exactly in the same direction as before, but the board having been moved, the line *d e* was now outside them. The change, however, did not at all discompose the ants; but instead of going, as before, through the tunnel and between the rows of bricks to *a*, they walked exactly along the old path to *e*.' Keeping the board steady, but moving the brick pathway to the left-hand corner of the board where the food was next placed (Fig. 5), had the effect of making the ant first go to the old position of the food at *a*, whence it veered to a new position, which we may call *x*. The bricks and food were then moved towards the right-hand corner of the board—*i.e.* over a distance of 8 inches (Fig. 6). The ant now first went to *a*,

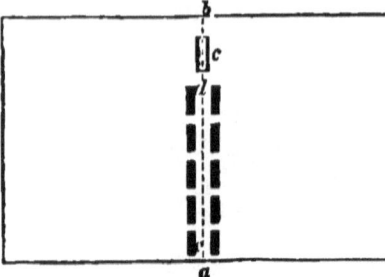

Fig. 3.

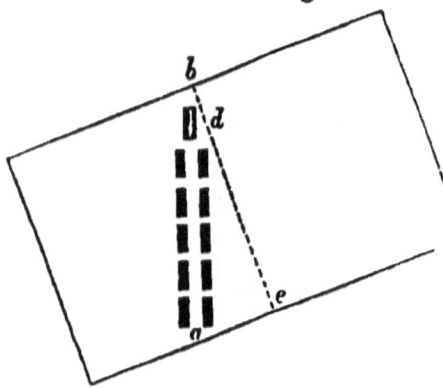

Fig. 4.

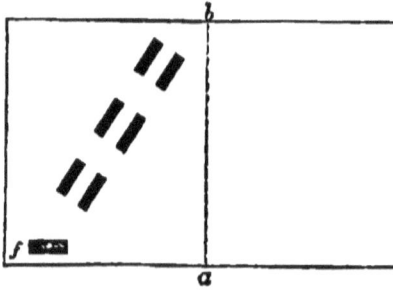

Fig. 5.

then to *x*, and not finding the food at either place, set to work to look for it at random, and was only successful after twenty-five minutes' wandering.

And, as evidence how much more dependence they place upon scent in finding their way than upon any other of their faculties, it is desirable to quote yet one further experiment, which is of great interest as showing that when their sense of smell is made to contradict their sense of direction, they follow the former, notwithstanding, as we shall presently see, the wonderful accuracy of the information which is supplied to them by the latter. 'If, when *F. niger* were carrying off larvæ placed in a cup on a piece of board, I turned the board round so that the side which had been turned towards the nest was away from it, and *vice versâ*, the ants always returned over the same track on the board, and, in consequence, directly away from home. If I moved my board to the other side of my artificial nest, the result was the same. Evidently they followed the road, not the direction.'

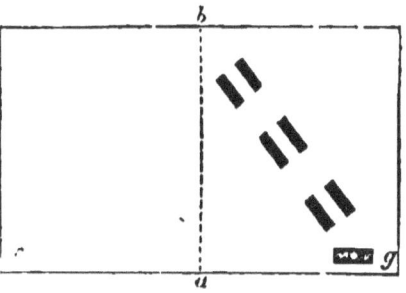

Fig. 6.

There can be little doubt that ants have a sense of taste, as they are so well able to distinguish sugary substances; and it is unquestionable that in their antennæ they possess highly elaborated organs of touch

Sense of Direction.

As evidence of the accuracy and importance of the sense of direction in the Hymenoptera, we must here adduce Sir John Lubbock's highly interesting experiments on ants—leaving his experiments in this connection on bees and wasps to be considered in the next chapter. He first accustomed some ants (*Lasius niger*) to go to and fro to food over a wooden bridge. When they had got quite accustomed to the way, he watched when an ant was upon a bridge which could be rotated, and while she was passing along it, he turned it round, so that end *b* was at *c*, and *c* at *b*. 'In most cases the ant immediately turned round also; but even if she went on to *b* or *c*, as the case might be, as soon as she came to the end of the bridge she turned round.' Next, between the nest and the food he placed a hat-box twelve inches in diameter and seven inches high, cutting two small holes, so that the ants in passing from the nest to the food had to pass in at one hole and out at the other. The box was fixed upon a central pivot, so as to admit of being rotated easily without much friction or disturbance. When the ants had well learnt their way, the box was turned half round as soon as an ant had entered it, 'but in every case the ant turned too, thus retaining her direction.' Lastly, Sir John took a disk of white paper, which he placed in the stead of the hat-box between the nest and the food. When an ant was on the disk making towards the food, he gently drew the disk to the other side of the food, so that the ant was conveyed by the moving surface in the same direction as that in which she was going, but *beyond* the point to which she intended to go. Under these circumstances ' the ant did not turn round, but went on ' to the further edge of the disk, when she seemed ' a good deal surprised at finding where she was.'

These experiments seem to show that the mysterious 'sense of direction,' and consequent faculty of 'homing,' are in ants, at all events, due to a process of registering, and, where desirable, immediately counteracting any change of direction, even when such change is gently made by a wholly closed chamber in which the animal is moving, and not by any muscular movements of the animal itself. And the fact that drawing the moving surface along in the same direction of advance as that which the insect is pursuing does not affect the movements of the latter, seems conclusively to show that the power of registration has reference only to *lateral* movements of the travelling surface; it has no reference to variations in the *velocity* of advance along the line in which the animal is progressing.[1]

[1] While this MS. is passing through the press Sir John Lubbock has

Powers of Memory.

Little need here be said to prove that ants display some powers of memory; for many of the observations and experiments already detailed constitute a sufficient demonstration of the statement that they do. Thus, for instance, the general fact that whenever an ant finds her way to a store of food or larvæ, she will return to it again and again in a more or less direct line from her nest, constitutes ample proof that the ant remembers the way to the store. It is of considerable interest, however, to note that the nature of this insect-memory appears to be, as far as it goes, precisely identical with that of memory in general. Thus, a new fact becomes *impressed* upon their memory by *repetition*, and the impression is liable to become effaced by lapse of time. More evidence on both these features of insect-memory will be adduced when we come to treat of the intelligence of bees; but meanwhile it is enough to refer to the fact that in his experiments on ants, Sir John Lubbock found it necessary to *teach* the insects by a repetition of several lessons their way to treasure, if that way was long or unusual.

With regard to the *duration* of memory, it does not appear that any experiments have been made; but the following observation by Mr. Belt on this point in the case of the leaf-cutting ant may here be stated. In June 1859 he found his garden invaded by these ants, and following up their paths he found their nest about a hundred yards

read another paper before the Linnæan Society, which contains some important additional matter concerning the sense of direction in ants. It seems that in the experiment above described, the hat-box was not provided with a cover or lid, *i.e.* was not a 'closed chamber,' and that Sir John now finds the ants to take their bearings from the direction in which they observe the light to fall upon them. For in the experiment with the uncovered hat-box, if the source of light (candle) is moved round together with the rotating table which supports the box, the ants continue their way without making compensating changes in their direction of advance. The same thing happens if the hat-box is covered, so as to make of it a dark chamber. Direction of light being the source of their information that their ground is being moved, we can understand why they do not know that it is being moved when it is moved in the direction of their advance, as in the experiment with the paper slip.

distant. He poured down their burrows a pint of common brown carbolic acid, mixed with four buckets of water. The marauding parties were at once drawn off from the garden to meet the danger at home, and the whole formicarium was disorganised, the ants running up and down again in the utmost perplexity. Next day he found them busily employed bringing up the ant-food from the old burrows, and carrying it to newly formed ones a few yards distant. These, however, turned out to be only intended as temporary repositories; for in a few days both the old and the new burrows were entirely deserted, so that he supposed all the ants to have died. Subsequently, however, he found that they had migrated to a new site, about two hundred yards from the old one, and there established themselves in a new nest. Twelve months later the ants again invaded his garden, and again he treated them to a strong dose of carbolic acid. The ants, as on the previous occasion, were at once withdrawn from the garden, and two days afterwards he found 'all the survivors at work on one track that led directly to the old nest of the year before, where they were busily employed in making fresh excavations. Many were bringing along pieces of ant-food' from the nest most recently deluged with carbolic acid to that which had been similarly deluged a year before, and from which all the carbolic acid had long ago disappeared. 'Others carried the undeveloped white pupæ and larvæ. It was a wholesale and entire migration;' and the next day the nest down which he had last poured the carbolic acid was entirely deserted. Mr. Belt adds: 'I afterwards found that when much disturbed, and many of the ants destroyed, the survivors migrate to a new locality. I do not doubt that some of the leading minds in this formicarium recollected the nest of the year before, and directed the migration to it.'

Now, I do not insist that the facts necessarily point to this conclusion; for it may have been that the leaders of the migration simply stumbled upon the old and vacant nest by accident, and finding it already prepared as a nest, forthwith proceeded to transfer the food and pupæ to it. Still, as the two nests were separated from one another by so considerable a distance, this hypothesis does not seem probable, and the only other one open to us is that the ants remembered the site of their former home for a period of twelve months. And this conclusion is rendered less improbable from a statement of Karl Vogt in his 'Thierstaaten,' to the effect that for several successive years ants from a certain nest used to go through certain inhabited streets to a chemist's shop 600 mètres distant, in order to obtain access to a vessel filled with syrup. As it cannot be supposed that this vessel was found in successive working seasons by as many successive accidents, it can only be concluded that the ants remembered the syrup store from season to season.

I shall now pass on to consider a class of highly remarkable facts, perhaps the most remarkable of the many remarkable facts connected with ant psychology.

It has been known since the observations of Huber that all the ants of the same nest or community recognise one another as friends, while an ant introduced from another nest, even though it be an ant of the same species, is known at once to be a foreigner, and is usually maltreated or put to death. Huber found that when he removed an ant from a nest and kept it away from its companions for a period of four months it was still recognised as a friend, and caressed by its previous fellow-citizens after the manner in which ants show friendship, viz., by stroking antennæ. Sir John Lubbock, after repeating and fully confirming these observations, extended them as follows. He first tried keeping the separated ant away from the nest for a still longer period than four months, and found that even after a separation of more than a year the animal was recognised as before. He repeated this experiment a number of times, and always with the same invariable difference between the reception accorded to a foreigner and a native—no matter, apparently, how long the native had been absent.

Considering the enormous number of ants that go to make a nest, it seems astonishing enough that they should be all personally known to one another, and still more astonishing that they should be able to recognise members of their community after so prolonged an absence. Thinking that the facts could only be explained, either by all the ants in the same nest having a peculiar smell, or by all the members of the same community having a particular pass-word or gesture-sign, Sir John Lubbock, with the view of testing this theory, separated some ants from a nest while still in the condition of pupæ, and, when they emerged from that state as perfect insects, transferred them back to the nest from which they had been taken as pupæ. Of course in this case the ants in the nest could never have *seen* those which had been removed, for a larval ant is as unlike the mature insect as a grub is unlike a beetle; neither can it be supposed that a larva, hatched out away from the nest, should retain, when a perfect insect, any smell belonging to its parent nest—more especially as it had been hatched out by ants in another nest;[1] nor, lastly, is it reasonable to imagine that the animal, while still a larval grub, can have been taught any gesture-signal used as a pass-word by the matured animals. Yet, although all these possible hypotheses seem to be thus fully excluded by the conditions of the experiment, the result showed unequivocally that the ants recognised their transformed larvæ as native-born members of their community.

Lastly, Sir John Lubbock tried the experiment of going still further back in the life-history of the ants before separating them from the nest. For in September he divided a nest into two halves, each having a queen. At this season there were neither larvæ nor eggs. The following April both the queens began to lay eggs, and in August—*i.e.* nearly a year after the original partitioning of the nest—he took some of the ants newly hatched from the pupæ in one division, and placed them in the other division, and *vice versâ*. In all cases these ants were received by the members of the other half of the divided nest as friends, although if a stranger were introduced into either half it was invariably killed. Yet the ants which

---

[1] It is to be noted that although ants will attack stranger ants introduced from other nests, they will carefully tend stranger larvæ similarly introduced.

were thus so certainly recognised by their kindred ants as friends had never, even in the state of an egg, been present in that division of the nest before. On this highly remarkable fact Sir John Lubbock says:—

These observations seem to me conclusive as far as they go, and they are very surprising. In my experiments of last year, though the results were similar, still the ants experimented with had been brought up in the nest, and were only removed after they had become pupæ. It might therefore be argued that the ants, having nursed them as larvæ, recognised them when they came to maturity; and though this would certainly be in the highest degree improbable, it could not be said to be impossible. In the present case, however, the old ants had absolutely never seen the young ones until the moment when, some days after arriving at maturity, they were introduced into the nest; and yet in all ten cases they were undoubtedly recognised as belonging to the community.

It seems to me, therefore, to be established by these experiments that the recognition of ants is not personal and individual; that their harmony is not due to the fact that each ant is individually acquainted with every other member of the community.

At the same time, the fact that they recognise their friends even when intoxicated, and that they know the young born in their own nest even when they have been brought out of the chrysalis by strangers, seems to indicate that the recognition is not effected by means of any sign or pass-word.

We must, therefore, conclude with reference to this subject that the mode whereby recognition is undoubtedly effected is as yet wholly unintelligible; and I have introduced these facts under the heading of memory only because this heading is not more inappropriate than any other that could be devised for their reception.

It ought here to be added also that the power of thus recognising members of their community is not confined by the limits of blood-relationship, for in an experiment made by Forel it was shown that Amazon ants recognised their own slaves almost instantaneously after an absence of four months.

Under this heading I may also adduce the evidence as to enormous masses, or, as we might say, a whole nation of ants recognising each other as belonging to the same nationality. New nests often spring up as offshoots from the older ones, and thus a nation of towns gradually spreads to an immense circumference around the original centre. Forel describes a colony of *F. exsecta* which comprised more than two hundred nests, and covered a space of nearly two hundred square mètres. 'All the members of such a colony, even those from the furthermost nest, recognise each other and admit no stranger.'

Similarly, MacCook describes an 'ant town' in the Alleghany Mountains of North America ('Trans. Amer. Entom. Soc.,' Nov. 1877) which was inhabited by *F. exsectoïdes*. It consists of 1,600 to 1,700 nests, which rise in cones to a height of from two to five feet. The ground below is riddled in every direction with subterranean passages of communication. The inhabitants are all on the most friendly terms, so that if any one nest is injured it is repaired by their united forces.

It remains to be added in connection with this subject that the recognition is not automatically invariable, but when 'ants are removed from a nest in the pupa state, tended by strangers, and then restored, some at least of their relatives are certainly puzzled, and in many cases doubt their claims to consanguinity. I say some, because while strangers under the circumstances would have been immediately attacked, these ants were in every case amicably received by the majority of the colony, and it was sometimes several hours before they came across one who did not recognise them.'

It may also be added that *Lasius flavus* behaves towards strangers quite differently and much more hospitably than is the case with *L. niger*. The stranger shows no alarm, but, on the contrary, will voluntarily enter the strange nest, and she is there received with kindness; although from the attention she excites, and the numerous communications which take place between her and her new friends, Sir John was 'satisfied that they knew she was not one of themselves. . . . Very different is the behaviour of *L. niger* under similar circumstances. I tried the same experiment with them. There was no communications with the antennæ, there was no cleaning, but every ant which the stranger approached flew at her like a little tigress. I tried this experiment four times; each stranger was killed and borne off to the nest.'

Emotions.

The pugnacity, valour, and rapacity of ants are too well and generally known to require the narration of special instances of their display. With regard to the tenderer emotions, however, there is a difference of opinion among observers. Before the researches of Sir John Lubbock it was the prevalent view that these insects display marked signs of affection towards one another, both by caressing movements of their antennæ, and by showing solicitude for friends in distress. Sir John, however, has found that the species of ants on which he has experimented are apparently deficient both in feelings of affection and of sympathy—or, at least, that such feelings are in these species much less strongly developed than the sterner passions.

He tried burying some specimens of *Lasius niger* beneath an ant-road; but none of the ants traversing the road made any attempt to release their imprisoned companions. He tried the same experiment with the same result on various other species. Even when the friends in difficulty are actually in sight, it by no means follows that their companions will assist them. Of this, he says, he could give almost any number of instances. Thus, when ants are entangled in honey, their companions devote themselves to the honey, and entirely neglect their friends in distress; and when partly drowned, their friends take no notice. When chloroformed or intoxicated their own companions either do not heed them, or else ' seem somewhat puzzled at finding their intoxicated fellow-creatures in such a condition, take them up, and carry them about for a time in a somewhat aimless manner.' Further experiments, however, on a larger scale, went to show that chloroformed ants were treated as dead, *i.e.* removed to the edge of the parade-board and dropped over into the surrounding moat of water; while intoxicated ants were generally carried into the nest, if they were ants belonging to that community; if not, they were thrown overboard. This care shown towards intoxicated friends appears to indicate a dim sense of sympathy towards afflicted individuals; but that this emotion or instinct does not in the case of these species extend to healthy individuals in distress seems to be proved, not only by the experiments of burying already described, but also by the following :—

On Sept. 2, therefore, I put two ants from one of my nests of *F. fusca* into a bottle, the end of which was tied up with muslin as described, and laid it down close to the nest. In a second bottle I put two ants from another nest of the same species. The ants which were at liberty took no notice of the bottle containing their imprisoned friends. The strangers in the other bottle, on the contrary, excited them considerably. The whole day one, two, or more ants stood sentry, as it were, over the bottle. In the evening no less than twelve were collected round it, a larger number than usually came out of the nest at any one time. The whole of the next two days, in the same way, there were more or less ants round the bottle containing the strangers; while, as far as we could see, no notice whatever was taken of the friends. On the 9th the ants had eaten through the muslin, and effected an entrance. We did not chance to be on the spot at the moment; but as I found two ants lying dead, one in the bottle and one just outside, I think there can be no doubt that the strangers were put to death. The friends throughout were quite neglected.

Sept. 21.—I then repeated the experiment, putting three ants from another nest in a bottle as before. The same scene was repeated. The friends were neglected. On the other hand, some of the ants were always watching over the bottle containing the strangers, and biting at the muslin which protected them. The next morning at 6 A.M. I found five ants thus occupied. One had caught hold of the leg of one of the strangers, which had unwarily been allowed to protrude through the meshes of the muslin. They worked and watched, though not, as far as I could see, with any system, till 7.30 in the evening, when they effected an entrance, and immediately attacked the strangers.

Sept. 24.—I repeated the same experiment with the same nest. Again the ants came and sat over the bottle containing the strangers, while no notice was taken of the friends.

The next morning again, when I got up, I found five ants round the bottle containing the strangers, none near the friends. As in the former case, one of the ants had seized a stranger by the leg, and was trying to drag her through the muslin. All day the ants clustered round the bottle, and bit perseveringly, though not systematically, at the muslin. The same thing happened all the following day.

On repeating these experiments with another species (viz., *Formica rufescens*) the ants took no notice of either bottle, and showed no sign either of affection or hatred. One is almost tempted to surmise that the spirit of these ants is broken by slavery [*i.e.* by the habit of keeping slaves]. But the experiments on *F. fusca* seem to show that in these curious insects hatred is a stronger passion than affection.

We must not, however, too readily assent to this general conclusion, that ants as a whole are deficient in the tenderer emotions; for although the case is doubtless so with the species which Sir John examined, it appears to be certainly otherwise with other species, as we shall presently see. But first it may be well to point out that even the hard-hearted species with which Sir John had to do seem not altogether devoid of sympathy with sick or mutilated friends, although they appear to be so towards healthy friends in distress. Thus the care shown to intoxicated friends seems to indicate, if not, as already observed, a dim sense of sympathy, at least an instinct to preserve the life of an ailing citizen for the future benefit of the community. Sir John also quotes some observations of Latreille showing that ants display sympathy with mutilated companions; and, lastly, mentions an instance which he has himself observed of the same thing. A specimen of *F. fusca* congenitally destitute of antennæ was attacked and injured by an ant of another species. When separated by Sir John, another ant of her own species came by. 'She examined the poor sufferer carefully, then picked her up tenderly, and carried her away into the nest. It would have been difficult for any one who witnessed this scene to have denied to this ant the possession of humane feelings.' Moggridge is also of opinion that the habit of throwing sick and apparently dead ants into the water, is 'in part to be rid of them, and partly, perhaps, with a view to effecting a possible cure; for I have seen one ant carry another down the twig which formed their path to the surface of the water, and, after dipping it in for a minute, carry it laboriously up again, and lay it in the sun to dry and recover.'

But that some species of ants display marked signs of what we may call sympathy even towards healthy companions in distress, is proved by the following observation of Mr. Belt. He writes:[1]—

One day, watching a small column of these ants (*i.e. Eciton humata*), I placed a little stone on one of them to secure it. The next that approached, as soon as it discovered its situation, ran backwards in an agitated manner, and soon communicated the intelligence to the others. They rushed to the rescue; some bit at the stone and tried to move it, others seized the prisoner by the legs and tugged with such force that I thought the legs would be pulled off, but they persevered until they got the captive free. I next covered one up with a piece of clay, leaving only the ends of its antennæ projecting. It was soon discovered by its fellows, which set to work immediately, and by biting off pieces of the clay soon liberated it. Another time I found a very few of them passing along at intervals. I confined one of these under a piece of clay at a little distance from the line, with his head projecting. Several ants passed it, but at last one discovered it and tried to pull it out, but could not. It immediately set off at a great rate, and I thought it had deserted its comrade, but it had only gone for assistance, for in a short time about a dozen ants came hurrying up, evidently fully informed of the circumstances of the case, for they made directly for their imprisoned comrade and soon set him free. I do not see how this action could be instinctive. It was sympathetic help, such as man only among the higher mammalia shows. The excitement and ardour with which they carried on their unflagging exertions for the rescue of their comrade could not have been greater if they had been human beings.

This observation seems unequivocal as proving fellow-feeling and sympathy, so far as we can trace any analogy between the emotions of the higher animals and those of

[1] *The Naturalist in Nicaragua*, 1874, p. 26.

insects. That insects with such highly organised social habits, and depending so greatly on the principles of co-operation, should manifest emotions or instincts of an incipiently altruistic character, is no more than we should antecedently expect on the general principle of survival of the fittest. Our only surprise should be that these emotions, or instincts, should appear to be so feebly developed in some species of ants, and, as we shall subsequently see, also of bees. But it may be worth while in this connection to point out that the valuable observation of Mr. Belt above quoted refers to the species of ant which, as we shall subsequently find, presents the most highly organised instincts of co-operation that are to be met with among ants, and therefore the greatest dependence of the welfare of the individual on that of the community. And the same remark is applicable to our native species, *F. sanguinea*, which the Rev. W. W. F. White has repeatedly seen rescuing buried companions very much in the manner described by Mr. Belt; and he does not appear to be acquainted with Mr. Belt's observations. He figures one case in which he saw three ants co-operating to dig out a buried comrade.[1]

Powers of Communication.

Huber, Kirby and Spence, Dugardin, Burmeister, Franklin, and other observers have all expressed themselves as more or less strongly of the opinion that members of the same community of ants, and other social Hymenoptera, are able to communicate information to one another by some system of language or signs. The facts, however, on which their opinion rests have not been stated with that degree of caution and detail which the acceptance of the conclusion requires. Thus, Kirby and Spence give only one instance of supposed communication between ants,[2] and even this one is inconclusive, as the facts described admit of being explained by supposing that the ants simply tracked one another by scent; while Huber

---

[1] See *Leisure Hour*, 1880, p. 390.
[2] *Introduction to Entomology*, vol. ii. p. 524.

merely deals in general statements as to 'contact of antennæ,' without narrating any particulars of his observations. Therefore, until within the last few years there was really no sufficient evidence to sustain the general opinion that ants are able to communicate with one another; but the observations which I shall now detail must be regarded as fully substantiating that general opinion by facts as abundant and conclusive as the most critical among us can desire. I shall first narrate in his own words the more important of Sir John Lubbock's experiments in this connection: —

I took three tapes, each about 2 feet 6 inches long, and arranged them parallel to one another and about 6 inches apart. An end of each I attached to one of the nests (*F. niger*), and at the other end I placed a glass. In the glass at the end of one tape I placed a considerable number (300 to 600) of larvæ. In the second I put two or three larvæ only, in the third none at all. The object of the last was to see whether many ants would come to the glasses under such circumstances by mere accident, and I may at once say that scarcely any did so. I then took two ants, and placed one of them to the glass with many larvæ, the other to that with two or three. Each of them took a larva and carried it to the nest, returning for another, and so on. After each journey I put another larva in the glass with only two or three larvæ, to replace that which had been removed. Now, if several ants came under the above circumstances as a mere matter of accident, or accompanying one another by chance, or if they simply saw the larvæ which were being brought, and consequently concluded that they might themselves find a larva in the same place, then the numbers going to the two glasses ought to be approximately equal. In each case the number of journeys made by the ants would be nearly the same; consequently, if it was a matter of scent, the two glasses would be in the same position. It would be impossible for an ant, seeing another in the act of bringing a larva, to judge for itself whether there were few or many left behind. On the other hand, if the strangers were brought, then it would be curious to see whether more were brought to the glass with many larvæ than to that which only contained two or three. I should also mention that every stranger was imprisoned until the end of the experiment.

The results of these experiments were that during

47½ hours the ants which had access to a glass containing numerous larvæ brought 257 friends to their assistance; while during an interval 5½ hours longer those which visited the glass with only two or three larvæ brought only 82 friends; and, as already mentioned, no single ant came to the glass which contained no larvæ. Now, as all the glasses were exposed to similar conditions, and as the roads to the first two must, in the first instance at all events, have been equally scented by the passage of ants over them, these results look very conclusive as proving some power of definite communication, not only that larvæ are to be found, but even where the largest store is to be met with.

To this interesting account Sir John Lubbock adds,—

One case of apparent communication struck me very much. I had had an ant (*F. niger*) under observation one day, during which she was occupied in carrying off larvæ to her nest. At night I imprisoned her in a small bottle; in the morning I let her out at 6.15, when she immediately resumed her occupation. Having to go to London, I imprisoned her again at 9 o'clock. When I returned at 4.40 I put her again to the larvæ. She examined them carefully, and went home without taking one. At this time no other ants were out of the nest. In less than a minute she came out again with eight friends, and the little heap made straight for the heap of larvæ. When they had gone two-thirds of the way I again imprisoned the marked ant; the others hesitated a few minutes, and then with curious quickness returned home. At 5.15 I put her again to the larvæ. She again went home *without a larva*, but after only a few seconds' stay in the nest, came out with no less than thirteen friends. They all went towards the larvæ, but when they had got about two-thirds of the way, although the marked ant had on the previous day passed over the ground about 150 times, and though she had just gone straight from the larvæ to the nest, she seemed to have forgotten her way, and considered; and after she had wandered about for half an hour, I put her to the larvæ. Now, in this case, the twenty-one ants must have been brought out by my marked one, for they came exactly with her, and there were no other ants out. Moreover, it would seem that they must have been told, because (which is very curious in itself) she did not in either case bring a larva, and consequently it cannot have been the mere sight of a larva which had induced them to follow her.

Further experiments proved, as we might have expected, that although an ant is able to communicate to her friends in the nest that she has found treasure somewhere outside, she is not able to describe to them its precise locality. Thus, having exposed larvæ and placed an ant upon them as before, Sir John watched every time she came out of the nest with friends to assist her, but instead of allowing her to pilot the way, he took her up and carried her to the larvæ, allowing her to return with a larva upon her own feet. Under these circumstances the friends, although evidently coming out with the intention of finding some treasure, were never able to find it; but wandered about in various directions for a while, and then returned to the nest. Thus, during two hours she brought out in her successive journeys altogether no less than 120 ants, of which number only 5 in their unguided wanderings happened to find the sought-for treasure. This result seems to prove, as we might have expected, that the communication is of the nature of some sign amounting to no more than a 'follow me.' Other experiments confirmed this result, and also brought out the fact that 'some species act much more in association than others—*Formica fusca*, for instance, much less than *Lasius niger*.' Thus Sir John Lubbock placed some honey before a marked specimen of the former species; but although she visited and revisited the honey during an entire day, she brought out no friends to share it; and although in her journeys to and from the nest she happened to pass and repass many other individuals, they took no notice of each other.

The obvious objection to these experiments, that an ant observing a friend bringing home food or a pupa might infer, without being told, that by accompanying the friend on the return journey she 'might participate in the good things,' has been partly met by the fact already stated, viz., that there is so very marked a difference in the result if, on experimenting on two ants, one had access to a large treasure and the other only to a small one. But to put this matter beyond question, Sir John Lubbock tried the experiment of pinning down a dead fly, so that the ant which found it was unable, with all her tugging, to move it towards the nest. At length she went back to the nest for assistance, and returned accompanied by seven friends. So great was her excitement, however, that she outran these friends, 'who seemed to have come out reluctantly, as if they had been asleep, and were only half awake;' and they failed to find the fly, slowly meandering about for twenty minutes. After again tugging for a time at the fly, the first ant returned a second time to the nest for assistance, and in less than a minute came out with eight friends. They were even less energetic than the first party, and having lost sight of their guide in the same manner as happened before, they all returned to the nest. Meanwhile several of the first party, which had all the while been meandering about, found the fly, and proceeded to dismember it, carrying the trophy to the nest, and calling out more friends in the ordinary way. This experiment was repeated several times and on different species, always with the same result. Now, as Sir John remarks, 'the two cases (*i.e.* those in which the ant brought out friends to her assistance even when she had no booty to show) surely indicate a distinct power of communication. . . . It is impossible to doubt that the friends were brought out by the first ant; and as she returned empty-handed to the nest, the others cannot have been induced to follow her by merely observing her proceedings. I conclude, therefore, that they possess the power of requesting their friends to come and help them.'

In order to ascertain whether the signs which communicating ants make to one another are made by means of sound, Sir John Lubbock placed near a nest of *Lasius flavus* six small upright pillars of wood about $1\frac{1}{2}$ inch high, and on one of these he put a drop of honey. 'I then put three ants to the honey, and when each had sufficiently fed, I imprisoned her, and put another; thus always keeping three ants at the honey, but not allowing them to go home. If, then, they could summon their friends by sound, there ought soon to be many ants at the honey.'

The result showed that the ants were not able thus to call to one another from a distance.

As additional proof of the general fact that at all events some ants have the power of communicating information to one another, it will be enough here to quote an exceedingly interesting observation of the distinguished geologist Hague. The quotations are taken from his letters written to Mr. Darwin, and published in *Nature*:[1]—

On the mantelshelf of our sitting-room my wife has the habit of keeping fresh flowers. A vase stands at each end, and near the middle a small tumbler, usually filled with violets. Some time ago I noticed a pile of very small red ants on the wall above the left-hand vase, passing upward and downward between the mantelshelf and a small hole near the ceiling, at a point where a picture nail had been driven. The ants, when first observed, were not very numerous, but gradually increased in number, until on some days the little creatures formed an almost unbroken procession, issuing from the hole at the nail, descending the wall, climbing the vase directly below the nail, satisfying their desire for water or perfume, and then returning. The other vase and tumbler were not visited at that time.

As I was just then recovering from a long illness it happened that I was confined to the house, and spent my days in the room where the operations of these insects attracted my attention. Their presence caused me some annoyance, but I knew of no effective means of getting rid of them. For several days in succession I frequently brushed the ants in great numbers from the wall down to the floor; but as they were not killed the result was that they soon formed a colony in the wall at the base of the mantel, ascending thence to the shelf, so that before long the vase was attacked from above and below.

One day I observed a number of ants, perhaps thirty or forty, on the shelf at the foot of the vase. Thinking to kill them, I struck them lightly with the end of my finger, killing some and disabling the rest. The effect of this was immediate and unexpected. As soon as those ants which were approaching arrived near to where their fellows lay dead and suffering, they turned and fled with all possible haste. In half an hour the wall above the mantelshelf was cleared of ants.

During the space of an hour or two the colony from below

[1] Vol. vii. pp. 443–4.

continued to ascend until reaching the lower bevelled edge of the shelf, at which point the more timid individuals, although unable to see the vase, somehow became aware of trouble, and turned about without further investigation, while the more daring advanced hesitatingly just to the upper edge of the shelf, when, extending their antennæ and stretching their necks, they seemed to peep cautiously over the edge until beholding their suffering companions, when they too turned and followed the others, expressing by their behaviour great excitement and terror. An hour or two later, the path or trail leading from the lower colony to the vase was almost entirely free from ants.

I killed one or two ants on their path, striking them with my finger, but leaving no visible trace. The effect of this was that as soon as an ant ascending towards the shelf reached the spot where one had been killed, it gave signs immediately of great disturbance, and returned directly at the highest possible speed.

A curious and invariable feature of their behaviour was that when such an ant, returning in fright, met another approaching, the two would always communicate, but each would pursue its own way, the second ant continuing its journey to the spot where the first had turned about, and then following that example.

For some days after this there were no ants visible on the wall, either above or below the shelf.

Then a few ants from the lower colony began to reappear, but instead of visiting the vase which had been the scene of the disaster, they avoided it altogether, and following the lower front edge of the shelf to the tumbler standing near the middle, made their attack upon that. I repeated the same experiment here with precisely the same result. Killing or maiming a few of the ants and leaving their bodies about the base of the tumbler, the others on approaching, and even before arriving at the upper surface of the shelf where their mutilated companions were visible, gave signs of intense emotion, some running away immediately, and others advancing to where they could survey the field and then hastening away precipitately.

Occasionally an ant would advance towards the tumbler until it found itself among the dead and dying; then it seemed to lose all self-possession, running hither and thither, making wide circuits about the scene of the trouble, stopping at times and elevating the antennæ with a movement suggestive of wringing them in despair, and finally taking flight. After this another interval of several days passed, during which no ants appeared. Now, three months later, the lower colony has been entirely abandoned. Occasionally, however, especially when fresh and fragrant violets have been placed on the shelf, a few 'prospectors' descend from the upper nail-hole, rarely, almost never, approaching the vase from which they were first driven away, but seeking to satisfy their desire at the tumbler. To turn back these stragglers and keep them out of sight for a number of days, sometimes for a fortnight, it is sufficient to kill one or two ants on the trail which they follow descending the wall. This I have recently done as high up as I can reach, three or four feet above the mantel. The moment this spot is reached, an ant turns abruptly and makes for home, and in a little while there is not an ant visible on the wall.

In a subsequent volume of 'Nature' (viii. p. 244), Mr. Darwin publishes another letter which he received from Mr. Hague upon the same subject. It seems that Mr. Moggridge suggested to Mr. Darwin that, as he and others had observed ants to be repelled by the mere scent of a finger drawn across their path, the observation of Mr. Hague might really resolve itself into a dislike on the part of the ants to cross a line over which a finger had been drawn, and have nothing to do with intelligent terror inspired by the sight of their slaughtered companions. The following is Mr. Hague's reply to Mr. Darwin's request for further experiments to test this point:—

Acting on Mr. M——'s suggestion, I first tried making simple finger-marks on their path (the mantel is of marble), and found just the results which he describes in his note as observed by himself at Mentone, that is, no marked symptoms of fear, but a dislike to the spot, and an effort to avoid it by going around it, or by turning back and only crossing it again after an interval of time. I then killed several ants on the path, using a smooth stone or piece of ivory, instead of my finger, to crush them. In this case the ants approaching all turned back as before, and with much greater exhibition of fear than when the simple finger-mark was made. This I did repeatedly. The final result was the same as obtained last winter. They persisted in coming for a week or two, during which I continued to kill them, and then they disappeared, and we have seen none since. It would appear from this that while the taint of the hand is sufficient to turn them back, the killing of their fellows with a stone or other material produces the effect described in my first note. This was made clear to me at that time, from the behaviour of the ants the first day I killed any, for on that occasion some of them approaching the vase from below, on reaching the upper edge of the mantel, peeped over, and drew back on seeing what had happened about the vase, then turned away a little, and after a moment tried again at another and another point along the edge, with the same result in the end. Moreover, those that found themselves among the dead and dying went from one writhing ant to another in great haste and excitement, exhibiting the signs of fright which I described.

I hardly hope that any will return again, but if they do, and give me an opportunity, I shall endeavour to act further on Mr. M——'s suggestion.

With this quotation I shall conclude the present division of the chapter; for, looking to all the other observations previously mentioned, there can be no question concerning the general fact that ants have the power of communicating with one another. And under subsequent headings abundant additional evidence on this point will be found implicated with the other facts detailed.

Habits General in Sundry Species.

Swarming.—The precise facts with regard to the swarming of ants are not yet certainly established. As regards some of the facts, however, there is no doubt. The winged males and females first quit the nest in enormous numbers, and choose some fine afternoon in July or August for their wedding flight. The entrances to the nest are widened by the workers and increased in number, and there is a great commotion on the surface of the nest. The swarm takes place as a thick cloud of all the male and female insects, rising together to a considerable height. The flight continues for several hours, usually circling round some tree or tower, and it is during the flight that fertilisation is effected. After it is effected, the swarm returns to the ground, when the males perish, either from falling a prey, in their shelterless condition, to birds or spiders, or, on account of not being able to feed themselves, from starvation. 'The workers, or neuter ants, of their own colony have lost all interest in them from the moment of their return, and trouble themselves no more about them, for they well know that the males have now fulfilled their vocation.' The great majority of the fertilised females share the same fate as the males. But a small proportion find concealment in holes, which they either dig for themselves, or happen to find ready made, and there found a new colony. The first thing they do is to pull off their now useless wings, by scratching and twisting them, one after the other, with the clawed ends of their feet. They then lay their eggs, and become the queens of new colonies.

Forel says that no fertilised female ever returns to her original home; but that the workers keep back a certain number of females which are fertilised before the swarming takes place; in this case the workers pull off the wings of the fertilised females. The majority of observers, however, maintain that some of the females composing the swarm return to their native home to become mothers where they had been children. Probably both statements are correct. A writer in the 'Groniger Deekblad' for June 16, 1877, observes that, looking to the injurious effects of in-breeding, the facts as related by Forel are less probable than those related by other observers, and that, if they actually occur, the females fertilised before flight are probably kept by the ants as a sort of 'reserve corps to which the workers resort only in case of need, and if they fail to secure any returning queens.'

Nursing.—The eggs will not develop into larvæ unless nursed. The nursing is effected by licking the surface of the eggs, which under the influence of this process increase in size, or grow. In about a fortnight, during which time the workers carry the eggs from higher to lower levels of the nest, and *vice versâ*, according to the circumstances of heat, moisture, &c., the larvæ are hatched out, and require no less careful nursing than the eggs. The workers feed them by placing mouths together and regurgitating food stored up in the crop or proventriculus into the intestinal tract of the young. The latter show their hunger by 'stretching out their little brown heads.

ANTS—NURSING AND EDUCATION. 59

Great care is also taken by the workers in cleaning the larvæ, as well as in carrying them up and down the chambers of the nest for warmth or shelter.

When fully grown the larvæ spin cocoons, and are then pupæ, or the 'ants' eggs' of bird-fanciers. These require no food, but still need incessant attention with reference to warmth, moisture, and cleanliness. When the time arrives for their emergence as perfect insects, the workers assist them to get out of their larval cases by biting through the walls of the latter. It is noticeable that in doing this the workers do not keep to any exact time, but free them sometimes earlier and sometimes later, in accordance with their rate of development. 'The little animal when freed from its chrysalis is still covered with a thin skin, like a little shirt, which has to be pulled off. When we see how neatly and gently this is done, and how the young creature is then washed, brushed, and fed, we are involuntarily reminded of the nursing of human babies. The empty cases, or cocoons, are carried outside the nest, and may be seen heaped together there for a long time. Some species carry them far away from the nest, or turn them into building materials for the dwelling.'[1]

Education.—The young ant does not appear to come into the world with a full instinctive knowledge of all its duties as a member of a social community. It is led about the nest, and 'trained to a knowledge of domestic duties, especially in the case of the larvæ.' Later on the young ants are taught to distinguish between friends and foes. When an ants' nest is attacked by foreign ants, the young ones never join in the fight, but confine themselves to removing the pupæ; and that the knowledge of hereditary enemies is not wholly instinctive in ants is proved by the following experiment, which we owe to Forel. He put young ants belonging to three different species into a glass case with pupæ of six other species—all the species being naturally hostile to one another. The young ants did not quarrel, but worked together to tend the pupæ. When the latter hatched out, an artificial colony was formed of

[1] Büchner, *Geistesleben der Thiere*, pp. 66-7.

a number of naturally hostile species all living together after the manner of the 'happy families' of the showmen.

Habit of keeping Aphides.—It is well known that various species of ants keep aphides, as men keep milch cows, to supply a nutritious secretion. Huber first observed this fact, and noticed that the ants collected the eggs of the aphides and treated them exactly as they treated their own, guarding and tending them with the utmost care. When these eggs hatch out the aphides are usually kept and fed by the ants, to whom they yield a sweet honey-like fluid, which they eject from the abdomen upon being stroked on this region by the antennæ of the ants. Mr. Darwin, who has watched the latter process, observes with regard to it,—

I removed all the ants from a group of about a dozen aphides on a dock plant, and prevented their attendance during several hours. After this interval, I felt sure that the aphides would want to excrete. I watched them for some time through a lens, but not one excreted; I then tickled them with a hair in the same manner, as well as I could, as the ants do with their antennæ; but not one excreted. Afterwards I allowed an ant to visit them, and it immediately seemed, by its eager way of running about, to be well aware what a rich flock it had discovered; it then began to play with its antennæ on the abdomen, first of one aphis and then of another; and each, as soon as it felt the antennæ, immediately lifted up its abdomen and excreted a limpid drop of sweet juice, which was eagerly devoured by the ant. Even quite young aphides behaved in this manner, showing that the action was instinctive, and not the result of experience.

The facts also show that the yielding of the secretion to the ants is, as it were, a voluntary act on the part of the aphides, or, perhaps more correctly, that the instinct to yield it has been developed in such a relation to the requirements of the ants, that the peculiar stimulation supplied by the antennæ of the latter is necessary to start the act of secretion; for in the absence of this particular stimulation the aphides will never excrete until compelled to do so by the superabundance of the accumulating secretion. The question, therefore, directly arises how, on evolutionary principles, such a class of facts is to be met; for it is certainly difficult to understand the manner in which this instinct, so beneficial to the ants, can have arisen in the aphides, to which it does not appear, at first sight, to offer any advantages. Mr. Darwin meets the difficulty thus: 'Although there is no evidence that any animal performs an action for the exclusive good of another species, yet each tries to take advantage of the instincts of others;' and 'as the secretion is extremely viscid, it is no doubt a convenience to the aphides to have it removed; therefore probably they do not excrete solely for the good of the ants.'[1]

Some ants which keep aphides build covered ways, or tunnels, to the trees or shrubs where the aphides live. Forel saw a tunnel of this kind which was taken up a wall and down again on the other side, in order to secure a safe covered way from the nest to the aphides. Occasionally such covered ways, or tubes, are continued so as to enclose the stems of the plants on which the aphides live. The latter are thus imprisoned by the walls of the tube, which, however, expand where they take on this additional function of stabling the aphides, so that these insects are really confined in tolerably large chambers. The doors of these chambers are too small to allow the aphides to escape, while large enough for the ants to pass in and out. Forel saw such a prison or stable shaped like a cocoon, and about a centimètre long, which was hanging on the branch of a tree, and contained aphides carefully tended by the ants. Huber records similar observations.

Sir John Lubbock has made an interesting addition to our knowledge respecting this habit as practised by a certain species of ant (*Lasius flavus*), which departs in a very remarkable manner from the habit as practised by other species. He says: 'The ants took the greatest care of these eggs, carrying them off to the lower chambers with the utmost haste when the nest was disturbed.' But the most interesting of Sir John Lubbock's observations in this connection is new, and reveals an astonishing

---

[1] *Origin of Species*, 6th ed. pp. 207-8.

amount of method shown by the ants in farming their aphides. He says:—

> When my eggs hatched I naturally thought that the aphides belonged to one of the species usually found on the roots of plants in the nests of *Lasius flavus*. To my surprise, however, the young creatures made the best of their way out of the nest, and, indeed, were sometimes brought out by the ants themselves. In vain I tried them with roots of grass, &c.; they wandered uneasily about, and eventually died. Moreover, they did not in any way resemble the subterranean species. In 1878 I again attempted to rear these young aphides; but though I hatched a great many eggs, I did not succeed. This year, however, I have been more fortunate. The eggs commenced to hatch the first week in March. Near one of my nests of *Lasius flavus*, in which I had placed some of the eggs in question, was a glass containing living specimens of several species of plants commonly found on or around ants' nests. To this some of the young aphides were brought by the ants. Shortly afterwards I observed on a plant of daisy, in the axils of the leaves, some small aphides, very much resembling those from my nest, though we had not actually traced them continuously. They seemed thriving, and remained stationary on the daisy. Moreover, whether they had sprung from the black eggs or not, the ants evidently valued them, for they built up a wall of earth round and over them. So things remained throughout the summer, but on October 9 I found that the aphides had laid some eggs exactly resembling those found in the ants' nests; and on examining daisy plants from outside, I found on many of them similar aphides, and more or less of the same eggs.
>
> I confess these observations surprised me very much. The statements of Huber have not, indeed, attracted so much notice as many of the other interesting facts which he has recorded, because if aphides are kept by ants in their nests, it seems only natural that their eggs should also occur. The above case, however, is much more remarkable. Here are aphides, not living in the ants' nests, but outside, on the leaf-stalks of plants. The eggs are laid early in October on the food-plant of the insect. They are of no direct use to the ants, yet they are not left where they are laid, where they would be exposed to the severity of the weather and to innumerable dangers, but brought into their nests by the ants, and tended by them with the utmost care through the long winter months until the following March, when the young ones are brought out and again placed

on the young shoots of the daisy. This seems to me a most remarkable case of prudence. Our ants may not perhaps lay up food for the winter, but they do more, for they keep during six months the eggs which will enable them to procure food during the following summer.

The following, which is taken from Büchner's 'Geistesleben der Thiere' is perhaps a still more striking performance of the same kind as that which Sir John Lubbock observed:—

The author is debtor to Herr Nottebohm, Inspector of Buildings at Karlsruhe, who related the following on May 24, 1876, under the title, 'Ants as Founders of Aphides' Colonies:'—'Of two equally strong young weeping ashes, which I planted in my garden at Kattowitz, in Upper Silesia, one succeeded well, and in about five or six years showed full foliage, while the other regularly every year was covered, when it began to bud, with millions of aphides, which destroyed the young leaves and sprouts, and thus completely delayed the development of the tree. As I perceived that the only reason for this was the action of the aphides, I determined to destroy them utterly. So in the March of the following year I took the trouble to clean and wash every bough, sprig, and bud before the bursting of the latter, with the greatest care, by means of a syringe. The result was that the tree developed perfectly healthy and vigorous leaves and young shoots, and remained quite free from the aphides until the end of May or the beginning of June. My joy was of short duration. One fine sunny morning I saw a surprising number of ants running quickly up and down the trunk of the tree; this aroused my attention, and led me to look more closely. To my great astonishment I then saw that many troops of ants were busied in carrying single aphides up the stem to the top, and that in this way many of the lower leaves had been planted with colonies of aphides. After some weeks the evil was as great as ever. The tree stood alone on the grass plot, and offered the only situation for an aphides' colony for the countless ants there present. I had destroyed this colony; but the ants replanted it by bringing new colonists from distant branches, and setting them on the young leaves.[1]

Again—

MacCook noticed, of the mound-making ants, that of the

[1] *Loc. cit.* p. 121.

workers returning to the nest from the tree on which the milking was going on, a far smaller number had distended abdomens than among those descending the tree itself. A closer investigation showed that at the roots of the trees, at the outlets of the subterranean galleries, a number of ants were assembled, which were fed by the returning ants after the fashion already described in feeding the larvæ, and which were distinguished by the observer as 'pensioners.' MacCook often observed the same fact later, among, with others, the already described Pennsylvanian wood-ant. Distinguished individuals in the body-guard of the queen were fed in like fashion. MacCook is inclined to think that the reason of this proceeding is to be found in the 'division of labour' so general in the ant republic, and that the members of the community which are employed in building and working within the nest, leave to the others the care of providing food for themselves as well as for the younger and helpless members; they thus have a claim to receive from time to time a reciprocal toll of gratitude, and take it, as is shown very clearly, in a way demanded by the welfare of the community.[1]

Aphides are not the only insects which ants employ as cows, several other insects which yield sweet secretions being similarly utilised in various parts of the world. Thus, gall insects and cocci are kept in just the same way as aphides; but MacCook observed that where aphides and cocci are kept by the same ants, they are kept in separate chambers, or stalls. The same observer saw caterpillars of the genus *Lycœna* kept by ants for the sake of a sweet secretion which they supply.

Habit of making Slaves.—This habit, or instinct, obtains among at least three species of ant, viz., *Formica rufescens*, *F. sanguinea*, and *strongylognathus*. It was originally observed by P. Huber in the first-named species. Here the species enslaved is *F. fusca*, which is appropriately coloured black. The slave-making ants attack a nest of *F. fusca* in a body; there is a great fight with much slaughter, and, if victorious, the slave-makers carry off the pupæ of the vanquished nest in order to hatch them out as slaves. Mr. Darwin gives an account of a battle which he himself observed.[2]

[1] *Loc. cit.* p. 123.
[2] *Origin of Species*, 6th ed. p. 218.

When the pupæ hatch out in the nest of their captors, the young slaves begin their life of work, and seem to regard their master's home as their own; for they never attempt to escape, and they fight no less keenly than their masters in defence of the nest. *F. sanguinea* content themselves with fewer slaves than do *F. rufescens*; and the work that devolves upon the slaves differs according to the species which has enslaved them. In the nests of *F. sanguinea* the comparatively few captives are kept as household slaves; they never either enter or leave the nest, and so are never seen unless the nest is opened. They are then very conspicuous from the contrast which their black colour and small size present to the red colour and much larger size of *F. rufescens*. As the slaves are by this species kept strictly indoors, all the outdoor work of foraging, slave-capturing, &c., is performed by the masters; and when for any reason a nest has to migrate, the masters carry their slaves in their jaws. *F. rufescens*, on the other hand, assigns a much larger share of labour to the slaves, which, as we have already seen, are present in much larger numbers to take it. In this species the males and fertile females do no work of any kind; and the workers, or sterile females, though most energetic in capturing slaves, do no other kind of work. Therefore the whole community is absolutely dependent upon its slaves. The masters are not able to make their own nests or to feed their own larvæ. When they migrate, it is the slaves that determine the migration, and, reversing the order of things that obtains in *F. sanguinea*, carry their masters in their jaws. Huber shut up thirty masters without a slave and with abundance of their favourite food, and also with their own larvæ and pupæ as a stimulus to work; but they could not feed even themselves, and many died of hunger. He then introduced a single slave, and she at once set to work, fed the surviving masters, attended to the larvæ, and made some cells.

In order to confirm this observation, Lespès placed a piece of sugar near a nest of slave-makers. It was soon found by one of the slaves, which gorged itself and returned to the nest. Other slaves then came out and did likewise. Then some of the masters came out, and, by pulling the legs of the feeding slaves, reminded them that they were neglecting their duty. The slaves then immediately began to serve their masters with the sugar. Forel also has confirmed all these observations of Huber. Indeed, in the case of *F. rufescens*, the structure of the animal is such as to render self-feeding physically impossible. Its long and narrow jaws, adapted to pierce the head of an enemy, do not admit of being used for feeding, unless liquid food is poured into them by the mouth of a slave. This fact shows of how ancient an origin the instinct of slave-making must be; it has altered in an important manner a structure which could not have been so altered prior to the establishment of the instinct in question.

Mr. Darwin thus sums up the differences in the offices of the slaves in the nests of *F. sanguinea* and *F. rufescens* respectively:—

The latter does not build its own nest, does not determine its own migrations, does not collect food for itself or for its fellows, and cannot even feed itself; it is absolutely dependent on its numerous slaves. *Formica sanguinea*, on the other hand, possesses much fewer slaves, and in the early part of the summer extremely few; the masters determine when and where a new nest shall be formed, and when they migrate, the masters carry the slaves. Both in Switzerland and England the slaves seem to have the exclusive care of the larvæ, and the masters alone go on slave-making expeditions. In Switzerland the slaves and masters work together, making and bringing materials for the nest; both, but chiefly the slaves, tend and milk, as it may be called, their aphides; and thus both collect food for the community. In England the masters alone usually leave the nest to collect building materials and food for themselves, their slaves and larvæ. So that the masters in this country receive much less service from their slaves than they do in Switzerland.

Mr. Darwin further observes that 'this difference in the usual habits of the masters and slaves in the two countries probably depends merely on the slaves being captured in greater numbers in Switzerland than in England;' and records that he has observed in a community of the English species having an unusually large stock of slaves that 'a few slaves mingled with their masters leaving the nest, and marched along the same road to a tall Scotch fir tree, twenty-five yards distant, which they ascended together, probably in search of aphides or cocci.' And, according to Huber, the principal office of the slaves in Switzerland is to search for aphides.

Mr. Darwin also made the following observation:— 'Desiring to ascertain whether *F. sanguinea* could distinguish the pupæ of *F. fusca*, which they habitually make into slaves, and which are an unwarlike species, from *F. flava*, which they rarely capture, and never without a severe fight,' he found 'it was evident that they did at once distinguish them;' for while 'they eagerly and instantly seized the pupæ of *F. fusca*, they were much terrified when they came across the pupæ, or even the earth from the nest, of *F. flava*, and quickly ran away; but in about a quarter of an hour, shortly after the little yellow ants had crawled away (from their nest having been disturbed by Mr. Darwin), they took heart and carried off the pupæ.'

Concerning the origin of this remarkable instinct, Mr. Darwin writes:—

As ants which are not slave-makers will, as I have seen, carry off pupæ of other species if scattered near their nests, it is possible that such pupæ originally stored as food might become developed, and the foreign ants thus unintentionally reared would then follow their proper instincts, and do what work they could. If their presence proved useful to the species which had seized them—if it were more advantageous to the species to capture workers than to procreate them—the habit of collecting pupæ, originally for food, might by natural selection be strengthened and rendered permanent for the very different purpose of raising slaves. When the instinct was once acquired, if carried out to a much less extent even than in our British *F. sanguinea*, which, as we have seen, is less aided by its slaves than the same species in Switzerland, natural selection might increase and modify the instinct, always supposing such modification to be of use to the species, until an ant was found as abjectly dependent on its slave as is the *Formica rufescens*.

Ants do not appear to be the only animals of which ants make slaves; for there seems to be at least one case in which these wonderful insects enslave insects of another species, which therefore may be said to stand to the ants in the relation of beasts of burden. The case to which I allude is one that is recorded in Perty's 'Intellectual Life of Animals' (2nd ed. p. 329), and is as follows:—

According to Audubon certain leaf-bugs are used as slaves by the ants in the Brazilian forests. When these ants want to bring home the leaves which they have bitten off the trees, they do it by means of a column of these bugs, which go in pairs, kept in order on either side by accompanying ants. They compel stragglers to re-enter the ranks, and laggards to keep up by biting them. After the work is done the bugs are shut up within the colony and scantily fed.

Wars.—On the wars of ants a great deal might be said, as the facts of interest in this connection are very numerous; but for the sake of brevity I shall confine myself to giving only a somewhat meagre account.

One great cause of war is the plundering of ants' nests by the slave-making species. Observers all agree that this plundering is effected by a united march of the whole army composing a nest of the slave-making species, directed against some particular nest of the species which they enslave. According to Lespès and Forel, single scouts or small companies are first sent out from the nest to explore in various directions for a suitable nest to attack. These scouts afterwards serve as guides to the marauding excursion. Forel saw several of these scouts of the species *F. rufescens* or Amazon carefully inspecting a nest of *F. fusca* which they had found, investigating especially the entrances. These are purposely made difficult to find by their architects, and it not unfrequently happens that after all precautions and inspections on the part of the invaders, an expedition fails on account of not finding the city gates.

When the scouts have been successful in discovering a suitable nest to plunder, and have completed their strategical investigations of the locality to their satisfaction, they return straight to their own nest or fortress. Forel has then seen them walking about on the surface of their nest for a long time, as if in consultation, or making up their minds. Then some of them entered the nest, soon after which hosts of warriors streamed out of the entrances, and ran about tapping each other with their heads and antennæ. They then formed into column and set out to pillage the nest of the slave ants. The following is the account which Lespès gives of such expeditions:—

They only take place towards the end of the summer and in autumn. At this time the winged members of the slave species (*F. fusca* and *F. cunicularia*) have left the nest, and the Amazons will not take the trouble to bring back useless consumers. When the sky is clear our robbers leave their town in the afternoon at about three or four o'clock. At first no order is perceptible in their movements, but when they are all gathered together they form a regular column, which then moves forward quickly, and each day in a different direction. They march closely pressed together, and the foremost always appear to be seeking for something on the ground. They are each moment overtaken by others, so that the head of the column is continually growing. They are in fact seeking the traces of the ants which they propose to plunder, and it is scent that guides them. They snuff over the ground like hounds following the track of a wild animal, and when they have found it they plunge headlong forward, and the whole column rushes on behind. The smallest armies I saw consisted of several hundred individuals, but I have also seen some four times as large. They then form columns which may be five mètres long, and as much as fifty centimètres wide. After a march, which often lasts a full hour, the column arrives at the nest of the slave species. The *F. cuniculariæ*, which are the strongest, offer keen opposition, but without much result. The Amazons soon penetrate within the nest, to come out again a moment later, while the assailed ants at the same time rush out in masses. During the whole time attention is directed solely to the larvæ and pupæ, which the Amazons steal while the others try to save as many as possible. They know very well that the Amazons cannot climb, so they fly with their precious burdens to the surrounding bushes or plants, whereto their enemies cannot follow them. They then pursue the retreating robbers and try to take away from them as much of their booty as possible. But the latter do not trouble themselves much about them, and hasten on home. On their return they do not follow the shortest road, but exactly the one by which they came, finding their way back by smell. Arrived at their nest, they immediately hand over their booty to the slaves, and trouble themselves no more about it. A few days afterwards the stolen pupæ or nymphæ emerge, without memory of their childhood, and immediately and without compulsion take part in all tasks.

According to Büchner's account,[1]—

From time to time the army makes a short halt, partly to let the rearguard close up, partly because different opinions arise as to the direction of the host, or because the place at which they are is unknown to them. Forel several times saw the army completely lose its way·—an incident only once observed by Huber. Forel puts the number of warriors in such an army at from one hundred to more than two thousand. Its speed is on an average a mètre per minute, but varies much according to circumstances, and is naturally least when returning laden with booty. If the distance be very great, such bodily fatigue may at last be felt that the whole attack on the hostile nest is given up, and a retreat is begun; Forel once saw this happen after they had passed over a distance of two hundred and forty yards. Sometimes it seems as though, on coming within sight of the hostile nest, a kind of discouragement took possession of them, and prevented their making the attack. If the nest cannot at once be found, the whole army halts, and some divisions are sent forward to search for it, and these are gradually seen returning towards the centre. Forel also saw such an army only searching the first day, advancing zigzag, and with frequent halts, whereas on the following day it went forward to its aim swiftly and without delay, having found out the road. It seems that a single ant, even if it knows the way and the place, is not able alone to lead a large army, but that a considerable number must be employed in this duty. Mistakes as to the road occur with special ease during the return journey, because the several ants are laden with booty and cannot readily understand each other. Individual ants are then seen to wander about in every direction often for a long time, until they at last reach a spot known to them, and then advance swiftly to their goal. Many never come back at all. These mistakes easily occur when the robbers which have passed into a hostile nest do not come out again at the same holes whereby they entered, but by others at some distance—for instance, by a subterranean canal. Coming out thus in a strange neighbourhood, they do not know which way to take, and only some chance to find the right road during their aimless wanderings about, and recognise and

---

[1] *Geistesleben der Thiere*, pp. 145-9.

follow it by smell. On the other hand, such mistakes scarcely ever happen to individuals in an unladen train, kept in good array. Other species of ants (*F. fusca, rufa, sanguinea*) know better how to manage under such circumstances than do the Amazons. The laden ones lay down their loads, first find where they are, and only take them up again after they have found their way. If the booty seized in the nest first attacked is too large to be all taken at once, the robbers return once, or oftener, so as to complete their work. . . . . The ants, as already said, have no regular leaders nor chiefs, yet it is certain that in each expedition, alteration of road, or other change, the decision during that event comes from a small knot of individuals, which have previously come to an understanding, and carry the rest and the undecided along with them. These do not always follow immediately, but only after they have received several taps on the head from the members of the 'ring.' The procession does not advance until the leaders have convinced themselves by their own eyesight that the main part of the army is following.

One day Forel saw some Amazons on the surface of a nest of the *F. fusca* seeking and sounding in all directions, without being able to find the entrance. At last one of them found a very little hole, hardly as large as a pin's head, through which the robbers penetrated. But since, owing to the smallness of the hole, the invasion went on slowly, the search was continued, and an entrance was found further off, through which the Amazon army gradually disappeared. All was quiet. About five minutes later Forel saw a booty-laden column emerge from each hole. Not a single ant was without a load. The two columns united outside and retreated together.

A marauding excursion of the Amazons against the *F. rufibarbis*, a sub-species of the *F. fusca*, or small black ants, took place as follows:—The vanguard of the robber army found that it had reached the neighbourhood of the hostile nest more quickly than it had expected; for it halted suddenly and decidedly, and sent a number of messengers which brought up the main body and the rearguard with incredible speed. In less than thirty seconds the whole army had closed up, and hurled itself in a mass on the dome of the hostile nest. This was the more necessary as the *rufibarbes* during the short halt had discovered the approach of the enemy, and had utilised the time to cover the dome with defenders. An indescribable struggle followed, but the superior numbers of the Amazons overcame, and they penetrated into the nest, while the defenders poured by thousands out of the same holes, with their larvæ and pupæ in their jaws, and escaped to the nearest plants and bushes, running over the heaps of their assailants. These looked on the matter as hopeless, and began to retreat. But the *rufibarbes*, furious at their proceedings, pursued them, and endeavoured to get away from them the few pupæ they had obtained, by trying to seize the Amazons' legs and to snatch away the pupæ. The Amazon lets its jaws slip slowly along the captive pupa, as far as the head of its opponent, and pierces it, if it does not, as generally happens, draw back. But it often manages to seize the pupa at the instant at which the Amazon lets it go and flies with it. This is managed yet more easily when a comrade holds the robber by the legs, and compels it to loose its prey in order to guard itself against its assailant. Sometimes the robbers seize empty cocoons and carry them away, but they leave them on the road when they have discovered their mistake. In the above case the strength of the *rufibarbes* proved at last so great that the rearguard of the retreating army was seriously pressed, and was obliged to give up its booty. A number of the Amazons also were overpowered and killed, but not without the *rufibarbes* also losing many people. None the less did some individuals, as though desperate, rush into the thickest hosts of the enemy, penetrated again into the nest, and carried off several pupæ by sheer audacity and skill. Most of them left their prey to go to the help of their comrades when assailed by the *rufibarbes*. Ten minutes after the commencement of the retreat all the Amazons had left the nest, and, being swifter than their opponents, they were only pursued for about halfway back. Their attack had failed on account of a short delay!

On another occasion observed by Forel, in which several fertile Amazons also took part and killed many enemies, the nest was thoroughly ravished, but the retreat was also in this case very much disturbed and harassed by the superior numbers of the enemy. There were many slain on both sides. That in spite of the above-mentioned unanimity different opinions among the members of an expedition sometimes hinder its conduct, the following observation seems to show :—An advancing column divided after it had gone about ten yards from the nest. Half turned back, while the other half went on, but after some time hesitated and also turned back. Arrived at home, it found those which had formerly turned back putting themselves in motion in a new direction. The newly returned followed them, and the reunited army, after various wheelings, halts, &c., at last turned home again by a long way round. The whole business looked like a promenade. But apparently different parties had different nests in view, while others were entirely against the expedition. Yet perhaps it was only a march for exercise. Outer obstacles do not, as a rule, hinder the Amazons when they are once on the march. Forel saw them wade through some shallow water, although many were drowned in it, and then march over a dusty high road, although the wind blew half of them away. As they returned, booty-laden, neither wind, nor dust, nor water could make them lay down their prey. They only got back with great trouble, and turned back again to bring fresh booty, although many lost their lives.

The following is also quoted from Büchner's excellent epitome of Forel's observations in this connection:—

The most terrible enemy of the Amazons is the sanguine ant (*F. sanguinea*), which also keeps slaves, and thereby often comes into collision with the Amazons on their marauding excursions. It is not equal to it in bodily strength or fighting capacity, but surpasses it in intelligence; according to Forel it is the most intelligent of all the species of ants. If Forel, for instance, poured out the contents of a sack filled with a nest of the slave species near an Amazon nest, the Amazons apparently generally regarded the tumbled together heap of ants, larvæ, pupæ, earth, building materials, &c., as the dome of a hostile nest, and took all imaginable but useless pains to find out the entrances thereinto, leaving on one side for this investigation their only object, the carrying off the pupæ; but the sanguine ants under similar circumstances did not allow themselves to be deceived, but at once ransacked the whole heap.

On another occasion, while a procession of Amazon ants was on its way to plunder a nest of *F. fusca*, before it arrived Forel poured out a sack-full of sanguine ants, and made a break in the nest:—

The sanguine ants pressed in, while the *fusca* came out to defend themselves. At this moment the first Amazons arrived. When they saw the sanguine ants they drew back and awaited the main army, which appeared much disturbed at the news. But once united, the bold robbers rushed at their foes. The latter gathered together and beat back the first attack, but the Amazons closed up their ranks and made a second assault, which carried them on to the dome and into the midst of the enemy. These were overthrown, as well as a number of *F. pratensis*, which Forel at this moment poured out on the nest. The conquerors delayed for a moment on the dome after their victory, and then entered the nest to bring out a little of the valuable booty. A few Amazons which were mad with anger did not return with the main army, but went on slaughtering blindly among the conquered and the fugitives of the three species, *fusca*, *pratensis*, and *sanguinea*.

The ravished *rufibarbes* once became so desperate at their overthrow that they followed the robbers to their own nest, and the latter had some trouble in defending it. The *rufibarbes* let themselves be killed in hundreds, and really seemed as though they courted death. A small number of the Amazons also sank under the bites of their enemies. The nest contained slaves of the *rufibarbis* species, which on this emergency fought actively against their own race. There were also slaves of the species *fusca*, so that the nest included three different species of ants.

The same nest is often revisited many times on the same day or at different periods, until either there is no more to steal, or the plundered folk have hit upon better mode of defence. A column which was in the act of going back to such a plundered nest turned when halfway there, and halted, apparently on no other ground than because it had met the rearguard of the army, and had learned that the nest was exhausted, and that there was nothing more to be had there. The robbers then went off to a *rufibarbis* nest which was in the neighbourhood, and killed half the inhabitants while plundering the nest. The surviving *rufibarbes* returned after the robbery and brought up new progeny; but thirteen days later the Amazons again reaped a rich harvest from the same nest. The Amazon army often severs itself into two separate divisions when there is not enough for both to do at the same spot. Sometimes one division finds something and the other nothing, and they then reunite. If any obstacle be placed in their way they try to overcome it, in doing which some leave the main army, lose themselves, and only find their way home again with difficulty. Forel has tried to establish the normal frequency of expeditions, and found that a colony watched by himself for a space of thirty days sent out no less than forty-four marauding excursions. Of these about eight-and-twenty were completely, nine partially, and the remainder not at all successful. He four times saw the army divide into two. Half the expeditions were levelled against the *rufibarbes*, half against the *fuscæ*. On an average a successful expedition would bring back to the colony a thousand pupæ or larvæ. On the whole, the number of future slaves stolen by a strong colony during a favourable summer may be reckoned at forty thousand!

The internecine battles which occasionally break out among the Amazons themselves are naturally the most cruel. They tear each other to pieces with incredible fury, and knots of five or six individuals which have pierced each other may be seen rolling over each other on the ground, it being impossible to distinguish between friend and foe. Civil wars among men are also known to be the most embittered and the most bloody.

The mode of attack practised by the other best known species of slave-making ant, *sanguinea*, is somewhat different:—

They march in small troops which, in case of need, summon reinforcements, and therefore as a rule only reach their goal slowly. Between the individual troops messengers or scouts run continually backwards and forwards. The first troop which arrives at the hostile nest does not rush at it, as do the Amazons, but contents itself with making provisional reconnaissances, wherein some of the assailants are generally made prisoners by the enemy, which have time to bethink and to collect themselves. Reinforcements are now brought up, and a regular siege of the nest begins. A sudden invasion, like that of the Amazons, is never seen. The besieging army forms a complete ring round the hostile nest, and the besiegers hold this with mandibles open and antennæ drawn back, without going nearer. In this position they beat off all assaults of the besieged, until they feel themselves strong enough to advance to the attack. This attack scarcely ever fails, and has for its chief object the mastering of the entrances and outlets of the nest. A special troop guards each opening, and only allows such of the besieged to pass out as carry no pupæ. This manœuvre gives rise to a number of comical and characteristic scenes. By this means the sanguine ants in a few minutes manage to have all the defenders out of the nests and the pupæ left behind. This is the case at least with the *rufibarbes*, while the rather less timid *fuscæ* try, even at the last moment when it is useless, to stop up or barricade the entrances. The sanguine ants do not indeed possess the terrible weapons and the warlike impetuosity of the Amazons, but they are stronger and larger. If a *fusca* or a *rufibarbis* fights with a sanguine ant for the possession of a pupa, it is generally very soon overcome. While the main part of the army is penetrating into the nest to steal the pupæ, some divisions pursue the fugitives, to take away from them the few pupæ which may chance to have been saved. They drive them even out of the cricket-holes in which they have meanwhile taken refuge. In short, it is a *razzia*, or sweeping burglary, as complete as can be imagined. In the retreat the robbers in no wise hurry themselves, for they know that they are threatened by no danger and no loss, and the complete emptying of a large and distant nest often takes several days in accomplishing. The ants which have been so thoroughly robbed scarcely every return to their former abode.

It must be admitted that a human army, robbing a foreign town or fortress, could not behave better or more prudently.

Huber gives the following account of a battle waged by sanguine ants:—

At ten, in a July morning, he noticed a small band of them emerge from their nest, and march rapidly towards a nest of negroes, around which it dispersed. A number of the blacks rushed out, gave battle, and succeeded in defeating their invaders, and in making several of them prisoners. Upon this, the remainder of the attacking force waited for a reinforcement. When this came up, they still declined further proceedings, and sent more aides-de-camp to their own nest. The result of these messages was a much larger reinforcement; but even yet the pirates appeared to shun the combat. At last, the negroes marched out from their nest in a phalanx of about two feet square, and a number of skirmishes began, which soon ended in a general *mêlée*. Long before the event seemed certain, the negroes carried off their pupæ to the most distant part of the nest; and when, after a longer encounter, they appeared to think further resistance vain, they retreated, attempting to take with them their young. In this, however, they were prevented, and the invaders obtained possession of their nest and the booty. When they had done this, they put in a garrison, and occupied the night and the succeeding day in carrying off their spoil.

Büchner says—

Battles between ants of the same species often end with a lasting alliance, especially when the number of the workers on both sides is comparatively small. The wise little animals under such circumstances discover, much more quickly and better than men, that they can only destroy each other by fighting, while union would benefit both parties. Sometimes they drive each other out of their nests in a quite friendly way. Forel laid on a table a piece of bark with a nest of the gentle *Leptothorax acervorum*, and then put on it the contents of another nest of the same species. The last comers were by far the more numerous, and soon possessed themselves of the nest, driving out the inmates. But the latter did not know whither to go, and turned back again. They were then seized by their opponents one after the other, carried away as far as possible from the nest, and there put down. The oftener they came back the further were they carried away. One of the carriers arrived in this fashion at the edge of the table, and after it had by means of its feelers convinced itself that it had reached the end of the world, mercilessly let its burden drop into the fathomless abyss. It waited a moment to see if it had attained its object, and then turned back to the nest. Forel picked up the ant which had fallen on the floor, and put it down right in front of the returning ant. The latter repeated the same manœuvre as at first, only stretching its neck further over the edge of the table. He several times reiterated his experiment, and always with the same result. Later the two colonies were shut up together in a glass case, and gradually learned to agree.

At other times, however, warlike ants show great and needless cruelty to one another:—

They slowly pull from their victim, that is rendered defenceless by wounds, exhaustion, or terror, first one feeler and then the other, then the legs one after another, until they at last kill it, or pull it in a completely mutilated and helpless condition to some out-of-the-way spot where it perishes miserably. Yet some compassionate hearts are to be found among the victors, which only pull the conquered to a distant place in order to get rid of them, and there let them go without injuring them.

The following account is also taken from Büchner's 'Mind in Animals,' p. 87:—

The doors are often guarded by special sentries, which fulfil their important duty in various ways. Forel saw a nest of the *Colobopsis truncata*, the two or three very small round openings of which were watched by soldiers, arranged so that their thick cylindrical heads stopped them up, just as a cork stops up the mouth of a bottle. The same observer saw the *Myrmecina Latreillei* defend themselves against the invasions of the slave-making *Strongylognathus*, by placing a worker at each of the little openings of the nest, which quite stops up the opening either with its head or abdomen. The *Campo-notus* species also defend their nests by stretching their heads in front of the openings, drawing back the antennæ. Each approaching enemy thus receives a sharp blow or bite delivered with the whole weight of the body. MacCook noticed in the nests of the soon to be described Pennsylvanian mound-building ants, the employment of special sentries, which lay watching within the nest entrances, and sprang out at the first sight of danger to attack the enemy; and it was wonderful to see with what swiftness the news of such an alarm spread through the nest, and how the inhabitants came out *en masse* to meet the enemy. The *Lasius* species defend their large, strong, and very extensive nests against hostile attack or sieges with equal courage and skill, while other timid species seek to fly as speedily as possible with their larvæ, pupæ, and fruitful queens. There is, as Forel tells us, a regular barricade fight. Passage after passage is stopped and defended to the uttermost, so that the assailants can only advance step and step. Unless the latter are in an enormous majority, the struggle may last a very long time with these tactics. During this time, other workers are busy preparing subterranean passages backwards for eventual flight. Generally such passages are already made, and during a fight a new dome of the *Lasius* may be seen rising at a distance, it not being difficult for them to make this with the help of their extended subterranean passages and communications.

The *F. exsecta* or *pressilabris* fights in a peculiar way, which is due to care of their small and very tender bodies. It avoids all single combats, and always fights in closed ranks. Only when it thinks victory secure does it spring on its enemy's back. But its chief strength lies in the fact that many together always attack a foe. They nail down their opponent by seizing its legs and holding them firmly to the ground, while a comrade springs on the back of the defenceless creature and tries to bite through its neck. But if threatened the holders sometimes take flight, and so it happens that in battles between the *exsectæ* and the much stronger *pratenses* not a few of the latter are seen running about with a small enemy clutching their shoulders, and making violent efforts to tear the neck of its foe. If the bearer is then seized with cramp, the nervous cord has been injured. On the other hand, if an *exsecta* is seized by the back by a *pratensis* it is at once lost.

The tactics of the turf ants resemble those of the *exsectæ*, three or four of them seizing an opponent and pulling off his legs. In similar fashion the attack of the *Lasius* species is chiefly directed against the legs of its enemies, three, four, or five uniting in the effort. They understand barricade fighting particularly well in their large well-built dwellings, and if it comes to the worst fly by subterranean passages. They are feared by most ants on account of their numerical superiority. Forel one day poured the contents of ten nests of *pratenses* in front of a tree trunk inhabited by *Lasius fuliginosus* (jet ant). The siege at once began; but the jet ants called in help from the nests connected with their colony, and thick black columns were at once seen coming out from the surrounding trees. The *pratenses* were obliged to fly, and left behind them a mass of dead as well as their pupæ, which last were carried off by the victors to their nests to be eaten.

Battles, however, are not confined to species of ants having warlike and slave-making habits. The agricultural ants likewise at times wage fierce wars with one another. The importance of seeds to these ants, and the consequent value which they set upon them, induce the animals, when supplies are scarce, to plunder each other's nests. Thus Moggridge says, —

By far the most savage and prolonged contests which I have witnessed were those in which the combatants belong to two different colonies of the same species. . . . The most singular contests are those which are waged for seeds by *A. barbara*, when one colony plunders the stores of an adjacent nest belonging to the same species, the weaker nest making prolonged though, for the most part, inefficient attempts to recover their property.

In the case of the other species of ant which I have watched fighting, the strife would last but a short time—a few hours or a day—but *A. barbara* will carry on the battle day after day and week after week. I was able to devote a good deal of time to watching the progress of a predatory war of this kind, waged by one nest of *barbara* against another, and which lasted for forty-six days, from January 18 to March 4!

I cannot of course declare positively that no cessation of hostilities may have taken place during the time, but I can affirm that whenever I visited the spot—and I did so on twelve days, or as nearly as possible twice a week—the scene was one of war and spoliation such as that which I shall now describe.

An active train of ants, nearly resembling an ordinary harvesting train, led from the entrance of one nest to that of another lower down the slope, and fifteen feet distant; but on

closer examination it appeared that though the great mass of
seed-bearers were travelling towards the upper nest, some few
were going in the opposite direction and making for the lower.
Besides this, at intervals, combats might be seen taking place,
one ant seizing the free end of a seed carried by another, and
endeavouring to wrench it away, and then frequently, as neither
would let go, the stronger ant would drag seed and opponent
towards its nest. At times other ants would interfere and seize
one of the combatants and endeavour to drag it away, this often
resulting in terrible mutilations, and especially in the loss of the
abdomen, which would be torn off while the jaws of the victim
retained their indomitable bull-dog grip upon the seed. Then
the victor might be seen dragging away his prize, while its adversary, though now little more than a head and legs, offered a
vigorous though of course ineffectual resistance. I frequently
observed that the ants during these conflicts would endeavour
to seize one another's antennæ, and that if this were effected,
the ant thus assaulted would instantly release his hold, whether
of seed or adversary, and appear utterly discomfited. No doubt
the antennæ are their most sensitive parts, and injuries inflicted
on these organs cause the greatest pain.

It was not until I had watched this scene for some days that
I apprehended its true meaning, and discovered that the ants of
the upper nest were robbing the granaries of the lower, while
the latter tried to recover the stolen seeds both by fighting
for them and by stealing seeds in their turn from the nest
of their oppressors. The thieves, however, were evidently
the stronger, and streams of ants laden with seeds arrived
safely at the upper nest, while close observation showed that
very few seeds were successfully carried on the reverse journey
into the lower and plundered nest.

Thus when I fixed my attention on one of these robbed ants
surreptitiously making its exit with the seed from the thieves'
nest, and having overcome the opposition and dangers met with
on its way, reaching, after a journey which took six minutes to
accomplish, the entrance to its own home, I saw that it was
violently deprived of its burden by a guard of ants stationed there
apparently for the purpose, one of whom instantly started off
and carried the seed all the way back again to the upper nest.

This I saw repeated several times.

After March 4 I never saw any acts of hostility between
these nests, though the robbed nest was not abandoned. In
another case of the same kind, however, where the struggle
lasted thirty-one days, the robbed nest was at length completely

abandoned, and on opening it I found all the granaries empty with one single exception, and this one was pierced by the matted roots of grasses and other plants, and must therefore have been long neglected by the ants. Strangely enough, not one of the seeds in this deserted granary showed traces of germination.

No doubt some very pressing need is the cause of these systematic raids in search of accumulations of seeds, and there can be little doubt that the requirements of distinct colonies of ants of the same species are often different even at the same season and date. Thus these warring colonies of ants were active on many days when the majority of the nests were completely closed; and I have even seen these robbers staggering along, enfeebled by the cold, and in wind and rain, when all other ants were safe below ground.

The agricultural ants of Texas do not appear to be less pugnacious than their European congeners. Thus MacCook says:—

A young community has sometimes to struggle into permanent prosperity through many perils. The following example is found in the unpublished Lincecum manuscripts. One day a new ant-city was observed to be located within ten or twelve yards of a long-established nest, a distance that the doctor thought would prove too near for peaceable possession—for the agriculturals seem to pre-empt a certain range of territory around their formicary as their own, within which no intrusion is allowed. He therefore concluded to keep these nests under close observation, and visited them frequently. Only a day or two had elapsed before he found that the inhabitants of the old city had made war upon the new. They had surrounded it in great numbers, and were entering, dragging out and killing the citizens. The young colonists, who seemed to be of less size than their adversaries, fought bravely, and, notwithstanding they were overwhelmed by superior numbers, killed and maimed many of their assailants. The parties were scattered in struggling pairs over a space ten or fifteen feet around the city gate, and the ground was strewed with many dead bodies. The new colonists aimed altogether at cutting off the legs of their larger foes, which they accomplished with much success. The old-city warriors, on the contrary, gnawed and clipped off the heads and abdomens of their enemies. Two days afterward the battle-field was revisited, and many ants were found lying dead tightly locked together by legs and mandibles, while hundreds of decapitated bodies and severed heads were strewed over the ground.

Another example, which is given in the published paper, is quite similar, and had like result. In forty-eight hours the old settlers had exterminated the new. The distance between the nests was about 20 feet. While the young colonists remained in concealment they were not disturbed, but as soon as they began to clear away their open disk war was declared.

MacCook, however, says that 'these ants are not always so jealous of territorial encroachment, or at least must have different standards of rights.' For he observed many cases of nests situated within twenty, and even ten feet of one another, without a battle ever occurring between members of the two communities. Therefore, without questioning the accuracy of Lincecum's observations—which, indeed, present no scope for inaccuracy—he adds, 'That neighbouring ants, like neighbouring nations of civilised men, will fall out and wage war Lincecum's examples show. Perhaps we should be quite as unsuccessful in case of these ants as of our human congeners, should we seek a sufficient reason for these wars, or satisfactory cause for these differences in dealing with neighbours which appear from the comparison of Lincecum's observations with mine.'

In connection with the wars of these ants, the following quotations may also be made from the same author:—

The erratic ants do not appear to be held as common enemies by the agriculturals, and they are even permitted to establish their formicaries within the limits of the open disk. Sometimes, however, the diminutive hillocks which mark the entrance to an erratic ant-nest multiply beyond the limit of the agriculturals' forbearance. But they do not declare war, nor resort to any personal violence. Nevertheless, they get rid of them, oddly enough, by a regular system of vexatious obstructions. They suddenly conclude that there is urgent demand for improving their public domain. Forthwith they sally forth in large numbers, fall eagerly to work gathering the little black balls which are thrown up by the earth-worms in great quantities everywhere in the prairie soil, which they bring and heap upon the paved disk until all the erratic ant-nests are covered ! The entire pavement is thus raised an inch or so, and pains are taken to deposit more balls upon and around the domiciles of their tiny neighbours than elsewhere. The erratics struggle vigorously against this Pompeian treatment; they bore through the avalanche of balls, only to find barriers laid in their way. The obstructions at length become so serious that it is impossible to keep the galleries open. The dwarfs cease to contend against destiny, and, gathering together their household stores, quietly evacuate the premises of the inhospitable giants. It is the triumph of the policy of obstruction, a bloodless but effectual opposition.

Lastly, MacCook records the history of an interesting engagement which he witnessed between two nests of *Tetramorium cæspitum*. It took place between Broad Street and Penn Square in Philadelphia, and lasted for nearly three weeks. Although all the combatants belonged to the same species, however great the confusion of the fight, friends were always distinguished from foes—apparently by contact of antennæ.

Habit of keeping Domestic Pets.—Many species of ants display the curious habit of keeping in their nests sundry kinds of other insects, which, so far as observation extends, are of no benefit to the ants, and which therefore have been regarded by observers as mere domestic pets. These 'pets' are for the most part species which occur nowhere else except in ants' nests, and each species of 'pet' is peculiar to certain species of ants. Thus Moggridge found 'a large number of a minute shining brown beetle moving about among the seeds' in the nests of the harvesting ant of the south of Europe, 'belonging to the scarce and very restricted genus *Colnocera*, called by Kraatz *C. attæ*, on account of its inhabiting the nests of ants belonging to the genus *Atta*.' He also observed inhabiting the same nests a minute cricket 'scarcely larger than a grain of wheat' (*Gryllus myrmecophilus*), which had been previously observed by Paolo Savi in the nests of several species of ants in Tuscany, where it lived on the best terms with its hosts, playing round the nests in warm weather, and retiring into them in stormy weather, while allowing the ants to carry it from place to place during migrations. Again, Mr. Bates observes that 'some of the most anomalous forms of coleopterous insects are those which live solely in the nests of ants.' Sir John Lubbock also, and other observers whom we need not wait to cite, mention similar facts. The Rev. Mr. White says that altogether 40 distinct species of Coleoptera, most of which he has in his own collection, are known to inhabit the nests of various species of ants, and to occur nowhere else.

As in all these cases the ants live on amicable terms with their guests, and in some cases even bestow labour upon them (as in carrying them from one nest to another during migration), it is evident that these insects are not only tolerated, but fostered by the ants. Moreover, as it seems absurd to credit ants with any mere fancy or caprice such as that of keeping pets, we can only conclude that these insects, like the aphides, are of some use to their hosts, although we are not yet in a position to surmise what this use can be.

Habits of Sleep and Cleanliness.—It is probable that all species of ants enjoy periods of true sleep alternating with those of activity; but actual observations on this subject have only been made on two or three species. The following is MacCook's account of these habits in the harvesting ant of Texas:—

The observation upon the ants now before me began at 8 o'clock; at 11 P.M. the cluster had nearly dissolved, only a few being asleep. To illustrate the soundness of this sleep I take the quill pen with which I write, and apply the feather end of it to an ant who is sleeping upon the soil. She has chosen a little oval depression in the surface, and lies with abdomen upon the raised edge, and face toward the lamp. Her legs are drawn up close to the body. She is perfectly still. I gently draw the feather tip along the body, stroking 'with the fur,' if I may so say. There is no motion. Again and again this action is repeated, the stroke gradually being made heavier, although always quite gentle. Still there is no change. The strokes are now directed upon the head, with the same result. Now the tip is applied to the neck, the point at which the head is united to the pro-thorax, with a waving motion intended to produce a sensation of tickling. The ant remains motionless. After continuing these experiments for several minutes, I arouse the sleeper by a sharp touch of the quill. She stretches out her head, then her legs, which she also shakes, steps nearer to the light, and begins to cleanse herself in the manner already described. This act invariably follows the waking of ants from sleep. The above description applies to the general habit of somnolence as observed upon the two named species of harvesting ants for nearly four months. I have often applied the quill, and even the point of a lead pencil, to the sleeping Floridians without breaking their slumber. There are some other details which have not appeared in the behaviour of the individual just put under observation.

Thus, I have several times seen the ants (*Cru lelis*) *yawning* after awaking. I use this word for lack of one which more accurately expresses the behaviour. The action is very like that of the human animal; the mandibles are thrown open with the peculiar muscular strain which is familiar to all readers; the tongue also is sometimes thrust out, and the limbs stretched with the appearance, at least, of that tension which accompanies the yawn in the genus *homo*. During sleep the antennæ have a gentle, quivering, apparently involuntary motion, which seemed to me, at times, to have the regularity of breathing. I also often noted an occasional regular lifting up and setting down of the fore-feet, one leg after another, with almost a rhythmic motion.

The length of time during which sleep is prolonged appears to vary according to circumstances and, perhaps, organism. The large head-soldiers of the Floridian harvesters appear to have a more sluggish nature than the smaller workers. Their sleep is longer and heavier. The former fact the watch readily determined. The latter appeared from the greater stolidity of the creatures under disturbance. While the ants of one group are taking sleep others may be busy at work, and these stalk among and over the sleepers, jostling them quite vigorously at times. Again, new members occasionally join the group, and, in their desire to get close up to the heat and light, crowd their drowsy comrades aside. I have seen ants who had been at work in the galleries drop their pellets, push thus into the cluster, and presently be apparently sound asleep. This rough treatment is invariably received with perfect good humour, as are like jostlings when the ants are awake. I have never seen the slightest display of anger or attempt to resent disturbance even under these circumstances, so peculiarly calculated to excite the utmost irritation in men. But of course some of the sleepers are aroused. They change position a little, or give themselves a brief combing, and then resume their nap, unless, indeed, they are satisfied. In watching these movements it was quite evident that the Florida soldiers were far less easily disturbed than their smaller fellows. They slept on stolidly while all the others were in agitation around them. Moreover, their very appearance, particularly when awaking out of sleep, indicated the greater sluggishness of their temperament in this respect.

The ordinary duration of sleep MacCook takes to be about three hours.

Ants, like many other insects, are in the habit of cleaning themselves, being, like them, provided by nature with combs and brushes, &c., for the purpose. But, unlike other insects, several species of ants are also in the habit of assisting each other in the performance of their toilet. The author last quoted gives the following account of this process in the genus *Atta* : —

We take a couple; the cleanser has begun at the face, which is licked thoroughly, even the mandibles being cared for, they being held apart for convenient manipulation. From the face the cleanser passes to the thorax, thence to the haunch, and so along the first leg, along the second and third in the same manner, around to the abdomen, and thence up the other side of the ant to the head. A third ant approaches and joins in the friendly task, but soon abandons the field to the original cleanser. The attitude of the cleansed all this while is one of intense satisfaction, quite resembling that of a family dog when one is scratching the back of his neck. The insect stretches out her limbs, and, as her friend takes them successively into hand, yields them limp and supple to her manipulation; she rolls gently over upon her side, even quite over upon her back, and with all her limbs relaxed presents a perfect picture of muscular surrender and ease. The pleasure which the creatures take in being thus 'combed' and 'sponged' is really enjoyable to the observer. I have seen an ant kneel down before another and thrust forward the head, drooping, quite under the face, and lie there motionless, thus expressing, as plainly as sign-language could, her desire to be cleansed. I at once understood the gesture, and so did the supplicated ant, for she at once went to work. If analogies in nature-studies were not so apt to be misleading, one might venture to suggest that our insect friends are thus in possession of a modified sort of Emmetonian Turkish bath.

The acrobatic skill of these ants, which has often furnished me amusement, and which I shall yet further illustrate, was fully shown one morning in these offices of ablution. The formicary was taken from the study, where the air had become chilled, and placed in an adjoining chamber upon the hearth, before an open-grate fire. The genial warmth was soon diffused throughout the nest, and aroused its occupants to unusual activity. A tuft of grass in the centre of the box was presently covered with them. They climbed to the very top of the spires, turned round and round, hanging by their paws, not unlike gymnasts performing upon a turning-bar. They hung or clung in various positions, grasping the grass blade with the third and fourth pairs of legs, which were spread out at length, cleansing their heads with the fore-legs or bending underneath to comb and lick the abdomen. Among these ants were several pairs, in one case a triplet, engaged in the cleansing operation just described. The cleanser clung to the grass, having a fore-leg on one side and a hind leg on the other side of the stem, stretched out at full length, while the cleansed hung in a like position below, and reached over and up, submitting herself to the pleasant process. As the progress of the act required a change of posture on the part of both insects, it was made with the utmost agility.

Similarly, Bates thus describes the cleansing process in another genus of ants (*Ecitons*):—

Here and there an ant was seen stretching forth first one leg and then another, to be brushed and washed by one or more of its comrades, who performed the task by passing the limb between the jaws and tongue, finishing by giving the antennæ a friendly wipe.

Habits of Play and Leisure.—The life of ants is not all work, or, at least, is not so in all species; for in some species, at any rate, periods of recreation are habitually indulged in.

Büchner ('Geistesleben der Thiere,' p. 163) gives the following abstract of Huber's celebrated observations in this connection:—

It was of the *pratensis* that Huber wrote the observations touching its gymnastic sports which became so famous. He saw these ants on a fine day assembled on the surface of their nest, and behaving in a way that he could only explain as simulating festival sports or other games. They raised themselves on their hind legs, embraced each other with their forelegs, seized each other by the antennæ, feet, or mandibles, and wrestled—but all in friendliest fashion. They then let go, ran after each other, and played hide-and-seek. When one was victorious, it seized all the others in the ring, and tumbled them over like ninepins.

This account of Huber's found its way into many popular books, but in spite of its clearness won little credence from the reading public. 'I found it hard to believe Huber's observation,' writes Forel, 'in spite of its exactness, until I myself had seen the same.' A colony of the *pratensis* several times gave him the opportunity when he approached it carefully. The players caught each other by the feet or jaws, rolled over each other on the ground like boys playing, pulled each other inside the entrances of their nest, only to come out again, and so on. All this was done without bad temper, or any spurting of poison, and it was clear that all the rivalry was friendly. The least breath from the side of the observer was enough to put an end to the games. 'I understand,' continues Forel, 'that the affair must seem marvellous to those who have not seen it, especially when we remember that sexual attraction can here play no part.'

MacCook also gives an account of habits of play as indulged in among ants of the other Hemisphere:—

At one formicary half a dozen or more young queens were out at the same time. They would climb up a large pebble near the gate, face the wind, and assume a rampant posture. Several having ascended the stone at one time, there ensued a little playful passage-at-arms as to position. They nipped each other gently with the mandibles, and chased one another from favourite spots. They, however, never nipped the workers. These latter evidently kept a watch upon the sportive princesses, occasionally saluted them with their antennæ in the usual way, or touched them at the abdomen, but apparently allowed them full liberty of action.

As to leisure, Bates writes:—

The life of these Ecitons is not all work, for I frequently saw them very leisurely employed in a way that looked like recreation. When this happened the place was always a sunny nook in the forest. The main column of the army and the branch columns, at these times, were in their ordinary relative positions; but instead of pressing forward eagerly and plundering right and left, they seemed to have been all smitten with a sudden fit of laziness. Some were walking sternly about, others were brushing their antennæ with their fore-feet; but the drollest sight was their cleaning each other. [Here follows the above-quoted passage.] The actions of these ants looked like simple indulgence in idle amusement. . . . . It is probable that these hours of relaxation and cleaning may be indispensable to the effective performance of their harder burdens; but whilst looking at them, the conclusion that the ants were engaged merely in play was irresistible.[1]

Funereal Habits.—In another connection it has already been stated that Sir John Lubbock found his ants to be very careful in disposing of the dead bodies of their comrades. This habit seems to be pretty general among many species of ants, and is no doubt due to sanitary requirements, thus becoming developed as a beneficial instinct by natural selection. The funereal habits of the agricultural ant are thus related by MacCook:[2]—

There is nothing which is apt to awaken deeper interest in the life-history of ants than what may properly be called their funereal habits. All species whose manners I have closely observed are quite alike in their mode of caring for their own dead, and for the dry carcasses of aliens. The former they appear to treat with some degree of reverence, at least to the extent of giving them a sort of sepulture without feeding upon them. The latter, after having exhausted the juices of the body, they usually deposit together in some spot removed from the nest. I did not see any of the 'cemeteries' of the agricultural ant upon the field, nor, indeed, observe any of their behaviour towards the dead, but my artificial nests gave me some insight of this. In the first colony had been placed eight agriculturals of another nest, which were literally cut to pieces. Very soon after the ants were comfortably established in their new home, a number of them laid hold upon these *disjecta membra*, and began carrying them back and forth around the formicarium. The next day this continued, and several of their own number

---

[1] *Loc. cit.* [2] *Loc. cit.* p. 337.

who had died were being treated in like manner. Back and forth, up and down, into every corner of the box the bearers wandered, the very embodiment of restlessness. For four days this conduct continued without any intermission. No sooner would a body or fragment thereof be dropped by one bearer than another would take it up and begin the restless circuit. The difficulty, I easily understood, was that there was no point to be found far enough removed from the living-rooms of the insects in which to inter these dead. Their desire to have their dead buried out of their sight was strong enough to keep them on this ceaseless round, apparently under the continuous influence of the hope that something might turn up to give them a more satisfactory burial-ground. It does not appear greatly to the credit of their wisdom that they were so long discovering that they were limited to a space beyond their power to enlarge. When, however, this fact was finally recognised they gave their habit its utmost bent, and began to deposit the carcasses in the extreme corner of the flat, as distant as possible from the galleries on the terrace above. Here a little hollow was made in the earth, quite up against the glass, wherein a number of bodies were laid. Portions of bodies were thrust into the chinks formed in the dry sod. This flat became the permanent charnel-house of the colony, and here, in corners, crevices, and holes, for the most part out of sight, but not always so, the dead were deposited. But the living never seemed quite reconciled to their presence. Occasionally, restless resurrectionists would disentomb the dead, shift them to another spot, or start them once more upon their unquiet wanderings. Even after the establishment of this cemetery, the creatures did not seem able to lay away their newly deceased comrades—for there were occasional deaths in the formicary—without first indulging in this funereal promenade.

In the formicaries established in glass jars, both of *barbatus* 'and *crudelis*, the same behaviour appeared. So great was the desire to get the dead outside the nest, that the bearers would climb up the smooth surface of the glass to the very top of the jar, laboriously carrying with them a dead ant. This was severe work, which was rarely undertaken except under the influence of this funereal enthusiasm. The jar was very smooth and quite high. Falls were frequent, but patiently the little 'undertaker' would follow the impulse of her instinct, and try and try again. Finally, as in the large box, the fact of a necessity seemed to dawn upon the ants, and a portion of the surface opposite from the entrance to the galleries, and close up against the glass, was used as burial-ground and sort of kitchen-midden, where all the refuse of the nest was deposited. Mrs. Treat has informed me that her artificial nests of *crudelis* behaved in precisely the same way.

An interesting fact in the funereal habits of *Formica sanguinea* was related to me by this lady. A visit was paid to a large colony of these slave-makers, which is established on the grounds adjoining her residence at Vineland, New Jersey. I noticed that a number of carcasses of one of the slave species, *Formica fusca*, were deposited together quite near the gates of the nest. These were probably chiefly the dry bodies of ants brought in from recent raids. It was noticed that the dead ants were all of one species, and thereupon Mrs. Treat informed me that the red slave-makers never deposited their dead with those of their black servitors, but always laid them by themselves, not in groups, but separately, and were careful to take them a considerable distance from the nest. One can hardly resist pointing here another likeness between the customs of these social hymenopters and those of human beings, certain of whom carry their distinctions of race, condition, or religious caste, even to the gates of the cemetery in which the poor body moulders into its mother dust!

It will be observed that none of these accounts furnish evidence of ants burying their dead, as Pliny asserts to have been the case with ants in the south of Europe. In the Proceedings of the Linnæan Society, however (1861), there is a very definite account of such a practice as obtaining among the ants of Sydney; and although it is from the pen of an observer not well known, the observation seems to have been one about which there could scarcely have been a mistake. The observer was Mrs. Hutton, and this is her account. Having killed a number of 'soldier ants,' and returning half an hour afterwards to the place where the dead bodies were lying, she says:

I saw a large number of ants surrounding the dead ones. I determined to watch their proceedings closely. I followed four or five that started off from the rest towards a hillock a short distance off, in which was an ants' nest. This they entered, and in about five minutes they reappeared, followed by others. All fell into rank, walking regularly and slowly two by two, until they arrived at the spot where lay the dead bodies of the soldier ants. In a few minutes two of the ants advanced and took up the dead body of one of their comrades; then two others, and so on, until all were ready to march. First walked two ants bearing a body, then two without a burden; then two others with another dead ant, and so on, until the line was extended to about forty pairs, and the procession now moved slowly onwards, followed by an irregular body of about two hundred ants. Occasionally the two laden ants stopped, and laying down the dead ant, it was taken up by the two walking unburdened behind them, and thus, by occasionally relieving each other, they arrived at a sandy spot near the sea. The body of ants now commenced digging with their jaws a number of holes in the ground, into each of which a dead ant was laid, where they now laboured on until they had filled up the ants' graves. This did not quite finish the remarkable circumstances attending this funeral of the ants. Some six or seven of the ants had attempted to run off without performing their share of the task of digging; these were caught and brought back, when they were at once attacked by the body of ants and killed upon the spot. A single grave was quickly dug, and they were all dropped into it.

The Rev. W. Farren White also, in his papers on ants published in the 'Leisure Hour' (1880), after alluding to the above case, corroborates it by some interesting observations of his own. He says:—

Several of the little sextons I observed with dead in their mandibles, and one in the act of burying a corpse. . . . I should mention that the dead are not interred without considerable difficulty, in consequence of the sides of the trays being almost perpendicular. The work of the sextons continued until no dead bodies remained upon the surface of the nest, but all were interred in the extramural cemeteries. Afterwards I removed the trays, and turned the contents of the formicarium upside down, and then I placed six trays on the surface of the earth, two of which I filled with sugar for food. All six were used freely as cemeteries, being crowded with the corpses of the little people and their young, the larvæ which had perished in the disruption of their home.

I have noticed in one of my formicaria a subterranean cemetery, where I have seen some ants burying their dead by placing earth above them. One ant was evidently much affected, and tried to exhume the bodies, but the united exertions of the yellow sextons were more than sufficient to neutralise the effort of the disconsolate mourner. The cemetery was now converted into a large vault, the chamber where the dead were placed, together with the passage which led to it, being completely covered in.

Habits Peculiar to Certain Species.

Leaf-cutting Ants of the Amazon (Œcodoma cephalotes).—The mode of working practised by these ants is thus described by Mr. Bates:—

They mount a tree in multitudes. . . . Each one places itself on the surface of a leaf, and cuts with its sharp scissor-like jaws a nearly semicircular incision on the upper side; it then takes the edge between its jaws, and by a sharp jerk detaches the piece. Sometimes they let the leaf drop to the ground, where a little heap accumulates, until carried off by another relay of workers; but generally each marches off with the piece it has operated on, and as all take the same road to the colony, the path they follow becomes in a short time smooth and bare, looking like the impression of a cart-wheel through the herbage.

Each ant carries its semicircular piece of leaf upright over its head, so that the home-returning train is rendered very conspicuous. Nearer observation shows that this home-returning or ladened train of workers keeps to one side of the road, while the outgoing or empty-handed train keeps to the other side; so that on every road there is a double train of ants going in opposite directions. When the leaves arrive at the nest they are received by a smaller kind of workers, whose duty it is to cut up the pieces of leaf into still smaller fragments, whereby the leaves seem to be better fitted for the purpose to which, as we shall presently see, they are put. These smaller workers never take any part in the outdoor labours; but they occasionally leave the nest, apparently for the sole purpose of obtaining air and exercise, for when they leave the nest they merely run about doing nothing, and frequently, as if in mere sport, mount some of the semicircular pieces of leaf which the carrier ants are taking to the nest, and so get a ride home.

From his continued observation of these ants, Bates concludes—and his opinion has been corroborated by that both of Belt and Müller—that the object of all this labour is highly interesting and remarkable. The leaves when gathered do not themselves appear to be of any service to the ants as food; but when cut into small fragments and stored away in the nests, they become suited as a nidus for the growth of a minute kind of fungus on which the ants feed. We may therefore call these insects the 'gardening ants,' inasmuch as all their labour is given to the rearing of nutritious vegetables on artificially prepared soil. They are not particular as to the material which they collect and store up for soil, provided that it is a material on which the fungus will grow. Thus they are very partial to the inside white rind of oranges, and will carry off the flowers of certain shrubs while leaving the leaves untouched. But, to quote again from Bates,—

They are very particular about the ventilation of their underground chambers, and have numerous holes leading up to the surface from them. These they open out or close up, apparently to keep up a regular degree of temperature below. The great care they take that the pieces of leaves they carry into the nest should be neither too dry nor too damp, is also consistent with the idea that the object is the growth of a fungus that requires particular conditions of temperature and moisture to ensure its vigorous growth. If a sudden shower should come on, the ants do not carry the wet pieces into the burrows, but throw them down near the entrances. Should the weather clear up again, these pieces are picked up when nearly dried, and taken inside: should the rain, however, continue, they get sodden down into the ground, and are left there. On the contrary, in dry and hot weather, when the leaves would get dried up before they could be conveyed to the nest, the ants, when in exposed situations, do not go out at all during the hot hours, but bring in their leafy burdens in the cool of the day and during the night. As soon as the pieces of leaves are carried in they must be cut up by the small class of workers into little pieces. Some of the ants make mistakes, and carry in unsuitable leaves. Thus grass is always rejected by them, but I have seen some ants, perhaps young ones, carrying leaves of grass; but after a while these pieces are always brought out again and thrown away. I can imagine a young ant getting a severe ear-wigging from one of the major-domos for its stupidity.

When a nest is disturbed and the masses of ant-food spread

about, the ants are in great concern to carry every morsel of it under shelter again; and sometimes, when I had dug into a nest, I found the next day all the earth thrown out filled with little pits, that the ants had dug into it to get out the covered-up food. When they migrate from one part to another, they also carry with them all the ant-food from their old habitations.

In Büchner's 'Geistesleben der Thiere' there is published an interesting description of the habits of these ants, which was communicated to the author by Dr. Fr. Ellendorf of Wiedenbrück, who has lived many years in Central America. Dr. Ellendorf says that—

It would be quite impossible for them to creep even through short grass with loads on their heads for miles. They therefore bite off the grass close to the ground for a breadth of about five inches, and throw it on one side. Thus a road is constructed, which is finally made quite smooth and even by the continual passing to and fro of millions upon millions night and day. . . . If the road is looked down upon from a height with these millions thickly pressed together, and all moving along with their green bannerets over their heads, it looks as though a giant green snake were gliding slowly along the ground; and this picture is all the more striking in that all these bannerets are swaying backwards and forwards.[1]

This observer made the experiment of interrupting the advance of a column of these ants, with the interesting result which he describes:—

I wished to see how they would manage if I put an obstacle in their way. Thick high grass stood on either side of their narrow road, so that they could not pass through it with the load on their heads. I placed a dry branch, nearly a foot in diameter, obliquely across their path, and pressed it down so tightly on the ground that they could not creep underneath. The first comers crawled beneath the branch as far as they could, and then tried to climb over, but failed owing to the weight on their heads. Meanwhile the unloaded ants from the other side came on, and when these succeeded in climbing over the bough there was such a crush that the unladen ants had to clamber over the laden, and the result was a terrible muddle. I now walked along the train, and found that all the ants with their bannerets on their heads were standing still,

[1] *Loc. cit.* p. 97.

thickly pressed together, awaiting the word of command from the front. When I turned back to the obstacle, I saw with astonishment that the loads had been laid aside by more than a foot's length of the column, one imitating the other. And now work began on both sides of the branch, and in about half an hour a tunnel was made beneath it. Each ant then took up its burden again, and the march was resumed in the most perfect order.

A migration of these ants is thus described by the same observer:—

The road led towards a cocoa plantation, and here I soon discovered the building which I afterwards visited daily. As I again went thither one day I was met, at a considerable distance from the nest, by a closely pressed column coming thence, and all the ants laden with leaves, beetles, pupæ, butterflies, &c.; the nearer I came to the nest, the greater was the activity. It was soon plain to me that the ants were in the act of leaving their dwelling, and I walked along the train to discover the new abode. They had gone for some distance along the old road, and had then made a new one through the grass to a cooler place, lying rather higher. The grass on the new road was all bitten off close to the ground, and thousands were busy carrying the path on to the new building. At the new home itself was an unusual stir of life. There were all sorts of labourers—architects, builders, carpenters, sappers, helpers. A number were busy digging a hole in the ground, and they carried out little pellets of earth and laid them together on end to make a wall. Others drew along little twigs, straws, and grass-stalks, and put them near the place of building. I was anxious to know why they had quitted their old home, and when the departure was complete, I dug it up with a spade. At a depth of about a foot and a half I found several tunnels of a large marmot species, the terror of cocoa planters, because in making their passages they gnaw off the thickest roots of the cocoa plants. The interior of the ant-hill had apparently fallen in through these mines. Unfortunately I was unable to follow further the progress of the new building, for I was obliged to leave the next day for San Juan del Sur. When I returned at the end of a week the building was finished, and the whole colony was again busy with the leaves of the coffee plants.

Harvesting Ants (Atta).—The ants which, so far as at

present known, practise the peculiar and distinctive habits to be described under this division belong for the most part to one genus, *Atta*, which, however, comprises a number of species distributed in localised areas over all the four quarters of the globe. Hitherto nineteen species have been detected as having the habits in question. These consist of gathering nutritious seeds of grasses during summer, and storing them in granaries for winter consumption. We owe our present knowledge concerning these insects to Mr. Moggridge,[1] who studied them in the south of Europe, Dr. Lincecum,[2] and Mr. MacCook,[3] who studied them in Texas, and Colonel Sykes[4] and Dr. Jerdon,[5] who made some observations upon them in India. They also occur scattered over a great part of Europe and in Palestine, where they were clearly known to Solomon and other classical writers of antiquity,[6] whose claim to accurate observation, although long disputed (owing to the authority of Huber), has now been amply vindicated.

Mr. Moggridge, who was a careful and industrious observer, found the following points of interest in the habits of the European harvesters. From the nest in various directions there proceed outgoing trains, which may be from twenty to thirty or more yards in length, and each consists of a double row of ants, moving, like the leaf-cutting ants, in opposite directions. Those in the outgoing row are empty-handed, while those in the incoming row are laden. But here the burdens are grass seeds. The roads terminate in the foraging ground, or ant-fields, and the insects composing the columns there become dispersed by hundreds among the seed-yielding grasses. The following is their method of collecting seeds; I quote from Moggridge:—

---

[1] *Harvesting Ants and Trap-door Spiders*, London, 1873 and Supplement, 1874.
[2] *Journal Linn Soc.*, vol. vi. p. 29, 1862.
[3] *Agricultural Ant of Texas*, Philadelphia, 1880.
[4] *Trans. Ent. Soc. Lond.*, i. 103, 1836.
[5] *Madras Journ. Lit. Sc.* 1851.
[6] For this see Moggridge, *loc. cit.* pp. 6-10, where, besides Prov. iv. 6-8, and xxx. 25, quotations are given from Horace, Virgil, Plautus, and others.

It is not a little surprising to see that the ants bring in not only seeds of large size and fallen grain, but also green capsules, the torn stalks of which show that they have been freshly gathered from the plant. The manner in which they accomplish this feat is as follows. An ant ascends the stem of a fruiting plant of shepherd's-purse (*Capsella bursa-pastoris*), let us say, and selects a well-filled but green pod about midway up the stem, those below being ready to shed their seeds at a touch. Then, seizing it in its jaws, and fixing its hind legs firmly as a pivot, it contrives to turn round and round, and so strain the fibres of the fruit-stalk that at length they snap. It then descends to the stem, patiently backing and turning upwards again as often as the clumsy and disproportionate burden becomes wedged between the thickly set stalks, and joins the line of its companions on their way to the nest. In this manner capsules of chickweed (*Stellaria media*) and entire calyces, containing the nutlets of calamint, are gathered; two ants also sometimes combine their efforts, when one stations itself near the base of the peduncle and gnaws it at the point of greatest tension, while the other hauls upon and twists it. I have never seen a capsule severed from its stalk by cutting alone, and the mandibles of this ant are perhaps incompetent to perform such a task. I have occasionally seen ants engaged in cutting the capsules of certain plants, drop them, and allow their companions below to carry them away; and this corresponds with the curious account given by Ælian of the manner in which the spikelets of corn are severed and thrown down 'to the people below,' τῷ δήμῳ τῷ κάτω.

The recognition of the principle of the division of labour which the latter observation supplies, is further proved by the following quotation from the same author. A dead grasshopper which was being carried into their nest was—

Too large to pass through the door, so they tried to dismember it. Failing in this, several ants drew the wings and legs as far back as possible, while others gnawed through the muscles where the strain was greatest. They succeeded at last in thus pulling it in.

The same thing is strikingly shown by the following quotation from Lespès:—

If the road from the place where they are gathering their

harvest to the nest is very long, they make regular depôts for their provisions under large leaves, stones, or other suitable places, and let certain workers have the duty of carrying them from depôt to depôt.

Büchner (*loc. cit.* p. 101) also makes the following references to the statements of previous observers:—

The subterranean workers of this remarkable genus are very clever. The Rev. H. Clark reports from Rio de Janeiro, that the *Sa-ubas* have made a regular tunnel under the bed of the river Parahyba, which is there as broad as the Thames at London, in order to reach a storehouse which is on the opposite bank. Bates tells us that close to the Magoary rice-mills, near Para, the ants bored through the dam of a large reservoir, and the water escaped before the mischief could be remedied. In the Para Botanical Gardens an enterprising French gardener did everything he could to drive the *Sa-ubas* away. He lit fires at the chief entrances of their nests, and blew sulphur vapour into their galleries by means of bellows. But how astonished was Bates when he saw the vapour come out at no less a distance than seventy yards! Such an extension have the subterranean passages of the *Sa-ubas*.

The recognition of the principle of the division of labour, which is shown by the above observations, is further corroborated by the following quotation from Belt:—

Between the old burrows and the new one was a steep slope. Instead of descending this with their burdens, they cast them down on the top of the slope, whence they rolled down to the bottom, where another relay of labourers picked them up and carried them to the new burrow. It was amusing to watch the ants hurrying out with bundles of food, dropping them over the slope, and rushing back immediately for more.

The same thing has been observed, as already stated, of the leaf-cutting ants—those engaged in cutting frequently throwing down the fragments of leaf which they cut to the carriers below. The prevalence of this habit among various species of ants therefore renders credible the following statements of Vincent Gredler of Botzen which are thus recorded in 'der Zool. Gart.,' xv. p. 434:—

In Herr Gredler's monastery one of the monks had been accustomed for some months to put food regularly on his window-sill for ants coming up from the garden. In consequence of Herr Gredler's communications he took it into his head to put the bait for the ants, pounded sugar, into an old inkstand, and hung this up by a string to the cross-piece of his window, and left it hanging freely. A few ants were in with the bait. These soon found their road out over the string with their grains of sugar, and so their way back to their friends. Before long a procession was arranged on the new road from the window-sill along the string to the spot where the sugar was, and so things went on for two days, nothing fresh occurring. But one day the procession stopped at the old feeding-place on the window-sill, and took the food thence, without going up to the pendent sugar-jar. Closer observation revealed that about a dozen of the rogues were in the jar above, and were busily and unwearyingly carrying the grains of sugar to the edge of the pot, and throwing them over to their comrades down below.

Many other instances of the division of labour might be given besides these, and those to be mentioned hereafter in other connections throughout the course of the present chapter; but enough has been said to show that the principle is unquestionably acted upon by sundry species of ants.

That ants are liable to make mistakes, and, when they do, that they profit by experience, is shown by the following experiment made by Moggridge; and many other instances might be given were it desirable:—

It sometimes happens that an ant has manifestly made a bad selection, and is told on its return that what it has brought home with much pains is no better than rubbish, and is hustled out of the nest, and forced to throw its burden away. In order to try whether these creatures were not fallible like other mortals, I one day took out with me a little packet of grey and white porcelain beads, and scattered these in the path of a harvesting train. They had scarcely lain a minute on the earth before one of the largest workers seized upon a bead, and with some difficulty clipped it with its mandibles and trotted back at a great pace to the nest. I waited for a little while, my attention being divided between the other ants who were vainly endeavouring to remove the beads, and the entrance down which the worker had disappeared, and then left the spot. On my return in an hour's time, I found the ants passing unconcernedly by and over the beads which lay where I had strewed them i..

apparently undiminished quantities; and I conclude from this that they had found out their mistake, and had wisely returned to their accustomed occupations.

When the grain is thus taken into the nest, it is stored in regular granaries, but not until it has been denuded of its 'husks' or 'chaff.' The denuding process is carried on below ground, and the chaff is brought up to the surface, where it is laid in heaps to be blown away by the wind.

It is a remarkable thing, and one not yet understood, why the seed, when thus stored in subterranean chambers just far enough below the surface to favour germination, does not germinate. Moggridge says that out of twenty-one nests and among many thousands of seeds that he examined, he only found twenty-seven cases of incipient germination. Moreover, all these cases occurred in months from November to February, while in the nests opened in October, March, April, and May, no sprouted seeds were discovered, though these are the months highly favourable to germination. He is at a loss to suggest the treatment to which the ants expose the seeds in order to prevent their sprouting. 'Apparently it is not that moisture or warmth or the influence of atmospheric air is denied to the seeds, for we find them in damp soil in genial weather, and often at but a trifling distance below the surface of the ground;' and he has proved that the vitality of the seeds is not impaired, for he succeeded in raising crops of young plants from seeds removed from the granaries.

He also says,—

By a fortunate chance I have been able to prove that the seeds will germinate in an undisturbed granary when the ants are prevented from obtaining access to it: and this goes to show not only that the structure and nature of the granary chamber is not sufficient of itself to prevent germination, but also that the presence of the ants is essential to secure the dormant condition of the seeds.

I discovered in two places portions of distinct nests of *Atta structor* which had been isolated owing to the destruction of the hollow wall behind which they lay, and then the granaries well filled up and literally choked with growing seeds, though the earth in which they lay completely enclosed and concealed them until by chance I laid them bare. In one case I knew that the destruction of the wall had only taken place ten days before, so that the seeds had sprouted in the interval.

My experiments also tend to confirm this, and to favour the belief that the non-germination of the seeds is due to some direct influence voluntarily exercised by the ants, and not merely to the conditions found in the nest, or to acid vapours which in certain cases are given off by the ants themselves.

These experiments consisted in confining a large number of harvesting ants with their queen and larvæ in a glass test-tube partly filled with damp soil and various seeds, the whole being closed with a cork in the mouth of the test-tube. Under these circumstances the seeds all sprouted, showing that mere confinement in an atmosphere of exhalations from the ants did not prevent germination. Another series of experiments, undertaken at the suggestion of Mr. Darwin, on the effects of an atmosphere of formic acid, showed that although this vapour was very injurious to the seeds, it did not prevent their incipient germination. Therefore it yet remains to be ascertained why the seeds do not germinate in the granaries of the ants.

But in whatever way the ants manage to prevent germination, it is certain that they are aware of the importance in this connection of keeping the seeds as dry as possible; for Moggridge repeatedly observed that when the seeds which had been stored proved over-moist, the ants again took them out and spread them in the sun to dry, to be again brought into the nest after a sufficient exposure.

Lastly, he also repeatedly observed the most surprising and interesting fact that when, as we have seen was occasionally the case, the seeds did begin to germinate in the nests, the ants knew the most effective method of preventing the germination from proceeding; for he found that in these cases the ants gnawed off the tips of the radicles. This fact deserves to be considered as one of the most remarkable among the many remarkable facts of ant-psychology.

Passing on now to the harvest'·g or agricultural ants

of Texas, attention was first called to the habits of this insect by Mr. Buckley in 1860,[1] and by Dr. Lincecum, who sent an account of his observations to Mr. Darwin, by whom they were communicated to the Linnæan Society in 1861. Five years later a paper was published in the Proceedings of the Academy of Natural Sciences of Philadelphia from the MS. of Dr. Lincecum. Lastly, in 1877 Mr. MacCook went to Texas expressly to study the habits of these insects, and he has recently embodied the results of his observations in a book of three hundred pages.[2] These observations are for the most part confirmatory of those of Lincecum, and for this as well as for reasons to be deduced from the work itself, they deserve to be accepted as trustworthy, notwithstanding that in some cases they are provokingly incomplete. The following is an epitome of these observations.

The ants clear away all the herbage above their nest in the form of a perfect circle, or 'disk,' 15 or 20 feet in diameter, by carefully felling every stalk of grass or weed that may be growing thereon. As the nests are placed in thickly grown localities, the effect of these bald or shaven disks is highly conspicuous and peculiar, exactly resembling in miniature the clearings which the settlers make in the American backwoods. The disk, however, is not merely cleared of herbage, but also carefully levelled, all inequalities of the surface being reduced by building pellets of soil into the hollows to an extent sufficient to make a uniformly flat surface. The action of rain and the constant motion of multitudes of ants cause this flat surface to become hard and smooth. In the centre of the disk is the gateway of the nest. This may be either a simple hole or a hollow cone.

From the disk in various directions there radiate ant-roads or avenues, which are cleared and smoothed like the disk itself, and which course through the thick surrounding grass, branching and narrowing as they go till they eventually taper away. These roads are usually three or four in number before they begin to branch, but may be

---

[1] Proc. Phil. Acad. Nat. Sci., xii. p. 445.
[2] Agricultural Ant of Texas (Lippincott & Co., Philadelphia, 1880).

as many as seven. They are usually two to three inches wide at their origin, but in large nests may be as much as five. MacCook found no road longer than sixty feet, but Lincecum describes one of three hundred feet. Along these hard and level roads there is always passing, during the daytime of the harvesting season, a constant stream of ants—those going from the nest being empty-handed, and those returning to it being laden with seeds. Of course the incoming ants, converging from all quarters upon the road, and therefore increasing in numbers as they approach the nest, require greater space for free locomotion; while the outgoing ants, diverging as they get further from home, also require greater proportional space the less their distance from the nest: hence the gradual swelling in the width of the roads as they approach the nests.

The manner of collecting the seeds in the jungle surrounding the roads is thus described by MacCook:—

At last a satisfactory seed is found. It is simply lifted from the ground, or, as often happens, has to be pulled out of the soil into which it has been tightly pressed by the rain or by passing feet. Now follows a movement which at first I thought to be a testing of the seed, and which, indeed, may be partially that; but finally I concluded that it was the adjusting of the burden for safe and convenient carriage. The ant pulls at the seed-husk with its mandibles, turning and pinching or 'feeling' it on all sides. If this does not satisfy, and commonly it does not, the body is raised by stiffening out the legs, the abdomen is curved underneath, and the apex applied to the seed. I suppose this to be simply a mechanical action for the better adjusting of the load. Now the worker starts homeward. It has not lost itself in the mazes of the grass forest. It turns directly towards the road with an unerring judgment. There are many obstacles to overcome. Pebbles, pellets of earth, bits of wood, obtruding rootlets, or bent-down spears of grass block up or hinder the way. These were scarcely noticed when the ant was empty-handed. But they are troublesome barriers now that she is burdened with a seed quite as thick, twice as wide, and half as long as herself. It is most interesting to see the skill, strength, and rapidity with which the little harvester swings her treasure over or around, or pushes it beneath these obstacles. Now the seed has caught against the herbage as the porter dodges under a too narrow opening. She backs out and tries another passage. Now the sharp points of the husk are entangled in the grass. She jerks or pulls the burden loose, and hurries on. The road is reached, and progress is comparatively easy. Holding the grain in her mandibles well above the surface, she breaks into what I may describe with sufficient accuracy as 'a trot,' and with little further interruption reaches the disk and disappears within the gate. There are variations from this behaviour, more or less marked, according to the nature of the grounds, the seeds, and (I suppose) the individuality of the harvesters; but the mode of ingathering the crop is substantially as above. Each ant operated independently. Once only did I see anything like an effort to extend sympathy and aid. A worker minor seeming to have difficulty in testing or adjusting a large seed of buffalo-grass, was assisted (apparently) by one worker major, and then by another, after which she went on her way.

But these ants do not confine their harvesting operations to gathering fallen seeds; they will, like the ants of Europe, also cut seeds from the stalk.

In order to test the disposition of *crudelis* to garner the seeds from the stem, bunches of millet were obtained from the North, and stalks eighteen inches high, crowned by the boll of close-set seeds, were stuck in the mound of an active formicary. The ants mounted the stems and set to work vigorously to secure the seeds, clusters of twenty or more being engaged at once upon one head. The seeds were carried off and stored within the nest. This experiment proved pretty conclusively that in the seeding season *crudelis* does not wait for the seeds to drop, but harvests them from the plant.

The 'granaries' into which the seeds are brought are kept distinct from the 'nurseries' for the pupæ. Their walls, floor, and roof are so hard and smooth, that MacCook thinks the insects must practise upon them 'some rude mason's craft.'

He traced these granaries to a depth of four feet below the surface of the ground, and believes, from the statements of a native peasant, that they, or at least the formicaries, extend to a depth of fifteen feet.

As regards the care that the ants take of the gathered grain, Lincecum describes the same habit as Moggridge and Sykes describe—viz., the sunning of wet seeds to dry. MacCook, however, neglected to make any experiments on this subject. Neither has he been able to throw any light upon the question as to why the stored seeds do not germinate, and is doubtful whether the habit of gnawing the radicle of sprouting seeds, which prevails in the European species, is likewise practised by the American. On two other points of importance MacCook's observations are also incomplete. One of these has reference to an alleged statement, which he is disposed to believe, that when some ants in a community have been killed by poison, the survivors avoid the poison: he, however, made no experiments to test this statement.

The other main point on which his observations are defective has reference to a remarkable statement made by Lincecum in the most emphatic terms. This statement is that upon the surface of their disk the ants sow the seeds of a certain plant, called ant-rice, for the purpose of subsequently reaping a harvest of the grain. There is no doubt that the ant-disks do very often support this peculiar kind of grass, and that the ants are particularly fond of its seed; but whether the plant is actually sown in these situations by the insects, or grows there on account of these situations being more open than the general surface of the ground—this question MacCook has failed to answer, or even to further. We are, therefore, still left with Dr. Lincecum's emphatic assurance that he has witnessed the fact. His account is that the seed of the ant-rice, which is a biennial plant, is sown in time for the autumnal rains to bring up. At the beginning of November a green row or ring of ant-rice, about four inches wide, is seen springing up round the circumference of the disk. In the vicinity of this circular ring the ants do not permit a single spire of any other grass or weed to remain a day, but leave the aristida, or ant-rice, untouched until it ripens, which occurs in June of the next year. After the maturing and harvesting of the seed, the dry stubble is cut away and removed from the pavement or disk, which is thus left unencumbered until the ensuing autumn, when the same species of grass again appears as before, and so on. Lincecum says he has seen the process go on year after year on the same ant-farms, and adds,—

> There can be no doubt of the fact that the particular species of grain-bearing grass mentioned above is intentionally planted. In farmer-like manner the ground upon which it stands is carefully divested of all other grasses and weeds during the time it is growing. When it is ripe the grain is taken care of, the dry stubble cut away and carried off, the paved area being left unencumbered until the ensuing autumn, when the same 'ant-rice' reappears within the same circle, and receives the same agricultural attention as was bestowed upon the previous crop—and so on year after year, as I *know* to be the case, in all situations when the ant's settlements are protected from graminivorous animals.

In a second letter Dr. Lincecum, in reply to an inquiry from Mr. Darwin whether he supposed that the ants plant seeds for the ensuing crop, says:—

> I have not the slightest doubt of it. And my conclusions have not been arrived at from hasty or careless observation, nor from seeing the ants do something that looked a little like it, and then guessing at the results. I have at all seasons watched the same ant-cities during the last twelve years, and I know that what I stated in my former letter is true. I visited the same cities yesterday, and found the crop of ant rice growing finely, and exhibiting also the signs of high cultivation, and not a blade of any other kind of grass or weed was to be seen within twelve inches of the circular row of ant-rice.—(*Journ. Linn. Soc.*, vol. vi. p. 30-1.)

Now, MacCook found the ant-rice growing as described, but only on some nests. Why it does not grow upon all the nests he does not understand. So far, then, as his observations go, they confirm those of Dr. Lincecum; but he does 'not believe that the ants deliberately sow a crop as Lincecum asserts;' he thinks 'that they have for some reason found it to their advantage to permit the aristida to grow upon their disks, while they clear off all other herbage;' but finally concludes 'that there is nothing unreasonable, nor beyond the probable capacity of the emmet intellect, in the supposition that the crop is actually sown. Simply, it is the Scotch verdict—" Not proven."'

The following facts with regard to 'modes of mining' are worth quoting from MacCook:—

In sinking the galleries the difficulty of carrying is not great in a moist or tough soil, which permits the ant to obtain goodly-sized pellets for portage. But when the soil is light and dry, so that it crumbles into dust as it is bitten off, the difficulty is greatly increased. It would be a very tedious task indeed to take out the diggings grain by grain. This difficulty the worker overcomes by balling the small particles against the surface of the gallery, the under side of the head, or within and against the mandibles. The fore-feet are used for this purpose, being pressed against the side face, turned under, and pushed upward with a motion similar to that of a man putting his hand upon his mouth. The abdomen is then swung underneath the body and the apex pressed against the little heap of grains of dirt massed against the under side of the mandibles, or between that and the smooth under surface of the head. Thus the dust is compressed into a ball which is of sufficient size to justify deportation.

The same operation is observed in the side-galleries, where the ants work very frequently upon their sides or backs, precisely as I have seen colliers do in Pennsylvania coal-mines.

The following is likewise worth quoting from the same author:—

Seeds are evidently not the only food of our agriculturals. When the ants at disk No. 2 had broken through the slight mud-sediment that sealed up their gate, as described above, they exhibited a peculiar behaviour. Instead of heading for the roads and pressing along them, they distributed themselves at once over the entire disk, radiating from the gate to all points in the circumference, from which they penetrated the jungle of grass beyond. In a moment a large number were returning across the roads, out of the grass, over the pavement toward the entrance. They bore in their mandibles objects which I presently found to be the males and females of white ants (*Termes flavipes*), which were filling the air, during and after the rain, in marriage flight. They had probably swarmed just before the shower. The agriculturals were under great excitement, and hurried forth and back at the top of their speed. The number of ants bearing termites was soon so great that the vestibule became choked, and a mass of struggling anthood was piled up around the gate. A stream of eager insects continually poured out of the door, pushing their way through the crowd that vainly but persistently endeavoured to get in with their burdens. The outcoming ants had the advantage, and succeeded in jostling through the quivering rosette of antennæ, legs, heads, and abdomens. Occasionally a worker gained an entrance by dint of sheer physical force and perseverance. Again and again would the crowd rush from all sides upon the gate, only to be pushed back by the issuing throng. In the meanwhile quite a heap of termites, a good handful at least, had been accumulated at one side of the gate, the ants having evidently dropped them, in despair of entrance, and hurried off to garner more.

In due time the pressure upon the vestibule diminished, the laden workers entered more freely, and in the end this heap was transferred to the interior. The rapidity with which the ants were distributed to all parts of their roads, after the first opening of the gates, was truly surprising. I was greatly puzzled, at the first, to know what the cause of such a rush might be. The whole behaviour was such as to carry the conviction that they knew accurately what effect the rain would have, had calculated upon it, and were acting in accordance with previous experience. I had no doubt at the time, and have none now, that the capturing of insects beaten down by the rain is one of the well-established customs of these ants. I saw a few other insects taken in, and one milliped, but chiefly the white ants.

That very afternoon I found in a formicary which I then opened several large colonies, or parts of one colony of termites, nested within the limits of the disk and quite at home. The next day numbers of the winged white ants were found stored within the granaries of a large formicary. There is no reason to doubt that these insects were intended for food, in accordance with the quite universal habit of the *Formicariæ*.

A curious habit has been noticed by most observers to occur in many species of ant, and it is one on which Mr. MacCook has a good deal to say. The habit in question consists in the ants transporting one another from place to place. The carrying ant seizes her comrade by the middle, and hurries along with it held aloft—the ant which is carried remaining quite motionless with all her legs drawn together. Huber supposed the process to be one enjoyable to both the insects concerned, and to be performed by mutual understanding and consent; but MacCook, in common with most other observers, supposes that it is merely a rough and primitive way of communicating to fellow-workers the locality where their services are required. He says:—

Keeping these facts in mind, we have a key to the solution of the press-gang operations which Lincecum observed among the agriculturals, and which have been fully described in other species. In the absence of any common head or directory, and of all executive officers, a change of location or any other concerted movement must be carried forward by the willing co-operation of individuals. At first sight, the act of seizing and carrying off workers does not appear like an appeal to free-will. It is indeed coercive, so far as the first act goes. But, in point of fact, the coercion ceases the moment the captive is set down within the precincts of the new movement. The carrier-ant has depended upon securing her consent and co-operation by thus bringing her within the circle of activity for which her service is sought. As a rule, no doubt, the deported ant at once yields to the influence around her, and drops into the current of fresh enterprise, in which she moves with as entire freedom and as independently as any other worker. But she is apparently under no restraint, and if she so please, may return to her former haunts.

Certain Ants of Africa.—Livingstone says of certain ants of Africa:—

They have established themselves on the plain where water stands so long annually as to allow the lotus and other aqueous plants to come to maturity. When all the ant-horizon is submerged a foot deep, they manage to exist by ascending to little houses built of black tenaceous loam on stalks of grass, and placed higher than the line of inundation. This must have been the result of experience, for, if they had waited till the water actually invaded their terrestrial habitations, they would not have been able to procure materials for their aërial quarters, unless they dived down to the bottom for every mouthful of clay.[1]

The Tree Ant of India and New South Wales.—These ants are remarkable from their habit of forming nests only in trees. According to Col. Sykes' account, the shape of the nest is more or less globular, and about ten inches in diameter. It is formed entirely of cow-dung, which the

[1] *Missionary Travels*, p. 328.

insects collect from the ground beneath, and work into the form of thin scales. These are then built together in an imbricated manner, like tiles or slates upon the roof of a house, the upper or outer scale, however, being one unbroken sheet, which covers the whole nest like a skull-cap. Below this the scales are placed one upon another in a wavy or scalloped manner, so that numerous little arched entrances are left, and yet, owing to the imbricated manner in which the scales are arranged, the interior of the nest is perfectly protected from rain. This interior consists of a number of irregular cells, the walls of which are formed by the same process as the exterior.

In New South Wales there is another species of ant which also frequents trees, but builds within the stem and branches. In the report of Captain Cook's expedition its habits are thus described:—' Their habitations are the insides of the branches of a tree, which they contrive to excavate, by working out the pith almost to the extremity of the slenderest twig ; the tree at the same time flourishing as if it had no such inmate.' On breaking one of the branches the ants swarm out in legions. Some of our native species also have the habit of excavating the interior of trees, though not on so extensive a scale.

Honey-making Ant (Myrmecocystus mexicanus).— This ant is found in Texas and New Mexico. Capt. W. B. Fleeson has observed its habits, and his observations have been communicated to the Californian Academy of Sciences, and also, by Mr. Henry Edwards, to Mr. Darwin. The following are the chief points of interest in Capt. Fleeson's results:—

The community appears to consist of three distinct kinds of ants, probably of two separate genera, whose offices in the general order of the nest would seem to be entirely apart from each other, and who perform the labour allotted to them without the least encroachment upon the duties of their fellows. These three kinds are—

  I. Yellow workers ; nurses and feeders of II.

  II. Yellow honey-makers ; sole function to secrete a kind of honey in their large globose abdomens, on which the other ants are supposed to feed. They never quit the nest, and are fed and tended by I.

III. Black workers, guards, and purveyors; surround the nest as guards or sentinels, in a manner presently to be described, and also forage for the food required for I. They are much larger and stronger insects than either I. or II., and are provided with very formidable mandibles.

The nest is placed in sandy soil in the neighbourhood of shrubs and flowers, is a perfect square, and occupies about four or five square feet of ground, the surface of which is kept almost unbroken. But the boundaries of the nest are rendered conspicuous by the guard of black workers (III.), which continuously parade round three of its sides in a close double line of defence, moving in opposite directions. In the accompanying diagram this sentry path is represented by the thick black lines. These always face the same points of the compass, and the direction in which the sentries march is one column from south-west to south-east, and the other column from south-east to south-west—each column, however, moving in regular order round three sides of a square. The southern side of the encampment is left unguarded; but if any enemy approaches on this or any other side, a number of the guards leave their stations, and sally forth to face the foe—raising themselves on their hind tarsi on meeting the enemy, and moving their large mandibles in defiance. Spiders, wasps, beetles, and other insects, if they venture too near the nest, are torn to pieces by the guard in a most merciless manner, and the dead body of the vanquished is speedily removed from the neighbourhood of the nest—the guard then marching back to resume their places in the line of defence, their object in destroying other insects being the defence of their encampment, and not the obtaining of food.

The object of leaving the southern side of the square encampment open is as follows. While some of the black workers are engaged on duty as guard, another and larger division are engaged on duty as purveyors. These enter and leave the quadrangle by its open or southern side along the dotted line marked a to the central point c. The incoming line is composed of individuals each bearing a burden of fragments of flowers or aromatic leaves. These are all deposited in the centre of the quadrangle *c*. Along the other diagonal *e* there is a no less incessantly moving double line of yellow workers (I.), whose office it is to convey the supplies deposited by the black workers at *c* to *b*, which is the gateway of the fortress. It is remarkable that no black ant is ever seen upon the line *e*,

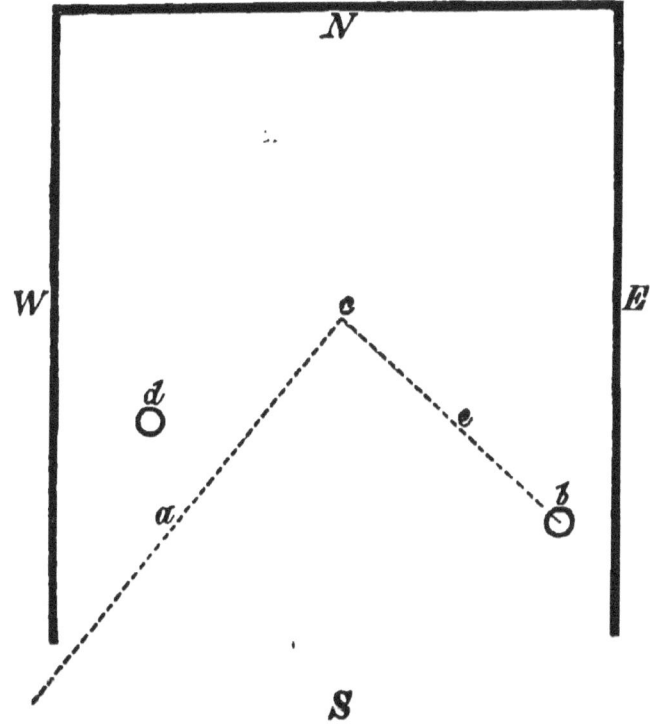

Fig. 7.

and no yellow one upon the line *a*; each keeps his own separate station, and follows his own particular duty with a steadfastness and apparent adherence to discipline that are most astonishing. The hole at *d* seems to be a ventilating shaft; it is never used as a gateway.

Section of the nest reveals, besides galleries, a small chamber about three feet below the surface, across which is spread, like a spider's web, a network of squares spun by the insects, the squares being about ¼ inch across, and the ends of the whole net being fastened to the earthen walls of the chamber. In each one of the squares, supported by the web, sits one of the honey-making ants (II.). Here these honey makers live in perpetual confinement, and receive a constant supply of flowers, pollen, &c., which is continually being brought them by (I.), and which, by a process analogous to that performed by the bee, they convert into honey.

Such is an epitome of the only account that the world has yet received of the habits and economy of this wonderful insect, whose instincts of military organization seem to be not less wonderful than those of the Ecitons, though in this case they are developed with reference to defence, and not to aggression. It is especially noteworthy that the black and yellow workers are believed to belong to 'two separate genera;' for if this is the case, it is the only one I can recall of two distinct species co-operating for a common end; for even the nearest parallel which we find supplied in other species of ants maintaining aphides, is not quite the same thing, seeing that the aphides are merely passive agents, like Class II., of the honey-making ant, and not actively co-operating members of the community, like Class I.

Ecitons.—We have next to consider the habits of the wonderful 'foraging,' or, as it might be more appropriately called, the military ant of the Amazon. These insects, which belong to several species of the same genus, have been carefully watched by Belt, Bates, and other naturalists. The following facts must therefore be accepted as fully established.

Eciton legionis moves in enormous armies, and everything that these insects do is done with the most perfect instinct of military organization. The army marches in the form of a rather broad and regular column, hundreds of yards in length. The object of the march is the capture and plunder of other insects, &c., for food, and as the well-organised host advances, its devastating legions set all other terrestrial life at defiance. From the main column there are sent out smaller lateral columns, the composing individuals of which play the part of scouts, branching off in various directions, and searching about with the utmost activity for insects, grubs, &c., over every log, under every fallen leaf, and in every nook and cranny where there is any chance of finding prey. When their errand is completed, they return into the main column. If the prey found is sufficiently small for the scouts themselves to manage, it is immediately seized, and carried back to the main column; but if the amount is too large for the scouts to deal with alone, messengers are sent back to the main column, whence there is immediately dispatched a detachment large enough to cope with the requirements. Insects which when killed are too large for single ants to carry, are torn in pieces, and the pieces conveyed back to the main army by different individuals. Many insects in trying to escape run up bushes and shrubs, where they are pursued from branch to branch and twig to twig by their remorseless enemies, until on arriving at some terminal ramification they must either submit to immediate capture by their pursuers, or drop down amid the murderous hosts beneath. As already stated, all the spoils that are taken by the scouts or by the detachments sent out in answer to their demands for assistance, are immediately taken back to the main column. When they arrive there, they are taken to the rear of that column by two smaller columns of carriers, which are constantly running, one on either side of the main column, with the supplies that are constantly pouring in from both sides. Each of these outside columns is a double line, the ants composing one of the two lines all running in the same direction as the main army, and the ants composing the other line all running in the opposite direction. The former are empty-handed carriers, which having deposited their burdens in the rear, are again advancing to the van for fresh burdens. Those composing the other line are all laden with the mangled remains of insects, pupæ of other ants, &c. On either side of the main column there are also constantly running up and down a few individuals of smaller size and lighter colour than the other ants, which seem to play the part of officers; for they never leave their stations, and while running up and down the outsides of the column, they every now and again stop to touch antennæ with some member of the rank and file, as if to give instructions. When the scouts discover a wasp's nest in a tree, a strong force is sent out from the main army, the nest is pulled to pieces, and all the larvæ carried to the rear of the army, while the wasps fly around defenceless against the invading multitude. Or, if the nest of any other species of ant is found, a similarly strong force, or perhaps the whole army is deflected towards it, and with the utmost energy the innumerable insects set to work to sink shafts and dig mines till the whole nest is rifled of its contents. In these mining operations the ants work with an extraordinary display of organized co-operation; for those low down in the shafts do not lose time by carrying up the earth which they excavate, but pass on the pellets to those above; and the ants on the surface, when they receive the pellets, carry them, 'with an appearance of forethought that quite staggered' Mr. Bates, only just far enough to ensure that they shall not roll back again into the shaft, and, after depositing them, immediately hurry back for more. But there is not a rigid division of labour, although the work ' seems to be performed by intelligent co-operation amongst a host of eager little creatures;' for some of them act ' sometimes as carriers of pellets, and at another as miners, and all shortly afterwards assume the office of conveyors of the spoil.' Again, as showing the instincts of co-operation, the following may also be quoted from Bates's account:—

On the following morning no trace of ants could be found near the place where I had seen them the preceding day, nor were there signs of insects of any description in the thicket; but at the distance of eighty or one hundred yards, I came upon the same army, engaged evidently on a razzia of a similar kind to that of the previous evening; but requiring other resources of their instinct, owing to the nature of the ground. They were eagerly occupied on the face of an inclined bank of light earth in excavating mines, whence, from a depth of eight or ten inches, they were extracting the bodies of a bulky species of ant of the genus Formica. It was curious to see them crowding round the orifices of the mines, some assisting their comrades to lift out the bodies of the Formicæ, and others tearing them in pieces, on account of their weight being too great for a single Eciton; a number of carriers seizing each a fragment, and carrying it off down the slope.

These Ecitons have no fixed nest themselves, but live, as it were, on a perpetual campaign. At night, however, they call a halt and pitch a camp. For this purpose they usually select a piece of broken ground, in the interstices of which they temporarily store their plunder. In the morning the army is again on the march, and before an hour or two has passed not a single ant is to be seen where the countless multitudes had previously covered the ground.

Another and larger species of Eciton (*E. humata*) hunts sometimes in dense armies, and sometimes in columns, according to the kind of prey of which they are in search. When in columns they are seeking for the nests of a certain species of ant which have their young in holes of rotten logs. These Ecitons when seeking for these nests hunt about, like those just described, in columns, which branch off in various directions. When a fallen log is reached, the column spreads over it, searching through all the holes and cracks. Mr. Belt says of them :—

The workers are of various sizes, and the smallest are here of use, for they squeeze themselves into the narrowest holes, and search out their prey in the furthest ramifications of the nests. When a nest of the *Hypoclinea* is attacked, the ants rush out, carrying the larvæ and pupæ in their jaws, but are immediately despoiled of them by the Ecitons, which are running about in every direction with great swiftness. Whenever they come across a *Hypoclinea* carrying a larva or pupa, they take it from it so quickly, that I could never ascertain exactly how it was done.

As soon as an Eciton gets hold of its prey, it rushes off back along the advancing column, which is composed of two sets, one hurrying forward, the other returning laden with their booty, but all and always in the greatest haste and apparent hurry. About the nest which they are harrying, all appears in confusion, Ecitons running here and there and everywhere in the greatest haste and disorder; but the result of all this apparent confusion is that scarcely a single *Hypoclinea* gets away with a pupa or larva. I never saw the Ecitons injure the Hypoclineas themselves, they were always contented with despoiling them of their young.

The columns of this species 'are composed almost entirely of workers of different sizes;' but, as in the species previously mentioned, 'at intervals of two or three yards there are larger and lighter coloured individuals that often stop, and sometimes run a little backward, stopping and touching some of the ants with their antennæ,' and looking 'like officers giving orders and directing the march of the column.'

Concerning the other habits of this species, the same author writes:—

The eyes in the Ecitons are very small, in some of the species imperfect, and in others entirely absent; in this they differ greatly from the *Pseudomyrma* ants, which hunt singly and which have the eyes greatly developed. The imperfection of eyesight in the Ecitons is an advantage to the community, and to their particular mode of hunting. It keeps them together, and prevents individual ants from starting off alone after objects that, if their eyesight was better, they might discover at a distance; the Ecitons and most other ants follow each other by scent, and, I believe, they can communicate the presence of danger, of booty, or other intelligence, to a distance by the different intensity or qualities of the odours given off. I one day saw a column of *Eciton hamata* running along the foot of a nearly perpendicular tramway cutting, the side of which was about six feet high. At one point I noticed a sort of assembly of about a dozen individuals that appeared in consultation. Suddenly one ant left the conclave, and ran with great speed up the perpendicular face of the cutting without stopping. It was followed by others, which, however, did not keep straight on like the first, but ran a short way, then returned, then again followed a little further than the first time. They were evidently scenting the trail of the pioneer, and making it permanently recognisable. These ants followed the exact line taken by the first one, although it was far out of sight. Wherever it had made a slight *détour* they did so likewise. I scraped with my knife a small portion of the clay on the trail, and the ants were completely at fault for a time which way to go. Those ascending and those descending stopped at the scraped portion, and made short circuits until they hit the scented trail again, when all their hesitation vanished, and they ran up and down it with the greatest confidence. On gaining the top of the cutting, the ants entered some brushwood suitable for hunting. In a very short space of time the information was communicated to the ants below, and a dense column rushed up to search for their prey. The Ecitons are singular amongst the ants in this respect, that they have no fixed habitations, but move on from one place to another, as they exhaust the hunting grounds around them. I think *Eciton hamata* does not stay more than four or five days in one place. I have sometimes come across the migratory columns; they may easily be known. Here and there one of the light-coloured officers moves backwards and forwards directing the columns. Such a column is of enormous length, and contains many thousands if not millions of individuals. I have sometimes followed them up for two or three hundred yards without getting to the end.

They make their temporary habitations in hollow trees, and sometimes underneath large fallen trucks that offer suitable hollows. A nest that I came across in the latter situation was open at one side. The ants were clustered together in a dense mass, like a great swarm of bees, hanging from the roof but reaching to the ground below. Their innumerable long legs looked like brown threads binding together the mass, which must have been at least a cubic yard in bulk, and contained hundreds of thousands of individuals, although many columns were outside, some bringing in the pupæ of ants, others the legs and dissected bodies of various insects. I was surprised to see in this living nest tubular passages leading down to the centre of the mass, kept open just as if it had been formed of inorganic materials. Down these holes the ants who were bringing in booty passed with their prey. I thrust a long stick down to the centre of the cluster, and brought out clinging to it many ants holding larvæ and pupæ, which probably were kept warm by the crowding together of the ants. Besides the common dark-coloured workers and light-coloured officers, I saw here many still larger individuals with enormous jaws. These they go about holding wide open in a threatening manner.

It was this ant which, as previously stated, showed sympathy and fellow-feeling with companions in difficulties.

The habits of *E. drepanophora* are closely similar to those of the species already described; and, indeed, except in matters of detail, all the species of Ecitons have much the same habits. Mr. Bates records an interesting observation which he made on one of the moving columns of this species. He says: 'When I interfered with the column or abstracted an individual from it, news of the disturbance was quickly communicated to a distance of several yards to the rear, and the column at that point commenced retreating.' The main column is in this species narrower, viz., 'from four to six deep,' but extends to a great length, viz., half a mile or more. It was this species of Eciton that the same naturalist describes as enjoying periods of leisure and recreation in the 'sunny nooks of the forest.'

Next we have to consider *E. prædator*, of which the same observer writes:—

This is a small dark reddish species, very similar to the common red stinging ant of England. It differs from all other Ecitons in its habit of hunting, not in columns, but in dense phalanxes consisting of myriads of individuals, and was first met with at Ega, where it is very common. Nothing in insect movements is more striking than the rapid march of these large and compact bodies. Wherever they pass, all the rest of the animal world is thrown into a state of alarm. They stream along the ground and climb to the summits of all the lower trees, searching every leaf to its apex, and whenever they encounter a mass of decaying vegetable matter, where booty is plentiful, they concentrate, like other Ecitons, all their forces upon it, the dense phalanx of shining and quickly-moving bodies, as it spreads over the surface, looking like a flood of dark red liquid. They soon penetrate every part of the confused heap, and then, gathering together again in marching order, onward they move. All soft-bodied and inactive insects fall an easy prey to them, and, like other Ecitons, they tear their victims in pieces for facility of carriage. A phalanx of this species, when passing over a tract of smooth ground, occupies a space of from four to six square yards; on examining the ants closely they are seen to move, not all together in one straightforward direction, but in variously spreading contiguous columns, now separating a little from the general mass, now reuniting with it. The margins of the phalanx spread out at times like a cloud of skirmishers from the flanks of an army. I was never able to find the hive of this species.

Lastly, there are two species of Eciton totally blind, and their habits differ from those of the species which we have hitherto considered. Bates writes of them:—

The armies of *E. vastator* and *E. erratica* move, as far as I could learn, wholly under covered roads, the ants constructing them gradually but rapidly as they advance. The column of foragers pushes forward step by step, under the protection of these covered passages, through the thickets, and on reaching a rotting log, or other promising hunting-ground, pour into the crevices in search of booty. I have traced their arcades, occasionally, for a distance of one or two hundred yards; the grains of earth are taken from the soil over which the column is passing, and are fitted together without cement. It is this last-mentioned feature that distinguishes them from the similar covered roads made by termites, who use their glutinous saliva to cement the grains together. The blind Ecitons, working in numbers, build up simultaneously the sides of their convex arcades, and contrive, in a surprising manner, to approximate them and fit in the key-stones without letting the loose uncemented structure fall to pieces. There was a very clear division of labour between the two classes of neuters in these blind species. The large-headed class, although not possessing monstrously lengthened jaws like the worker-majors in *E. hamata* and *E. drepanophora*, are rigidly defined in structure from the small-headed class, and act as soldiers, defending the working community (like soldier termites) against all comers. Whenever I made a breach in one of their covered ways, all the ants underneath were set in commotion, but the worker-minors remained behind to repair the damage, whilst the large-heads issued forth in a most menacing manner, rearing their heads and snapping their jaws with an expression of the fiercest rage and defiance.

Annornia arcens.—This is the so-called 'driver' or 'marching' ant of West Africa, which in habits and intelligence closely resembles the military ants of the other hemisphere. I shall therefore not wait again to describe these habits in detail. Like the Ecitons, the marching ants of Africa have no fixed nest, but make temporary halts in the shade of hollow trees, overhanging rocks, &c. They march in large armies, and, like the Ecitons, always in the form of a long close column; but in this case the relative position of the carriers of spoil and larvæ is reversed, for while these occupy the middle place the soldiers and officers march on either side. These have large heads armed with powerful jaws, and never take part in carrying; their function is to maintain order, act as scouts, and attack prey. The habits of these ants resemble most closely those of the blind Ecitons in that they very frequently, and indeed generally, build covered ways; they do so apparently in order to protect themselves from the heat of the African sun. Their line of march is therefore marked by a continuous arch or tunnel, which is always being constructed by the van of the column. The structure is made of earth moulded together by saliva, and is very quickly built. But it is only built in places where the line of march is exposed to the sunlight; at night, or in the shadow of trees or long grass, it is not made. If their camp is flooded by a tropical rainstorm, the ants congregate in a close mass, with the younger ants in the centre; they thus form a floating island.

It is remarkable that ants of different hemispheres should manifest so close a similarity with respect to all these wonderful habits. The Chasseur ants of Trinidad, and, according to Madame Merian, the ants of visitation of Cayenne, also display habits of the same kind.

General Intelligence of Various Species.

Many of the foregoing facts display an astonishing degree of intelligence as obtaining among ants; for I think that however much latitude we may be inclined to allow to ' blind instinct ' in the way of imitating actions elsewhere due to conscious purpose, some at least of these foregoing facts can only be fairly reconciled with the view that the insects know what they are doing and why they are doing it. But as I am myself well aware of the difficulty that arises in all such cases of drawing the line between purposeless instinct and purposive intelligence, I have thought it desirable to reserve for this concluding division of the present chapter several isolated facts which have been observed among sundry species of ants, and which do not

ANTS—GENERAL INTELLIGENCE.

seem to admit of being reasonably comprised under the category of instinctive action, if by the latter we mean action pursued without knowledge of the relation between the means adopted and the ends attained.

It will be remembered that our test of instinctive as distinguished from truly intelligent action is simply whether all individuals of a species perform similar adaptive movements under the stimulus supplied by similar and habitual circumstances, or whether they manifest individual and peculiar adaptive movements to meet the exigencies of novel and peculiar circumstances. The importance of this distinction may be rendered manifest by the following illustrations.

We have already seen that the ants which Sir John Lubbock observed display many and complex instincts, which together might seem to justify us in anticipating that animals which present such wonderful instincts must also present sufficient general intelligence to meet simple though novel exigencies by such simple adaptations as the unfamiliar circumstances require. Yet experiments which

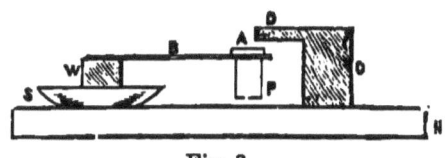

Fig. 8.

he made in this connection seem to show that such is not the case, but that these ants, with all their wealth of instinctive endowments, are utterly destitute of intelligent resources; they have abundance of common and detailed knowledge (supposing the adaptations to be made consciously) how to act under certain complex though familiar circumstances, but appear quite unable to originate any adaptive action to obviate even the simplest conceivable difficulty, if this is of a kind which they have not been previously accustomed to meet. Thus, on a horizontal rod B supported in a saucer of water S, and therefore inaccessible to the ants from beneath, he placed some larvæ A. On the nest N he then placed a block of wood C D, constructed so that the portion D should touch the larvæ at A. When the ants had made a number of journeys over C D A and back again, he raised the block C D so that there was an interval $\frac{3}{10}$ of an inch between the end of the block D and the larvæ at A.

The ants kept on coming, and tried hard to reach down from D to A, which was only just out of their reach. . . . After a while they all gave up their efforts and went away, losing their prize in spite of most earnest efforts, because it did not occur to them to drop $\frac{3}{10}$ of an inch. At the moment when the separation was made there were fifteen ants on the larvæ. These could, of course, have returned if one had stood still and allowed the others to get on its back. This, however, did not occur to them; nor did they think of letting themselves drop from the bottom of the paper (P) on to the nest. Two or three, indeed, fell down, I have no doubt by accident; but the remainder wandered about, until at length most of them got into the water.

In another experiment he interposed a light straw bridge on the way between the nest and the larvæ, and when the ants had well learnt the way, he drew the bridge a short distance towards the nest, so that a small chasm was made in the road. The ants tried hard and ineffectually to reach across it, but it did not occur to them to *push* the straw into its original position.

The following experiment is still more illustrative of the absence of intelligence, because the adjustive action required would not demand the exercise of such high powers of imagination and abstraction as would have been required for the moving forwards of the paper drawbridge.

To test their intelligence I made the following experiments: I suspended some honey over a nest of *Lasius flavus* at a height of about $\frac{1}{2}$ an inch, and accessible only by a paper bridge more than 10 feet long. Under the glass I then placed a small heap of earth. The ants soon swarmed over the earth on to the glass, and began feeding on the honey. I then removed a little of the earth, so that there was an interval of about $\frac{1}{3}$ of an inch between the glass and the earth; but though the distance was so small, they would not jump down, but preferred to go round by the long bridge. They tried in vain to stretch up from the earth to the glass, which, however, was just out of their reach, though they could touch it with their antennæ; but it did not occur to them to heap the earth up a little, though if they had moved only half a dozen particles of earth they would have secured for themselves direct access to the food. This, however, never occurred to them. At length they gave up all attempts to reach up to the glass, and went round by the paper bridge. I left the arrangement for several weeks, but they continued to go round by the long paper bridge.

Another and somewhat similar experiment consisted in placing an upright stick A, supporting at an angle another stick B, which nearly but not quite touched the ground at C. At the end of the stick B there were placed some larvæ in a horizontal glass cell at D. Into this cell were also placed a number of ants along with the larvæ. The drop from D to C was only ½ an inch; 'still, though the ants reached over and showed a great anxiety to take this short cut home, they none of them faced the leap, but all went round by the sticks, a distance of nearly 7 feet.' Sir John then reduced the interruption to ⅔ of an inch, so that the ants could even touch the glass cell with their antennæ; yet all day long the ants continued to go the long way round rather than face the drop. Next, therefore, he took still longer sticks and tapes, and arranged them as before, only horizontally instead of vertically. He also placed some fine earth under the glass cell containing the larvæ. The ants as before continued to go the long way round (16 feet), though the drop could not have hurt either themselves or the larvæ, and though even this drop might have been obviated by heaping up the fine earth into a little mound ⅛ of an inch high, so as to touch the glass cell.

It is desirable, however, here to state that all species of ants do not show this aversion to allowing themselves to drop through short distances; for Moggridge describes the harvesting ants of Europe as seeming rather to enjoy acrobatic performances of this kind; and the same fact is recorded by Belt of the leaf-cutting ants of the Amazons. Dr. Bastian, in his work on 'Brain as an Organ of Mind,' suggests that the 'seeming lack of intelligence betrayed by our English ants, from their disinclination to take a small leap, may be due simply to their defective sight.'

(pp. 241-2). But even this consideration does not extenuate the stupidity of the ants which failed to heap up the fine earth to reach the glass cell which they were able to touch with their antennæ.

That the species of ants on which Sir John Lubbock experimented were not, however, quite destitute of intelligence is proved by the result of the following experiment :—

I put some provisions in a shallow box with a glass top and a single hole in one side ; I then put some specimens of *Lasius niger* to the food, and soon a stream of ants was at work busily carrying supplies off to the nest. When they had got to know their way thoroughly, and from thirty to forty were so occupied, I poured some fine mould in front of the hole, so as to cover it to a depth of about $\frac{1}{2}$ an inch. I then took out the ants which were actually in the box. As soon as the ants had recovered from the shock of this unexpected proceeding on my part, they began to run all round and about the box, looking for some other place of entrance. Finding none, however, they began digging down into the earth just over the hole, carrying off the grains of earth one by one and depositing them without any order all round at a distance of from $\frac{1}{2}$ to 6 inches, until they had excavated down to the doorway, when they again began carrying off the food as before.

This experiment was several times repeated on *L. niger* and on *L. flavus*, always with the same result.

Thus, then, we may conclude that the reasoning power of these ants, although shown by the first experiments to be almost *nil*, is shown by this experiment to be not quite *nil* ; for the attempt to meet the exigencies of the case by first going round the box to seek another entrance, before taking the labour to remove the earth from the known entrance, implies a certain rudimentary degree of adaptive capacity which belongs to the category of the rational.

Another point of considerable interest, as bearing on the general intelligence of ants, is one that was brought out as the result of a laborious series of hourly observations, extending without intermission from 6.30 A.M. to 10 P.M. for a period of three months. The object of these observations was to ascertain whether the principle of the division of labour is practised by the ants. The result of these observations was to show that during the wintertime, when the ants are not active, certain individuals are told off to forage for supplies, and that when any casualty overtakes these individuals, others are told off to supply their places. Thus, in the words of Sir John Lubbock's analysis of his lengthy tables,—

The feeders at the beginning of the experiment were those known to us as Nos. 5, 6, and 7. On the 22nd of November a friend, registered as No. 8, came to the honey, and again on the 11th December; but with these two exceptions the whole of the supplies were carried in by Nos. 5 and 6, with a little help from No. 7. Thinking now it might be alleged that possibly these were merely unusually active or greedy individuals, I imprisoned No. 6 when she came out to feed on the 5th. As will be seen from the table, no other ant had been out to the honey for some days; and it could therefore hardly be accidental that on that very evening another ant (then registered as No. 9) came out for food. This ant, as will be seen from the table, then took the place of No. 6 (No. 5 being imprisoned). On the 11th January No. 9 took in all the supplies, again with a little help from No. 7. So matters continued until the 17th, when I imprisoned No. 9, and then again, *i.e.* on the 19th, another ant (No. 10) came out for the food, aided, on and after the 22nd, by another (No.11). This seems to me very curious. From the 1st November to the 5th January, with two or three casual exceptions, the whole of the supplies were carried in by three ants, one of whom, however, did comparatively little. The other two are imprisoned, and then, but not till then, a fresh ant appears on the scene. She carries in the food for a week, and then she being imprisoned, two others undertake the task. On the other hand, in nest 1, when the first foragers were not imprisoned, they continued during the whole time to carry in the necessary supplies.

The facts, therefore, certainly seem to indicate that certain ants are told off as foragers, and that during winter, when but little food is required, two or three such foragers are sufficient to provide it.

Although Sir John Lubbock's ants showed such meagre resources of intelligent adjustment, other species of ants, which we have already had occasion to consider, appear to be as remarkable in this respect as they are in respect of their instinctive adjustments. Unfortunately observations on this subject are very sparse, but such as they are they hold out a strong inducement for any one who has the opportunity to experiment with the view of testing the intelligence of those species in connection with which the following observations have been made.

Réaumur states that ants will make no attempt to enter an inhabited beehive to get at the contained honey, knowing that the bees will slaughter them if they do so. But if the hive is uninhabited, or the bees all dead, the ants will swarm into the hive as long as any honey is to be found there.

P. Huber records that a wall which had been partly erected by ants was observed by him—

As though it were intended to support the still unfinished arched roof of a large room, which was being built from the opposite side. But the workers which had begun the arch had given it too low an elevation for the wall on which it was to rest, and if it had been continued on the same lines it would have met the partition wall halfway up, and this was to be avoided. I had just made this criticism to myself, when a new arrival, after looking at the work, came to the same conclusion. For it began at once to destroy what had been done, and to heighten the wall on which it was supported, and to make a new arch with the materials of the old one under my very eyes. When the ants begin an undertaking it seems exactly as if an idea slowly ripened into execution in their minds. Thus if one of them finds two stalks lying crosswise on the nest, which make possible the formation of a room, or some little rafters which suggest the walls and the corners, it first observes the various parts accurately, and then quickly and neatly heaps little pellets of earth in the interspaces and alongside the stalks. It brings from every side materials that seem appropriate, and sometimes takes such from the uncompleted works of its companions, so much is it urged on by the idea which it has once conceived, and by the desire to execute it. It goes and comes and turns back again, until its plan is recognisable by the others.

Ebrard, in his 'Etudes de Mœurs' (p. 3), gives the following remarkable instance of the display of intelligence of *F. fusca* :—

The earth was damp and the workers were in full swing.

It was a constant coming and going of ants, coming forth from their underground dwelling, and carrying back little pellets of earth for building. In order to concentrate my attention I fixed my gaze on the largest of the rooms which were being built, wherein several ants were busy. The work had made considerable progress; but although a projection could be plainly seen along the upper edge of the wall, there remained an interspace of about twelve or fifteen millimetres to fill in. Here would have been the place, in order to support the earth still to be brought in, to have had recourse to those pillars, buttresses, or fragments of dried leaves, which many ants are wont to use in building. But the use of this expedient is not customary with the ants I was observing (*F. fusca*). Our ants, however, were sufficient for the occasion. For a moment they seemed inclined to leave their work, but soon turned instead to a grass-plant growing near, the long narrow leaves of which ran close together. They chose the nearest, and weighted its distal end with damp earth, until its apex just bent down to the space to be covered. Unfortunately the bend was too close to the extremity, and it threatened to break. To prevent this misfortune, the ants gnawed at the base of the leaf until it bent along its whole length and covered the space required. But as this did not seem to be quite enough, they heaped damp earth between the base of the plant and that of the leaf, until the latter was sufficiently bent. After they had thus attained their object, they heaped on the buttressing leaf the materials required for building the arched roof.

The characteristic *trait* of the building of ants, says Forel, is the almost complete absence of an unchangeable model, peculiar to each species, such as is found in wasps, bees, and others. The ants know how to suit their indeed little perfect work to circumstances, and to take advantage of each situation. Besides, each works for itself and on a given plan, and is only occasionally aided by others when these understand its plan. Naturally many collisions occur, and some destroy that which others have made. This also gives the key to understanding the labyrinth of the dwelling. For the rest, it is always those workers which have discovered the most advantageous method, or which have shown the most patience, which win over to their plan the majority of their comrades and at last the whole colony, although not without many fights for supremacy. But if one succeeds in obtaining a second to follow it, and this second draws the others after it, the first is soon lost again in the crowd.

Espinas also observed ('Thierischen Gesellschaften,' German translation, 1879, p. 371) that each single ant made its own plan and followed it until a comrade, which had caught the idea, joined it, and then they worked together in the execution of the same plan.

Moggridge says of the harvesters of Europe,—

I have observed on more than one occasion that when in digging into an ants' nest I have thrown out an *elater* larva, the ants would cluster round it and direct it towards some small opening in the soil, which it would quickly enlarge and disappear down. At other times, however, the ants would take no notice of the *elater*, and it is my belief that the attentions paid to it on former occasions were purely selfish, and that they intended to avail themselves of the tunnel thus made down into the soil, with the view of reopening communications with the galleries and granaries concealed below, the approaches to which had been covered up. I have frequently watched the ants make use of these passages mined by the *elater* on these occasions.

And again, as showing apparently intelligent adaptation of their usual habits to altered circumstances, he gives an account of the behaviour of these ants when a great crowd of them were confined by him in a glass jar containing earth. He says:—

On the following morning the openings were ten in number, and the greatly increased heaps of excavated earth showed that they must probably have been at work all night. The amount of work done in this short time was truly surprising, for it must be remembered that, eighteen hours before, the earth presented a perfectly level surface, and the larvæ and ants, now housed below, found themselves prisoners in a strange place, bounded by glass walls, and with no exit possible.

It seems to me that the ants displayed extraordinary intelligence in having thus at a moment's notice devised a plan by which the superabundant number of workers could be employed at one time without coming in one another's way. The soil contained in the jar was of course less than a tenth part of that comprised within the limits of an ordinary nest, while the number of workers was probably more than a third of the total number belonging to the colony. If therefore but one or two entrances had been pierced in the soil, the workers would have been for ever running against one another, and a great number could never have got below to help in the all-important task of preparing passages and chambers for the accommodation of the larvæ. These numerous and funnel-shaped entrances admitted of the simultaneous descent and ascent of large numbers of ants, and the work progressed with proportionate rapidity. After a few days only three entrances, and eventually only one, remained open.

Concerning the harvesting ant of Texas, the following quotation may be made, under the present head, from MacCook. After remarking that these ants always select sunny places wherein to build their nests, or disks, he goes on to say that within a few paces of his tent—

A nest was made which was partly shaded by a small mesquite tree that stood just beyond the margin of the clearing. The sapling had probably grown up after the location of the community, and for some reason had been permitted to remain until too old to kill off. The shadow thrown upon the pavement was very slight; nevertheless, fifteen feet distant a new formicary was being established. The path from the ranch to the spring ran between this new hill and the old one, and ants were in communication between the two. An opening had been made in the ground, and the beginnings of a new formicary were quite apparent. This is the only instance observed of what seemed an attempt at colonising or removing, and I associated it with the presence of the small but growing shadow of the young tree.

He also gives us a still more remarkable observation, which indeed, I must candidly say, does not appear to me credible. I am, therefore, glad to add that it does not appear very distinctly from the account whether the author himself made the observation, or had it narrated to him by his guide. But here is the observation in his own words:—

While studying the habits of the cutting ant I was tempted to make a night visit to a farm some distance from camp, by the farmer's story of depredations made by these insects upon certain plants and vegetables. A long, dark tramp, a blind and vain search among the fields, compelled us at last to call out the countryman from his bed. He led us directly to one of the cutting ants' nests, which was overshadowed by a young peach tree. 'There they be, sir,' cried he triumphantly.

They were agriculturals! So also were the other nests shown. The reason for this confounding of the two ants on the part of the people hereabouts, and the reason for the 'cutting' operations of our harvesters, will be explained farther on. It is only in point here to say that the farmer affirmed that the ants under the peach tree had stripped off the first tender leaves last spring, so that scarcely one had been left upon the limbs. I am convinced that the reason for this onslaught was the desire to be rid of the obnoxious shade, and open the formicary to the full light of the sun.

From this account it is not very clear whether the writer himself saw evidence of the former denudation of the tree, and if so whether there was any indication, other than the word of the farmer, that the denudation had been effected by the ants. To make this conclusion credible the best conceivable evidence would be required, and this, unfortunately, is just what we find wanting. Somewhat the same remarks may be made on the following quotation from the same writer, though in this case his view is to some extent supported by an observation of Moggridge, as well as by that of Ebrard already quoted :—

Here I observed what appeared to be a new mode of operation. The workers, in several cases, left the point at which they had begun a cutting, ascended the blade, and passed as far out toward the point as possible. The blade was thus borne downward, and as the ant swayed up and down it really seemed that she was taking advantage of the leverage thus gained, and was bringing the augmented force to bear upon the fracture. In two or three cases there appeared to be a division of labour; that is to say, while the cutter at the roots kept on with her work, another ant climbed the grass blade and applied the power at the opposite end of the lever. This position may have been quite accidental, but it certainly had the appearance of a voluntary co-operation. I was sorry not to be able to establish this last inference by a series of observations, as the facts were only observed in this one nest.

The observation of Moggridge, to which I have alluded as in some measure rendering support to the foregoing, is as follows. Speaking of European harvesters which he kept in an artificial nest for the purposes of close observation, he says :-

I was also in this way able to see for myself much that I otherwise could not have seen. Thus I was able to watch the operation of removing roots which had pierced through their galleries, belonging to seedling plants growing on the surface, and which was performed by two ants, one pulling at the free end of the root, and the other gnawing at its fibres where the strain was greatest, until at length it gave way.

And again,—

Two ants sometimes combine their efforts, when one stations itself near the base of the peduncle, and gnaws it at the point of greatest tension, while the other hauls upon and twists it. . . . I have occasionally seen ants engaged in cutting the capsules of certain plants, drop them, and allow their companions below to carry them away.

Lastly, the statements of these three observers taken together serve to render credible the following quotation from Bingley,[1] who says that in Captain Cook's expedition in New South Wales ants were seen by Sir Joseph Banks and others—

As green as a leaf, which live upon trees and build their nests of various sizes, between that of a man's head and his fist. These nests are of a very curious structure: they are formed by bending down several of the leaves, each of which is as broad as a man's hand, and gluing the points of them together so as to form a purse. The viscous matter used for this purpose is an animal juice. . . . Their method of bending down leaves we had no opportunity to observe; but we saw thousands uniting all their strength to hold them in this position, while other busy multitudes were employed within, in applying this gluten, that was to prevent their returning back. To satisfy ourselves that the leaves were bent and held down by the efforts of these diminutive artificers, we disturbed them in their work; and as soon as they were driven from their station, the leaves on which they were employed sprang up with a force much greater than we could have thought them able to conquer by any combination of their strength.

This remarkable fact also seems to be corroborated by the following independent observation of Sir E. Tennent:—

[1] *Animal Biography,* 'Ants.'

The most formidable of all is the great red ant, or Dimiya. It is particularly abundant in gardens and on fruit-trees; it constructs its dwellings by gluing the leaves of such species as are suitable from their shape and pliancy into hollow balls, and these it lines with a kind of transparent paper, like that manufactured by the wasp. I have watched them at the interesting operation of forming these dwellings;—a line of ants standing on the edge of one leaf bring another into contact with it, and hold both together with their mandibles till their companions within attach them firmly by means of their adhesive paper, the assistants outside moving along as the work proceeds. If it be necessary to draw closer a leaf too distant to be laid hold of by the immediate workers, they form a chain by depending one from the other till the object is reached, when it is at length brought into contact, and made fast by cement.

I shall now pass on to the remarkable observation communicated to Kirby by Colonel Sykes, F.R.S., and which is thus narrated by Kirby in his 'History, Habits, and Instincts of Animals:'—

When resident at Poona, the dessert, consisting of fruits, cakes, and various preserves, always remained upon a small side table, in a verandah of the dining-room. To guard against inroads, the legs of the table were immersed in four basins filled with water; it was removed an inch from the wall, and, to keep off dust from open windows, was covered with a tablecloth. At first the ants did not attempt to cross the water, but as the strait was very narrow, from an inch to an inch and a half, and the sweets very tempting, they appear, at length, to have braved all risks, to have committed themselves to the deep, to have scrambled across the channel, and to have reached the objects of their desires, for hundreds were found every morning revelling in enjoyment: daily vengeance was executed upon them without lessening their numbers; at last the legs of the table were painted, just above the water, with a circle of turpentine. This at first seemed to prove an effectual barrier, and for some days the sweets were unmolested, after which they were again attacked by these resolute plunderers; but how they got at them seemed totally unaccountable, till Colonel Sykes, who often passed the table, was surprised to see an ant drop from the wall, about a foot above the table, upon the cloth that covered it; another and another succeeded. So that though the turpentine and the distance from the wall appeared effectual barriers, still the resources of the animal, when determined to carry its point, were not exhausted, and by ascending the wall to a certain height, with a slight effort against it, in falling it managed to land in safety upon the table.

Colonel Sykes was a good observer, so that this statement, standing upon his authority, ought not, perhaps, to be questioned. But in all cases of remarkable intelligence displayed by animals, we naturally and properly desire corroboration, however good the authority may be on which the statement of such cases may rest. I will, therefore, add the following instances of the ingenious and determined manner in which ants overcome obstacles, and which so far lend confirmation to the above account.

Professor Leuckart placed round the trunk of a tree, which was visited by ants as a pasture for aphides, a broad cloth soaked in tobacco-water. When the ants returning home down the trunk of the tree arrived at the soaked cloth, they turned round, went up the tree again to some of the overhanging branches, and allowed themselves to drop clear of the obnoxious barrier. On the other hand, the ants which desired to mount the tree first examined the nature of the barrier, then turned back and procured from a distance little pellets of earth, which they carried in their jaws and deposited one after another upon the tobacco-cloth till a road of earth was made across it, over which the ants passed to and fro with impunity.

This interesting, and indeed surprising observation of Leuckart's is, in turn, a corroboration of an almost identical one made more than a century ago by Cardinal Fleury, and communicated by him to Réaumur, who published it in his 'l'Histoire des Insectes' (1734). The Cardinal smeared the trunk of a tree with birdlime in order to prevent the ants from ascending it; but the insects overcame the obstacle by making a road of earth, small stones, &c., as in the case just mentioned. In another instance the Cardinal saw a number of ants make a bridge across a vessel of water surrounding the bottom of an orange-tree tub. They did so by conveying a number of little pieces of *wood*, the choice of which material instead of earth or stones, as in the previous case, seems to betoken no small knowledge of practical engineering.

Büchner, after quoting these cases, proceeds to say (*loc. cit.*, p. 120),—

The ants behaved in yet more ingenious fashion under the following very similar circumstances. Herr G. Theuerkauf, the painter (Wasserthorstr. 49, Berlin), writes to the author, November 18, 1875 : 'A maple tree standing on the ground of the manufacturer, Vollbaum, of Elbing (now of Dantzic), swarmed with aphides and ants. In order to check the mischief, the proprietor smeared about a foot width of the ground round the tree with tar. The first ants who wanted to cross naturally stuck fast. But what did the next? They turned back to the tree and carried down aphides, which they stuck down on the tar one after another until they had made a bridge over which they could cross the tarring without danger. The above-named merchant, Vollbaum, is the guarantor of this story, which I received from his own mouth on the very spot whereat it occurred.

Büchner also gives the following case on the authority of Karl Vogt (*loc. cit.*, p. 128). An apiary of a friend was invaded by ants :—

To make this impossible for the future, the four legs of the beehive-stand were put into small, shallow bowls filled with water, as is often done with food in ant-infested places. The ants soon found a way out of this, or rather a way into their beloved honey, and that over an iron staple with which the stand was attached to a neighbouring wall. The staple was removed, but the ants did not allow themselves to be defeated. They climbed into some linden trees standing near, the branches of which hung over the stand, and then dropped upon it from the branches, doing just the same as their comrades do with respect to food surrounded by water, when they drop upon it from the ceiling of the room. In order to make this impossible, the boughs were cut away. But once more the ants were found in the stand, and closer investigation showed that one of the bowls was dried up, and that a crowd of ants had gathered in it. But they found themselves puzzled how to go on with their robbery, for the leg did not, by chance, rest on the bottom of the bowl, but was about half an inch from it. The ants were seen rapidly touching each other with their antennæ, or carrying on a consultation, until at last a rather larger ant came forward and put an end to the difficulty. It rose to its full height on its hind legs, and struggled until at last it seized a rather projecting splinter of the wooden leg, and managed to take hold of it. As soon as this was done other ants ran on to it, strengthened the hold by clinging, and so made a little living bridge, over which the others could easily pass.

The same author publishes the following very remarkable observation, quoted from a letter to him by Dr. Ellendorf:—

It is a hard matter to protect any eatables from these creatures, let the custody be ever so close. The legs of cupboards and tables in or on which eatables are kept are placed in vessels of water. I myself did this, but I none the less found thousands of ants in the cupboard next morning. It was a puzzle to me how they crossed the water, but the puzzle was soon solved; for I found a straw in one of the saucers, which lay obliquely across the edge of the pan and touched the leg of the press: this they had used for a bridge. Hundreds were drowned in the water, apparently because disorder had reigned at first, those coming down with booty meeting those going up. But now there was perfect order; the descending stream used one side of the straw, the ascending the other. I now pushed the straw about an inch away from the cupboard leg; a terrible confusion arose. In a moment the leg immediately over the water was covered with hundreds of ants, feeling for the bridge in every direction with their antennæ, running back again and coming in ever larger swarms, as though they had communicated to their comrades within the cupboard the fearful misfortune that had taken place. Meanwhile the new-comers continued to run along the straw, and not finding the leg of the cupboard the greatest perplexity arose. They hurried round the edge of the pan, and soon found out where the fault lay. With united forces they quickly pulled and pushed at the straw, until it again came into contact with the wood, and the communication was again restored.

This observation is strikingly, though unconsciously, confirmed by a recent writer in the *Leisure Hour* (1880, pp. 718-19), who having been much troubled by small red ants in the tropics swarming over his provisions, placed the latter in a meat-safe detached from the wall and standing on four legs, each of which was placed in a little tin vessel containing water. Eight or ten days afterwards he found his provisions in the safe swarming with ants as before, and on investigating their mode of access to them found—

Proceeding along the whitewashed wall a string of ants going and coming from the outer door to a height of four feet on my wall, and corresponding with that of the safe; and looking between it and the wall, I discovered the secret—the bridge which these persevering little insects had made. It consisted of a broken bit of straw, which rested with one end on a mud buttress fixed to the wall, and the other on the overhanging or projecting top of the safe, which came within an inch and a half of the wall. So they must have carried the straw up from the floor, and resting their end of it on the support they had prepared, let it fall until its other end reached the safe, and then crossed and completed the structure, for it was fastened at both ends with the mortar composed of their saliva and fine earth. Ruthlessly I destroyed the bridge, and moving the safe farther from the wall, managed to prevent their inroads for that season at least. Since then I have frequently seen short bridges, composed entirely of the concrete or mortar which the white ants use to cover up their workings, extending from a damp earthen wall to anything not more than three-quarters of an inch from it.

Of the Ecitons Mr. Belt says:—

I shall relate two more instances of the use of a reasoning faculty in these ants. I once saw a wide column trying to pass along a crumbling, nearly perpendicular slope. They would have got very slowly over it, and many of them would have fallen, but a number having secured their hold, and reaching to each other, remained stationary, and over them the main column passed. Another time they were crossing a watercourse along a small branch, not thicker than a goose-quill. They widened this natural bridge to three times its width by a number of ants clinging to it and to each other on each side, over which the column passed three or four deep; whereas excepting for this expedient they would have had to pass over in single file, and treble the time would have been consumed. Can it be contended that such insects are not able to determine by reasoning powers which is the best way of doing a thing?

Another observer, writing from the same part of the world to Büchner, gives a still more wonderful account of the ingenuity of Ecitons in crossing water. This observer is Herr H. Kreplin, of Heidemühl (Station Ducherom), 'who lived for nearly twenty years in South America as an engineer, and had often the opportunity of seeing the driver ants in the forests there.' He writes to Büchner, under date May 10, 1876, as follows:—

On both sides of the train, at about 10 mm. distance from each other, stronger ants are to be seen, distinguishable from the others by their foxy colour and very thick heads with gigantic mandibles. These 'thickheads' play the same rôle in the ant-state for which they are cast in cultured communities. They look after the order of the march, and allow none to turn either to the right or left. The least confusion in the regularity of the march makes them turn round and put things straight again. While the procession of the brown workers streams on unceasingly with a swarming motion, the 'officers,' as the natives call these thickheads, run constantly backwards and forwards, ready to take the command on meeting any difficulty. The crossing of streams by these creatures is the most interesting point. If the watercourse be narrow, the thickheads soon find trees, the branches of which meet on the bank on either side, and after a short halt the column set themselves in motion over these bridges, rearranging themselves in the narrow train with marvellous quickness on reaching the further side. But if no natural bridge be available for the passage, they travel along the bank of the river until they arrive at a flat sandy shore. Each ant now seizes a bit of dry wood, pulls it into the water, and mounts thereupon. The hinder rows push the front ones even further out, holding on to the wood with their feet and to their comrades with their jaws. In a short time the water is covered with ants, and when the raft has grown too large to be held together by the small creatures' strength, a part breaks itself off and begins the journey across, while the ants left on the bank busily pull their bits of wood into the water, and work at enlarging the ferry-boat until it again breaks. This is repeated as long as an ant remains on shore. I had often heard described this method of crossing rivers, but in the year 1859 I had the opportunity of seeing it for myself.

It is remarkable that the military or driving ants of Africa exhibit precisely similar devices for the bridging of streams, namely, by forming a chain of individuals over which the others pass. By means of similar chains they also let themselves down from trees. It must be observed, however, that these and all the above observations, being independently made and separately recorded, serve to corroborate one another so strongly that we can entertain no reasonable doubt concerning the wonderful facts which they convey.

I shall now bring these numerous instances to a close with a quotation from Mr. Belt, which reveals in the most unequivocal manner surprising powers of observation and rational action on the part of the leaf-cutting ants of South America, whose general habits we have already considered:—

A nest was made near one of our tramways, and to get to the trees the ants had to cross the rails, over which the waggons were continually passing and repassing. Every time they came along a number of ants were crushed to death. They persevered in crossing for some time, but at last set to work and tunnelled underneath each rail. One day, when the waggons were not running, I stopped up the tunnels with stones; but although great numbers carrying leaves were thus cut off from the nest, they would not cross the rails, but set to work making fresh tunnels underneath them.

Anatomy and Physiology of Nerve-centres and Sense-organs.

The foregoing facts concerning the intelligence of ants fully justifies Mr. Darwin's observation that 'the brain of an ant is one of the most marvellous atoms of matter in the world, perhaps more so than the brain of a man.' It may therefore be interesting in this particular case to depart from the lines otherwise laid down throughout the present work, and to devote a short section to the anatomy and physiology of this nerve-centre with its appended organs of sense.

The brain of an ant, then, is proportionally larger than that of any other insect. (See Titus Graber, 'Insects,' vol. i. p. 255.) In structure, also, the brain of an ant is in advance of that of other insects, its nearest analogue being the brain of a bee. The superiority of development is particularly remarkable with reference to the 'stalked bodies' of Dujardin; and these are largest in neuter

ANTS—GENERAL INTELLIGENCE.

workers, which are the most intelligent members of the community.

Injury of the brain causes, as in higher animals, tetanic spasms and involuntary reflex movements, followed by stupefaction.

An ant, whose brain has been perforated by the pointed mandibles of an amazon, remains as though nailed to its place; a shudder runs from time to time through its body, and one of its legs is lifted at regular intervals. It occasionally makes a short and quick step, as though driven by an unseen spring, but, like that of an automaton, aimless and objectless. If it is pulled, it makes a movement of avoidance, but falls back into its stupefied condition as soon as it is released. It is no longer capable of action consciously directed to a given object; it neither tries to escape, nor to attack, nor to go back to its home, nor to rejoin its companions, nor to walk away; it feels neither heat nor cold, it knows neither fear nor desire for food. It is merely an automatic and reflex machine, and is exactly similar to one of those pigeons from which Flourens removed the hemispheres of the cerebrum. Just in the same way behaves the body of an ant from which the head has been taken away. In the numerous fights between amazons and other ants, countless cases have been observed of slight injury to the brain, which have caused the most remarkable phenomena. Many of the wounded were seized with a mad rage, and flung themselves at every one that came in their way, whether friend or foe. Others assumed an appearance of indifference, and walked serenely about in the midst of the fighting. Others exhibited a sudden failure of strength; but they still recognised their enemies, approached them, and tried to bite them in cold blood, in a way quite foreign to the behaviour of healthy ants. They were also often observed to run round and round in a circle, the motion resembling the *manège*, or riding-school action of mammals, when one of the crura cerebri has been removed.

If an ant is cut in half through the thorax, so that the great nerve ganglia of the pro-thorax remain untouched, the behaviour of the head shows that intelligence also remains untouched. Ants mutilated in this way try to go forwards with their two remaining legs, and beg with their antennæ for their companions' aid. If one of these latter lets itself be stopped, then we observe a lively interchange of thanks and sympathy expressed by the actively moving antennæ. Forel placed near to each other two such mutilated bodies of the *F. rufibarbis*. They conversed with each other in the above-described way, and appeared each to beg for help. But when he put in some similarly mutilated ants of a hostile species, *F sanguinea*, the picture was changed; war broke out between these cripples just in the same way and with the same fury as between perfect ants.[1]

The antennæ appear to be the most important of the sense-organs, as their removal produces an extraordinary disturbance in the intelligence of the animal. An ant so mutilated can no longer find its way or recognise companions, and therefore is unable to distinguish between friends and foes. It is also unable to find food, ceases to engage in any labour, and loses all its regard for larvæ, remaining permanently quiet and almost motionless. A somewhat similar disturbance, or rather destruction, of the mental faculties is observable as a result of the same mutilation in the case of bees.[2]

[1] Büchner, *Geistesleben der Thiere*, English translation, p. 49.
[2] While this work is passing through the press, an interesting Essay has been published by Mr. MacCook on the Honey-making Ant. I am not here able to refer to this Essay at greater length, but have done so in a review in *Nature* (March 2, 1882.)—G. J. R.

CHAPTER IV.

BEES AND WASPS.

ARRANGING this chapter under the same general headings as the one on ants, we shall consider first—

Powers of Special Sense.

Bees and wasps have much greater powers of sight than ants. They not only perceive objects at a greater distance, but are also able to distinguish their colours. This was proved by Sir John Lubbock, who placed honey on slips of paper similarly formed, but of different colours; when a bee had repeatedly visited a slip of one colour (A), he transposed the slips during the absence of the bee; on its return the insect did not fly to slip B, although this now occupied the *position* which had been previously occupied by slip A, but again visited slip A, although this now occupied the position which had been previously occupied by slip B. Therefore, as these experiments were again and again repeated both on bees and wasps with uniform results, there can be no question that the insects by their first visits to slip A established an association between the colour of A and the honey upon it, such that, when they again returned and found B in the place of A, they were guided by their memory of the colour rather than by their memory of the position. It was thus shown that the insects could distinguish green, red, yellow, and blue. These experiments also brought out the further fact that both bees and wasps exhibit a marked preference for some colours over others. Thus, in a series of black, white, yellow, orange, green, blue, and red slips, two or three bees paid twenty-one visits to the orange and yellow, and only four to all the other slips. The slips were then moved, after which, out of thirty-two visits, twenty-two were to the orange and yellow. Another colour to which a similar preference is shown is blue.

As regards scent, Sir John found that on putting a few drops of eau de Cologne at the entrance of a beehive, 'immediately a number (about 15) came out to see what was the matter.' Other scents had a similar effect; but on repetition several times the bees became accustomed to the scent, and no longer came out.

As in ants, so in bees, Sir John's experiments failed to yield any evidence of a sense of hearing. But in this connection we must not forget the well-known fact, first observed by Huber, that the queen bee will answer by a certain sound the peculiar piping of a pupa queen; and again, by making a certain cry or humming noise, will strike consternation suddenly on all the bees in the hive—these remaining for a long time motionless as if stupefied.

Sense of Direction.

The following are Sir John Lubbock's observations upon this subject in the case of bees and wasps:—

Every one has heard of a 'bee-line.' It would be no less correct to speak of a wasp-line. On August 6 I marked a wasp, the nest of which was round the corner of the house, so that her direct way home was not out at the window by which she had entered, but in the opposite direction, across the room to a window which was closed. I watched her for some hours, during which time she constantly went to the wrong window, and lost much time in buzzing about at it. For ten consecutive days this wasp paid numerous visits, coming in at the open window, and always trying, though always unsuccessfully, to return to her nest in the 'wasp-line' of the closed window— buzzing about that window for hours at a time, though eventually on finding it closed she returned and went round through the open window by which she always entered.

This observation shows how strong must be the instinct in a wasp to take the shortest way home, and how much the insect depends upon its sense of direction in so doing. It also shows how long a time it requires to learn by individual experience the properties of a previously unknown substance such as glass. But to this latter point we shall presently have occasion to return.

Next we must adduce evidence to show that in way-finding the 'sense of direction' in bees appears to be largely supplemented by observation of particular objects.

Sir John Lubbock observes: 'I never found bees to return if brought any considerable distance at once. By taking them, however, some twenty yards each time they came to the honey, I at length *trained* them to come to my room;' that is to say, bees require to *learn* their way little by little before they can return to a store of honey which they may have been fortunate enough to find; their general sense of direction is not in itself a sufficient guide. This, at least, is the case where, as in the experiments in question, the bees are *carried* from the hive to the store of honey (here a distance of less than 200 yards): possibly if they had found the honey by themselves flying towards it, and so probably taking note of objects by the way, one journey might have proved sufficient to teach them the way. But, whether or not this would have been the case, the fact that when carried they required also to be taught the way piece by piece, is conclusive proof that their sense of direction *alone* is not sufficient to enable them to traverse a route of 200 yards a second time.

The same result is brought out by other experiments conducted on a different plan, though not apparently with this object. 'My room is square, with two windows on the south-west side, where the hive was placed, and one on the south-east.' Besides the ordinary entrance from outside, the hive had a small postern door opening into the room.

At 6.50 a bee came out through the little postern door. After she had fed, she evidently did not know her way home; so I put her back.
At 7.10 she came out again. I again fed her and put her back.
At 10.15 she came out a third time; and again I had to put her back.

At 10.55 she came out again, and still did not remember the door. Though I was satisfied that she really wished to return, and was not voluntarily remaining outside; still, to make the matter clear, I turned her out of a side window into the garden, when she at once returned to the hive.

At 11.15 she came out again; and again I had to show her the way back.

At 11.20 she came out again; and again I had to show her the way back (this makes five times); when, however,—

At 11.30 she came out again after feeding, she returned straight to the hive.

At 11.40 she came out, fed, and returned straight to the hive.

At 11.50 she came out, fed, and returned straight to the hive; she then stayed in for some time.

At 12.30 she came out again, but seemed to have forgotten the way back; after some time, however, she found the door and went in.

Again:—August 24 at 7.20 a bee came through the postern: I fed her; and though she was not frightened or disturbed, when she had finished her meal she flew to the window and had evidently lost her way; so at 8 o'clock I in pity put her back myself.

August 29.—A bee came out to the honey at 10.10; at 10.12 she flew to the window, and remained buzzing about till 11.12, when, being satisfied that she could not find her way, I put her in.

Nay, even those who seemed to know the postern, if taken near the other window, flew to it, and seemed to have lost themselves.

This cost me a great many bees. Those which got into my room by accident continually died on the floor near the window.

These observations show that even when a bee is not *carried* from the hive to the honey, but herself *flies* to it, her sense of direction is not alone sufficient to enable her to find the way back to the hive—or, rather, to the unaccustomed entrance to the hive from which she had come out. Probably if the side window had been open, the bee would have returned to the hive round the corner of the house, and through the entrance to which she was most accustomed. But as it was she had to *learn*, by five or six journeys, the way between the postern entrance and the food.

But the following observation on a wasp is in this connection the most conclusive.

A marked wasp visited honey exposed in the room before mentioned. 'The next morning she came—

At 7.25, and fed till 7.28, when she began flying about the room and even into the next; so I thought it well to put her out of the window, when she flew straight away to her nest. My room, as already mentioned, had windows on two sides; and the nest was in the direction of a closed window, so that the wasp had to go out of her way in going out through the open one.

At 7.45 she came back. I had moved the glass containing the honey about two yards; and though it stood conspicuously, the wasp seemed to have much difficulty in finding it. Again she flew to the window in the direction of her nest, and I had to put her out, which I did at 8.2.

At 8.15 she returned to the honey almost straight. 8.21, she flew again to the closed window, and apparently could not find her way; so at 8.35 I put her out again. It seems obvious from this that wasps have a sense of direction, and do not find their way merely by sight.

At 8.50 back to honey, and 8.54 again to wrong window; but finding it closed, she took two or three turns round the room, and then flew out through the open window.

At 9.24 back to the honey; and 9.27 away, first, however, paying a visit to the wrong window, but without alighting.

At 9.36 back to the honey, and 9.39 away, but, as before, going first to wrong window.

She was away therefore 9 minutes.

9.50	,,	9.53	away, this time straight. ,,	11	,,
10	,,	10. 7	,,	,, 11	,,
10.19	,,	10.22	,,	,, 12	,,
10.35	,,	10.39	,,	,, 13	,,
10.47	,,	10.50	,,	,, 9	,,
11. 4	,,	11. 7	,,	,, 14	,,
11.21	,,	11.24	,,	,, 14	,,
11.34	,,	11.37	,,	,, 10	,,
11.49	,,	11.52	,,	,, 1	,,

12. 3 back to the honey,		12. 5 away.		Away therefore	11	minutes.
12.13	,,	12.15½	,,	,,	8	,,
12.25	,,	12.28	,,	,,	10	,,
12.39	,,	12.43	,,	,,	11	,,
12.54	,,	12.57	,,	,,	11	,,
1.15	,,	1.19	,,	,,	18	,,
1.27	,,	1.30	,,	,,	8	minutes,'

&c., &c., the way being now clearly well learnt.

But that the sense of direction is of much service to bees in finding the locality of their hives seems to be indicated by the following observation thus narrated, on the authority of the authors themselves, by Messrs. Kirby and Spence :—

In vain, during my stay at St. Nicholas, I sallied out at every outlet to try to gain some idea of the extent and form of the town. Trees, trees, trees, still met me, and intercepted the view in every direction; and I defy any inhabitant bee of this rural metropolis, after once quitting its hive, ever to gain a glimpse of it again until nearly perpendicularly over it. The bees, therefore, ... must be led to their abodes by instinct, &c.

The observation, however, is not so conclusive as its authors suppose; for there is nothing to show that the bees did not take note of particular objects on their accustomed routes, and so learn these routes by stages. It would be worth while in this connection to try the effect of hooding the eyes of a bee, or, if this were deemed too disturbing an experiment, removing the hive bodily to a distance from its accustomed site, and observing whether the bees start away boldly as before for long flights, or learn their new routes by stages.

In this connection I may quote the following.

Mr. John Topham, of Marlborough House, Torquay, writing to 'Nature,'[1] says :—

On October 29, 1873, I removed a hive of bees in my garden, after it was quite dark, for a distance of 12 yards from the place in which it had stood for several months; and between its original situation and the new one there was a bushy evergreen tree, so that all sight of its former place was

[1] Vol. ix. p. 484.

obstructed to a person looking from the new situation of the hive.

Notwithstanding this change, the bees every day flew to the locality where they formerly lived, and continued flying around the site of what had been their home until, as night came on, they many of them sank upon the grass exhausted and chilled by the cold. Numbers, however, returned alive to their new position, after having looked in vain for their hive in its old place. At night I picked the exhausted bees up, and having restored warmth to them (by leaving them for a time on my coat-sleeve), I returned them to their companions.

Here was an illustration that the faculty of memory was superior to that of observation; but that was not all. Nearly every bee which I picked up during the 23 days through which this effort of memory lasted was an *old one*, as was easily deduced from observing the worn edges of the wings; showing that whilst the young insects were quick in receiving new impressions and in correcting errors, the nervous system of the old bees continued *acting in the direction which early habit had effected*. So true it is that 'one touch of nature makes the whole world kin.'

A closely similar observation has been told me by a friend, Mr. George Turner. He found that when he removed a beehive only a yard or two from its accustomed site, the bees, on returning home, flew in swarms around the latter, and for a long time were unable to find the hive. And several other similar cases might be adduced. Lastly, Thompson says:—

It is highly remarkable that they [bees] know their hive more from its locality than from its appearance, for if it be removed during their absence and a similar one be substituted, they enter the strange one. If the position of a hive be changed, the bees for the first day take no distant flight till they have thoroughly scrutinised every object in its neighbourhood.[1]

On the other hand, the writer of the article on 'Bees' in the 'Encyclopædia Britannica' says that in certain parts of France it is the habit of bee-keepers to place a number of hives upon a boat, which, in charge of a man, floats slowly down a river. The bees are thus continuously changing their pasture-ground, and yet do not lose their locomotive hives.

[1] *Passions of Animals*, p. 53.

It may be here worth while to add, parenthetically, as the only authentic observation with which I am acquainted concerning the distance that bees are accustomed to forage, the following statement of Prof. Hugh Blackburn. Writing from Glasgow University to 'Nature,'[1] he says that bees are found in a certain peach-house every spring at the time of blossom, although, so far as he can ascertain, the beehives nearest to the peach-house in question are his own, and these are at a distance of ten miles.

On the whole, then, and in the absence of further experiments, we must conclude it to be probable that the sense of direction with which hymenopterous insects are, as shown by some of Sir John Lubbock's experiments, unquestionably endowed, is of no small use to them in finding their way from home to food and *vice versâ*; although it appears certain, from other of his experiments, that this sense of direction is not in all cases a sufficient guide, and therefore requires to be supplemented by the definite observation of landmarks.

But the most conclusive evidence on this latter point is afforded by a highly interesting observation of Mr. Bates on the sand-wasps at Santurem, which may here be suitably introduced, as the insects are not distantly allied. He describes these animals as always taking a few turns in the air round the hole they had made in the sand before leaving to seek for flies in the forest, apparently in order to mark well the position of the burrow, so that on their return they might find it without difficulty. This observation has been since confirmed in a striking manner by Mr. Belt, who found that the sand-wasp takes the most precise bearings of an object the position of which she desires to remember. This observation is so interesting that it deserves to be rendered *in extenso* :—

A specimen of *Polistes carnifex* (*i.e.* the sand-wasp noticed by Mr. Bates) was hunting about for caterpillars in my garden. I found one about an inch long, and held it out towards it on the point of a stick. It seized it immediately, and commenced biting it from head to tail, soon reducing the soft body to a mass of pulp. It rolled up about one-half of it into a ball, and pro-

[1] Vol. xii. p. 68.

pared to carry it off. Being at the time amidst a thick mass of a fine-leaved climbing plant, it proceeded, before flying away, to take note of the place where it was leaving the other half. To do this, it hovered in front of it for a few seconds, then took small circles in front of it, then larger ones round the whole plant. I thought it had gone, but it returned again, and had another lock at the opening in the dense foliage down which the other half of the caterpillar lay. It then flew away, but must have left its burden for distribution with its comrades at the nest, for it returned in less than two minutes, and making one circle around the bush, descended to the opening, alighted on a leaf, and ran inside. The green remnant of the caterpillar was lying on another leaf inside, but not connected with the one on which the wasp alighted, so that in running in it missed it, and soon got hopelessly lost in the thick foliage. Coming out again, it took another circle, and pounced down on the same spot again, as soon as it came opposite to it. Three small seed-pods, which here grew close together, formed the marks that I had myself taken to note the place, and these the wasp seemed also to have taken as its guide, for it flew directly down to them, and ran inside; but the small leaf on which the fragment of caterpillar lay not being directly connected with any on the outside, it again missed it, and again got far away from the object of its search. It then flew out again, and the same process was repeated again and again. Always when in circling round it came in sight of the seed-pods down it pounced. alighted near them, and recommenced its quest on foot. I was surprised at its perseverance, and thought it would have given up the search; but not so, it returned at least half-a-dozen times, and seemed to get angry, hurrying about with buzzing wings. At last it stumbled across its prey, seized it eagerly, and as there was nothing more to come back for, flew straight off to its nest, without taking any further note of the locality. Such an action is not the result of blind instinct, but of a thinking mind; and it is wonderful to see an insect so differently constructed using a mental process similar to that of man.

Memory.

We may here first allude to an observation of Sir John Lubbock already quoted in another connexion (see p. 147). It is here evident that the wasp, after finding the store of honey in the room, and after finding the window closed in the 'wasp-line' direction to its nest, required three repeated *lessons* from Sir John before she *learnt* that the window on the other side of the room, and away from the direction of her nest, afforded no obstacle to her exit. Having learnt this, the fourth time she came she again flew to the closed window as before, and then, as if but dimly remembering that there was another opening somewhere that offered no such mysterious resistance to her passage, 'she took two or three turns round the room, and then flew out through the open window.' Having now taken the bearings of all the room upon her own wings, and having again found the difference between the two windows in respect of resistance, although in all other respects so much alike, the next time she came she made in the first instance as it were an experimental flight towards the closed window, but clearly had the alternative of going to the open one in her memory; for on finding the window closed as before, she did not alight, but flew straight from the closed to the open window. The same thing happened once again, but now, with the distinction between the two windows thus fully learnt, and with it the perception that in this case 'the shortest cut was the longest way round,' she never again flew to the closed window; in the forty successive visits which she paid through the remainder of that day, and the hundred visits or so which she made during the two following days, she seems to have uniformly flown to the open window.

As evidence of *forgetfulness*, it will be enough to refer to the case of another wasp which, under precisely similar circumstances to those just detailed, learnt her way out of the open window one day, having made fifty passages through it in five hours. Yet Sir John remarks,—

It struck me as curious that on the following day this wasp seemed by no means so sure of her way, but over and over again went to the closed window.

It is further of interest to note, as showing the similarity of the memory displayed by these insects with that of the higher animals, that there are considerable individual differences to be found in the degree of its manifestation.

In this respect they certainly differ considerably. Some of the bees which came out of the little postern door (already described) were able to find their way back after it had been shown to them a few times. Others were much more stupid; thus one bee came out on the 9th, 10th, 11th, 12th, 14th, 15th, 16th, 17th, 18th, and 19th, and came to the honey; but though I repeatedly put her back through the postern, she was never able to find her way for herself.

I often found that if bees which were brought to honey did not return at once, still they would do so a day or two afterwards. For instance, on July 11, 1874, a hot thundery day, and when the bees were much out of humour, I brought twelve bees to some honey; only one came back, and that one only twice; but on the following day several of them returned.

This latter observation is important, as proving that bees can remember for at least a whole day the locality where they have found honey only once before, and that they so far think about their past experiences as to return to that locality when foraging.

As the association of ideas by contiguity is the principle which forms the basis of all psychology, it is desirable to consider still more attentively this the earliest manifestation that we have of it in the memory of the Hymenoptera. That it is not exercised with exclusive reference to *locality* is proved by the following observation of Sir John Lubbock:—

I kept a specimen of *Polistes Gallica* for no less than nine months.[1] . . . I had no difficulty in inducing her to feed on my hand; but at first she was shy and nervous. She kept her sting in constant readiness. . . . Gradually she became quite used to me, and when I took her on my hand apparently expected to be fed. She even allowed me to stroke her without any appearance of fear, and for some months I never saw her sting.

One other observation which goes to prove that other things besides locality are noted and remembered by bees may here be quoted. Sir John placed a bee in a bell jar, the closed end of which he held towards a window. The bee buzzed about at that end trying to

---

[1] 'Three months' in the Journal of the Linnæan Society, but Sir John Lubbock informs me that this is a misprint.

make for the open air. He then showed her the way out of the open end of the jar, and after having thus learnt it, she was able to find the way out herself. This seems to show that the bee, like the wasp on the closed windowpane, was able to appreciate and to remember the difference between the quality of glass as resisting and air as permeable, although to her sense of vision the difference must have been very slight. In other words, the bee must have remembered that by first flying *away* from the window, round the edge of the jar, and then *towards* the window, she could surmount the transparent obstacle; and this implies a somewhat different act of memory from that of associating a particular object—such as honey—with a particular locality. It is noteworthy that a fly under similar circumstances did not require to be taught to find its way out of the jar, but spontaneously found its own way out. This, however, may be explained by the fact that flies do not always direct their flight towards windows, and therefore the escape of this one was probably not due to any act of intelligence.

While upon the subject of memory in the Hymenoptera, it is indispensable that we should again refer to the observation of Messrs. Belt and Bates already alluded to on pages 150–51. For it is from that observation rendered evident that these sand-wasps took definite pains, as it were, to *teach themselves* the localities to which they desired to return. Mr. Bates further observed that after thus taking a careful mental note of the place, they would return to it without a moment's hesitation after an absence of an hour. The observation of Mr. Belt, already quoted *in extenso*, proves that these mental notes may be taken with the utmost minuteness, so that even in the most intricate places the insect, on its return, is perfectly confident that it has not made a mistake.

With regard to the duration of memory, Stickney relates a case in which some bees took possession of a hollow place beneath a roof, and having been then removed into a hive, continued for several years to return and occupy the same hole with their successive swarms.[1]

[1] See Kirby and Spence, vol. ii. p. 591.

Similarly Huber relates an observation of his own showing the duration of memory in bees. One autumn he put some honey in a window, which the bees visited in large numbers. During the winter the honey was taken away and the shutters shut. When they were again opened in the spring the bees returned, although there was no honey in the window.

These two cases amply prove that the memory of bees is comparable with that of ants, which, as we have seen from analogous facts, also extends at least over a period of many months.

Emotions.

Sir John Lubbock's experiments on this head go to show that the social sympathies of bees are even less developed than he found them to be in certain species of ants. Thus he says:—

I have already mentioned with reference to the attachment which bees have been said to show for one another, that though I have repeatedly seen them lick a bee which had smeared herself in honey, I never observed them show the slightest attention to any of their comrades who had been drowned in water. Far, indeed, from having been able to discover any evidence of affection among them, they appear to be thoroughly callous and utterly indifferent to one another. As already mentioned, it was necessary for me occasionally to kill a bee; but I never found that the others took the slightest notice. Thus on the 11th of October I crushed a bee close to one which was feeding —in fact, so close that their wings touched; yet the survivor took no notice whatever of the death of her sister, but went on feeding with every appearance of composure and enjoyment, just as if nothing had happened. When the pressure was removed, she remained by the side of the corpse without the slightest appearance of apprehension, sorrow, or recognition. It was, of course, impossible for her to understand my reason for killing her companion; yet neither did she feel the slightest emotion at her sister's death, nor did she show any alarm lest the same fate should befall her also. In a second case exactly the same occurred. Again, I have several times, while a bee has been feeding, held a second bee by the leg close to her; the prisoner, of course, struggled to escape, and buzzed as loudly as she could; yet the selfish eater took no notice whatever. So far, therefore, from being at all affectionate, I doubt whether bees are in the least fond of one another.

Réaumur, however ('Insects,' vol. v., p. 265), narrates a case in which a hive-bee was partly drowned and so rendered insensible; the others in the hive carefully licked and otherwise tended her till she recovered. This seems to show that bees, like ants, are more apt to have their sympathies aroused by the sight of ailing or injured companions than by that of healthy companions in distress; but Sir John Lubbock's observations above quoted go to prove that even in this case display of sympathy is certainly not the rule.

Powers of Communication.

Huber says that when one wasp finds a store of honey 'it returns to its nest, and brings off in a short time a hundred other wasps;' and this statement is confirmed by Dujardin, who witnessed a somewhat similar performance in the case of bees—the individual which first found a concealed store informing other individuals of the fact, and so on till numberless individuals had found it.

Although the systematic experiments of Sir John Lubbock have not tended to confirm these observations with regard to bees and wasps, we must not too readily allow his negative results to discredit these positive observations—more especially as we have seen that his *later* experiments have fully confirmed the opinion of these previous authors with respect to ants. His experiments on bees and wasps consisted in exposing honey in a hidden situation, marking a bee or wasp that came to it, and observing whether it afterwards brought any companions to share the booty. He found that although the same insect would return over and over again, strangers came so rarely that their visits could only be attributed to accidental and independent discovery. Only if the honey were in an exposed situation, where the insects could *see* one another feeding, would one follow the other to the food.

But we have the more reason not to accept unreservedly

the conclusion to which these experiments in themselves might lead, because the very able observer F. Müller states an observation of his own which must be considered as alone sufficient to prove that bees are able to communicate information to one another:—

Once (he says[1]) I assisted at a curious contest, which took place between the queen and the other bees in one of my hives, which throws some light on the intellectual faculties of these animals. A set of forty-seven cells have been filled, eight on a newly completed comb, thirty-five on the following, and four around the first cell of a new comb. When the queen had laid eggs in all the cells of the two older combs she went several times round their circumference (as she always does, in order to ascertain whether she has not forgotten any cell), and then prepared to retreat into the lower part of the breeding-room. But as she had overlooked the four cells of the new comb, the workers ran impatiently from this part to the queen, pushing her, in an odd manner, with their heads, as they did also other workers they met with. In consequence the queen began again to go around on the two older combs; but as she did not find any cell wanting an egg she tried to descend, but everywhere she was pushed back by the workers. This contest lasted for a rather long while, till the queen escaped without having completed her work. Thus the workers knew how to advise the queen that something was as yet to be done, but they knew not how to show her *where* it had to be done.

Again, Mr. Josiah Emery, writing to 'Nature,'[2] with reference to Sir John Lubbock's experiments, says that the faculty of communication which bees possess is so well and generally known to the 'bee-hunters' of America, that the recognised method of finding a bees' nest is to act upon the faculty in question:—

Going to a field or wood at a distance from tame bees, with their box of honey they gather up from the flowers and imprison one or more bees, and after they have become sufficiently gorged, let them out to return to their home with their easily gotten load. Waiting patiently a longer or shorter time, according to the distance of the bee-tree, the hunter scarcely ever fails to see the bee or bees return accompanied with other bees, which are in like manner imprisoned till they in turn are

[1] Letter to Mr. Darwin, published in *Nature*, vol. x., p. 102.
[2] Vol. xii., pp. 25-6.

filled, when one or more are let out at places distant from each other, and the direction in each case in which the bee flies noted, and thus, by a kind of triangulation, the position of the bee-tree proximately ascertained.

Those who have stored honey in their houses understand very well how important it is to prevent a single bee from discovering its location. Such discovery is sure to be followed by a general onslaught from the hive unless all means of access is prevented. It is possible that our American are more intelligent than European bees, but hardly probable; and I certainly shall not ask an Englishman to admit it. Those in America who are in the habit of playing first, second, and third fiddle to instinct will probably attribute this seeming intelligence to that principle.

According to De Fravière, bees have a number of different notes or tones which they emit from the stigmata of the thorax and abdomen, and by which they communicate information. He says:—

As soon as a bee arrives with important news, it is at once surrounded, emits two or three shrill notes, and taps a comrade with its long, flexible, and very slender feelers, or antennæ. The friend passes on the news in similar fashion, and the intelligence soon traverses the whole hive. If it is of an agreeable kind—if, for instance, it concerns the discovery of a store of sugar or of honey, or of a flowering meadow—all remains orderly. But, on the other hand, great excitement arises if the news presages some threatened danger, or if strange animals are threatening invasion of the hive. It seems that such intelligence is conveyed first to the queen, as the most important person in the state.

This account, which is quoted from Büchner, no doubt bears indications of imaginative colouring; but if the observation as to the emission of sounds is correct—and, as we shall see, this point is well confirmed by other observers—it is most likely concerned in communicating by tone a general idea of good or harm: probably in the former case it acts as a sign, 'follow me;' and in the latter as a signal of danger. Büchner further says that, according to Landois, if a saucer of honey is placed before a hive, a few bees come out, which emit a cry of tut, tut, tut. This note is rather shrill, and resembles the cry of an attacked bee. Hereupon a large number of bees come out of the hive to collect the offered honey.

Again,—

The best way to observe the power of communication possessed by bees by means of their interchange of touches, is to take away the queen from a hive. In a little time, about an hour afterwards, the sad event will be noticed by a small part of the community, and these will stop working and run hastily about over the comb. But this only concerns part of the hive, and the side of a single comb. The excited bees, however, soon leave the little circle in which they at first revolved, and when they meet their comrades they cross their antennæ and lightly touch the others with them. The bees which have received some impression from this touch now become uneasy in their turn, and convey their uneasiness and distress in the same way to the other parts of the dwelling. The disorder increases rapidly, spreads to the other side of the comb, and at last to all the people. Then arises the general confusion before described.

Huber tested this communication by the antennæ by a striking experiment. He divided a hive into two quite separate parts by a partition wall, whereupon great excitement arose in the division in which there was no queen, and this was only quieted when some workers began to build royal cells.

He then divided a hive in similar fashion by a trellis, through which the bees could pass their feelers. In this case all remained quiet, and no attempt was made to build royal cells: the queen could also be clearly seen crossing her antennæ with the workers on the other side of the trellis.

Apparently the feelers are also connected with the exceedingly fine scent of the bees, which enables them, wonderful as it may seem, to distinguish friend and foe, and to recognise the members of their own hive among the thousands and thousands of bees swarming around, and to drive back from the entrance stranger or robber bees. The bee-masters, therefore, when they want two separate colonies or the members of them to unite in one hive, sprinkle water over the bees, or stupefy them with some fumigating substance, so as to make them to a certain extent insensible to smell, in order to attain their object. It is always possible to unite colonies by making the bees smell of some strong-smelling stuff, such as musk.[1]

Lastly, under the present heading I shall quote one o her observation, for which I am also indebted to

[1] *Loc. cit.*

Büchner's very admirable collection of facts relating to the psychology of Hymenoptera :—

Herr L. Brofft relates, in 'der Zoologische Garten' (XVIII. Year, No. 1, p. 67), that a poor and a rich hive stood next each other on his father's bee stand, and the latter suddenly lost its queen. Before the owner had come to a decision thereupon the bees of the two hives came to a mutual understanding as to the condition of their two states. The dwellers in the queenless hive, with their stores of provisions, went over into the less populous or poorer hive, after they had assured themselves, by many influential deputations, as to the state of the interior of the poor hive, and, as appeared, especially as to the presence of an egg-laying queen!

General Habits.

The active life of bees is divided between collecting food and rearing young. We shall therefore consider these two functions separately.

The food collected consists of two kinds, honey (which, although stored in the 'crop' for the purpose of carriage from the flowers to the cells, appears to be but the condensed nectar of flowers) and so-called 'bee-bread.' This consists of the pollen of flowers, which is worked into a kind of paste by the bees and stored in their cells till it is required to serve as food for their larvæ. It is then partly digested by the nurses with honey, so that a sort of chyle is formed. It is observable that in each flight the 'carrier bees' collect only one kind of pollen, so that it is possible for the 'house bees' (which, by the way, are the younger bees left at home to discharge domestic duties with only a small proportion of older ones, left probably to direct the more inexperienced young) to sort it for storage in different cells. In the result there are several different kinds of bee-bread, some being more stimulating or nutritious than others. The most nutritious has the effect, when given to any female larva, of developing that larva into a queen or fertile female. This fact is well known to the bees, who only feed a small number of larvæ in this manner, and the larvæ which they select so to feed they place in larger or 'royal' cells, with an obvious foreknowledge of the increased dimensions to which the animal will grow under the influence of this food. Only one queen is required for a single hive; but the bees always raise several, so that if any mishap should occur to one, other larvæ may be ready to fall back upon.

Besides honey and bee-bread two other substances are found in beehives. These are propolis and beeswax. The former is a kind of sticky resin collected for the most part from coniferous trees. This is used as mortar in building, &c. It adheres so strongly to the legs of the bee which has gathered it, that it can only be detached by the help of comrades. For this purpose the loaded bee presents her legs to her fellow-workers, who clean it off with their jaws, and while it is still ductile, apply it round the inside of the hive. According to Huber, who made this observation, the propolis is applied also to the insides of the cells. The workers first planed the surfaces with their mandibles, and one of them then pulled out a thread of propolis from the heap deposited by the carrier bees, severed it by a sudden throwing back of the head, and returned with it to the cell which it had previously been planing. It then laid the thread between the two walls which it had planed; but, proving too long, a portion of the thread was bitten off. The properly measured portion was then forced into the angle of the cell by the fore-feet and mandibles. The thread, now converted into a narrow ribbon, was next found to be too broad. It was therefore gnawed down to the proper width. Other bees then completed the work which this one had begun, till all the walls of the cells were framed with bands of propolis. The object of the propolis here seems to be that of giving strength to the cells.

The wax is a secretion which proceeds from between the segments of the abdomen. Having ingested a large meal of honey, the bees hang in a thick cluster from the top of their hive in order to secrete the wax. When it begins to exude, the bees, assisted by their companions, rub it off into heaps, and when a sufficient quantity of the material has been thus collected, the work begins of building the cells. As the cells are used both for storing food and rearing young, I shall consider them later on. Now we have to pass to the labours incidental to propagation.

All the eggs are laid by one queen, who requires during this season a large amount of nourishment, so much, indeed, that ten or twelve working bees (*i.e.* sterile females) are set apart as her feeders. Leaving the 'royal cell,' she walks over the nursery-combs attended by a retinue of workers, and drops a single egg into each open cell. It is a highly remarkable fact that the queen is able to control the sex of the eggs which she lays, and only deposits drone or male eggs in the drone cells, and worker or female eggs in the worker cells—the cells prepared for the reception of drone larvæ being larger than those required for the worker larvæ. Young queens lay more worker eggs than old queens, and when a queen, from increasing age or any other cause, lays too large a proportion of drone eggs, she is expelled from the community or put to death. It is remarkable, also, under these circumstances, that the queen herself seems to know that she has become useless, for she loses her propensity to attack other queens, and so does not run the risk of making the hive virtually queenless. There is now no doubt at all that the determining cause of an egg turning out male or female is that which Dzierzon has shown, namely, the absence or presence of fertilisation—unfertilised eggs always developing into males, and fertilised ones into females. The manner, therefore, in which a queen controls the sex of her eggs must depend on some power that she has of controlling their fertilisation.

The eggs hatch out into larvæ, which require constant attention from the workers, who feed them with the chyle or bee-bread already mentioned. In three weeks from the time that the egg is deposited, the white worm-like larva has passed through its last metamorphosis. When it has emancipated itself its nurses assemble round it to wash and caress it, as well as to supply it with food. They then clean out the cell which it has left.

When so large a number of the larvæ hatch out as to overcrowd the hive, it is the function of the queen to lead forth a swarm. Meanwhile several larval queens have been in course of development, and matters are so arranged by the foresight of the bees, that one or more young queens are ready to emerge at a time when otherwise the hive would be left queenless. But the young queen or queens, although perfectly formed, must not escape from their royal prison-houses until the swarm has fairly taken place; the worker bees will even strengthen the coverings of these prison-houses if, owing to bad weather or other causes, swarming is delayed. The prisoner queens, which are fed through a small hole in the roof of their cells, now continually give vent to a plaintive cry, called by the bee-keepers 'piping,' and this is answered by the mother queen. The tones of the piping vary. The reason why the young queens are kept such close prisoners till after the departure of the mother queen with her swarm, is simply that the mother queen would destroy all the younger ones, could she get the chance, by stinging them. The workers, therefore, never allow the old queen to approach the prisons of the younger ones. They establish a guard all round these prisons or royal cells, and beat off the old queen whenever she endeavours to approach. But if the swarming season is over, or anything should prevent a further swarm from being sent out, the worker bees offer no further resistance to the jealousy of the mother queen, but allow her in cold blood to sting to death all the young queens in their nursery prisons. As soon as the old queen leaves with a swarm, the young queens are liberated in succession; but at intervals of a few days; for if they were all liberated at once they would fall upon and destroy one another. Each young queen as it is liberated goes off with another swarm, and those which remain unliberated are as carefully guarded from the liberated sister queen as they were previously guarded from the mother queen. When the season is too late for swarming the remaining young queens are liberated simultaneously, and are then allowed to fight to the death, the survivor being received as sovereign.

The bees, far from seeking to prevent these battles, appear to excite the combatants against each other, surrounding and bringing them back to the charge when they are disposed to recede from each other; and when either of the queens shows a disposition to approach her antagonist, all the bees forming the cluster instantly give way to allow her full liberty of attack. The first use which the conquering queen makes of her victory is to secure herself against fresh dangers by destroying all her future rivals in the royal cells; while the other bees, which are spectators of the carnage, share in the spoil, greedily devouring any food which may be found at the bottom of the cells, and even sucking the fluid from the abdomen of the pupæ before they toss out the carcasses.[1]

Similarly, when a strange queen is put into a hive already provided with a queen—

A circle of bees instinctively crowd around the invader, not, however, to attack her—for a worker never assaults a queen—but to respectfully prevent her escape, in order that a combat may take place between her and their reigning monarch. The lawful possessor then advances towards the part of the comb where the invader has established herself, the attendant workers clear a space for the encounter, and, without interfering, wait the result. A fearful encounter then ensues, in which one is stung to death, the survivor mounting the throne. Although the workers of a *de facto* monarch will not fight for her defence, yet, if they perceive a strange queen *attempting* to enter the hive, they will surround her, and hold her until she is starved to death; but such is their respect for royalty that they never attempt to sting her.[2]

All these facts display a wonderful amount of apparently sagacious purpose on the part of the workers, although they may not seem to reflect much credit on the intelligence of the queens. But in this connection we must remember the observation of F. Huber, who saw two queens, which were the only ones left in the hive, engaged in mortal combat; and when an opportunity arose for each to sting the other simultaneously, they simultaneously released each other's grasp, as if in horror of a situation that might have ended in leaving the hive queenless. This, then, is the calamity to avert which all

[1] Art. 'Bees,' *Encycl. Brit.*
[2] Dr. Kemp, *Indications of Instinct.*

the instincts both of workers and queens are directed. And that these instincts are controlled by intelligence is suggested, if not proved, by the adaptations which they show to special circumstances. Thus, for instance, F. Huber smoked a hive so that the queen and older bees effected their escape, and took up their quarters a short distance away. The bees which remained behind set about constructing three royal cells for the purpose of rearing a new queen. Huber now carried back the old queen and ensconced her in the hive. Immediately the bees set about carrying away all the food from the royal cells, in order to prevent the larvæ contained therein from developing into queens. Again, if a strange queen is presented to a hive already provided with one, the workers do not wait for their own queen to destroy the pretender, but themselves sting or smother her to death. When, on the other hand, a queen is presented to a hive which is without one, the bees adopt her, although it is often necessary for the bee-master to protect her for a day or two in a trellis cage, until her subjects have become acquainted with her. When a hive is queenless, the bees stop all work, become restless, and make a dull complaining noise. This, however, is only the case if there is likewise a total absence of royal pupæ, and of ordinary pupæ under three days of age—*i.e.* the age during which it is possible to rear an ordinary larva into a queen.

As soon as the queen has been fertilised, and the services of the drones therefore no longer required, the worker bees fall upon their unfortunate and defenceless brothers to kill them, either by direct stinging or by throwing them out of the hive to perish in the cold. The drones' cells are then torn down, and any remaining drone eggs or pupæ destroyed. Generally all the drones—which may number more than a thousand—are slaughtered in the course of a single day. Evidently the object of this massacre is that of getting rid of useless mouths; but there is a more difficult question as to why these useless mouths ever came into existence. It has been suggested that the enormous disproportion between the present number of males and the single fertile female refers to a time before the social instincts became so complex or consolidated, and when, therefore, bees lived in lesser communities. Probably this is the explanation, although I think we might still have expected that before this period in their evolution had arrived bees might have developed a compensating instinct, either not to allow the queen to lay so many drone eggs, or else to massacre the drones while still in the larval state. But here we must remember that among the wasps the males do work (chiefly domestic work, for which they are fed by their foraging sisters); so it is possible that in the hive-bee the drones were originally useful members of the community, and that they have lost their primitively useful instincts. But whatever the explanation, it is very curious that here, among the animals which are justly regarded as exhibiting the highest perfection of instinct, we meet with perhaps the most flagrant instance in the animal kingdom of instinct unperfected. It is the more remarkable that the drone-killing instinct should not have been better developed in the direction of killing the drones at the most profitable time—namely, in their larval or oval state—from the fact that in many respects it seems to have been advanced to a high degree of discriminative refinement. Thus, to quote Büchner,—

> That the massacre of the drones is not performed entirely from an instinctive impulse, but in full consciousness of the object to be gained, is proved by the circumstance that it is carried out the more completely and mercilessly the more fertile the queen shows herself to be. But in cases where this fertility is subject to serious doubt, or when the queen has been fertilised too late or not at all, and therefore only lays drones' eggs, or when the queen is barren, and new queens, to be fertilised later, have to be brought up from working-bee larvæ, then all or some of the drones are left alive, in the clear prevision that their services will be required later. . . . This wise calculation of consequences is further exemplified in that sometimes the massacre of the drones takes place before the time for swarming, as, for instance, when long-continued unfavourable weather succeeds a favourable beginning of spring, and makes the bees anxious for their own welfare. If, however, the weather breaks, and work again becomes possible, so that the

bees take courage anew, they then bring up new drones, and prepare them in time for the swarming. This killing of drones is distinguished from the regular drone massacre by the fact that the bees then only kill the developed drones, and leave the drone larvæ, save when absolute hunger compels their destruction. Not less can it be regarded as a prudent calculation of circumstances when the bees of a hive, brought from our temperate climate to a more southern country, where the time of collecting lasts longer, do not kill the drones in August, as usual, but at a later period, suitable to the new conditions.

But the philosophy of drone-killing is, I think, even more difficult in the case of the wasps than in that of the bees. For, unlike the bees, whose communities live from year to year, the wasps all perish at the end of autumn, with the exception of a very few fertilised females. As this season of universal calamity approaches, the workers destroy all the larval grubs—a proceeding which, in the opinion of some writers, strikingly exemplifies the beneficence of the Deity! Now, it does not appear to me easy to understand how the presence of such an instinct in this case is to be explained. For, on the one hand, the individual females which are destined to live through the winter cannot be conspicuously benefited by this slaughter of grubs; and, on the other hand, the rest of the community is so soon about to perish, that one fails to see of what advantage it can be to it to get rid of the grubs. If the whole human race, with the exception of a few women, were to perish periodically once in a thousand years, the race would profit nothing by destroying, a few months before the end of each millennium, all sick persons, lunatics, and other 'useless mouths.' I have not seen this difficulty with regard to the massacring instinct in wasps mentioned before, and I only mention it now in order to draw attention to the fact that there seems to be a more puzzling problem presented here than in the case of the analogous instinct as exhibited by bees. The only solution which has suggested itself to my mind is the possibility that in earlier times, or in other climates, wasps may have resembled bees in living through the winter, and that the grub-slaying instinct is in them a survival of one which was then, as in the case of the bees now, a clearly beneficial instinct.

. For some days before swarming begins, there is a great excitement and buzzing in the hive, the temperature of which rises from 92° to 104°. Scouts having been previously sent out to explore for suitable quarters wherein to plant the new colony, these now act as guides. The swarm leaves the hive with their queen. The bees which remain behind busy themselves in rearing out the pupæ, which soon arriving at maturity, also quit the hive in successive swarms. According to Büchner, 'secondary swarms with young queens send out no scouts, but fly at random through the air. They clearly lack the experience and prudence of the older bees.' And, regarding the behaviour of the scouts sent out by primary swarms, this author says :—

M. de Fravière had the opportunity of observing the manner in which such an examination is carried on, and with what prudence and accuracy. He placed an empty beehive, made in a new style, in front of his house, so that he could exactly watch from his own window what went on inside and out without disturbance to himself or to the bees. A single bee came and examined the building, flying all round it and touching it. It then let itself down on the board, and walked carefully and thoroughly over the interior, touching it continually with its antennæ so as to subject it on all sides to a thorough investigation. The result of its examination must have been satisfactory, for after it had gone away it returned accompanied by a crowd of some fifty friends, which now together went through the same process as their guide. This new trial must also have had a good result, for soon a whole swarm came, evidently from a distant spot, and took possession. Still more remarkable is the behaviour of the scouts when they take possession of a satisfactory hive or box for an imminent or approaching swarm. Although it is not yet inhabited they regard it as their property, watch it and guard it against stranger bees or other assailants, and busy themselves earnestly in the most careful cleansing of it, so far as this cleansing is impossible to the setter up of the hive. Such a taking possession sometimes occurs eight days before the entrance of the swarm.

Wars.—As with ants, so with bees, the great cause of

war is plunder; and facts now well substantiated by numberless observers concerning 'robber-bees' indicate a large measure of intelligence. These aim at lessening their labour in collecting honey by plundering the store of other hives. The robberies may be conducted singly or in concert. When the thieving propensity is developed only in individual cases, the thieves cannot rely on force in plundering a foreign state, and so resort to cautious stealth. 'They show by their whole behaviour—creeping into the hive with careful vigilance—that they are perfectly conscious of their bad conduct; whereas the workers belonging to the hive fly in quickly and openly, and in full consciousness of their right.' If such solitary burglars are successful in obtaining plunder, their bad example leads other members of their own community to imitate them; thus it is that the whole bee-nation may develop marauding habits, and when they do this they act in concert to rob by force. In this case an army of bees precipitates itself upon the foreign hive, a battle ensues, and if successful in overcoming resistance, the invaders first of all search out the queen-bee and put her to death, whereby they disorganise their enemies and plunder the hive with ease. It is observed that when this policy is once successful, the spirit of aggrandisement is encouraged, so that the robber-bees 'find more pleasure in robbery than in their own work, and become at last formidable robber-states.' When an invaded hive is fairly overcome by the invaders killing the queen, the owners of the hive, finding that all is lost, not only abandon further resistance, but very often reverse their policy and join the ranks of their conquerors. They assist in the tearing down of their cells, and in the conveyance of the honey to the hive of their invaders. 'When the assailed hive is emptied, the next ones are attacked, and if no effective resistance is offered, are robbed in similar fashion, so that in this way a whole bee-stand may be gradually destroyed.' Siebold observed the same facts in the case of wasps (*Polistes gallica*). If, however, the battle turns in favour of the defenders, they pursue the flying legions of their enemies to a distance from their home. It sometimes happens that the plundered hive offers no resistance at all, owing to the robbers having visited the same flowers as the robbed, and so probably (having much the same smell) not being recognised as belonging to a different community. The thieves, when they find such to be the case, may become so bold as to stop the bees that are returning to the hive with their loads, of which they deprive them at the entrance of the hive. This is done by a process which one observer, Weygandt,[1] calls 'milking,' and it seems that the milking bee attains the double advantage of securing the honey from the milked one and disarming suspicion of the other bees by contracting its smell and entering the hive loaded, into which it is admitted without opposition to continue its plunder.

Sometimes robber-bees attack their victims in the fields at a distance from the hives. This sort of highway robbery is generally conducted by a gang of four or five robber-bees which set upon a single honest bee, 'hold him by the legs, and pinch him until he unfolds his tongue, which is sucked in succession by his assailants, who then suffer him to depart in peace.'

It is strange that hive-bees of dishonest temperaments seem able to coax or wheedle humble-bees into the voluntary yielding of honey. 'Humble-bees have been known to permit hive-bees to take the whole honey that they have collected, and to go on gathering more, and handing it over, for three weeks, although they refuse to part with it, or seek refuge in flight, when wasps make similar overtures.'[2]

Besides theft and plunder, there are other causes of warfare among bees, which, however, are only apparent in their effects. Thus, for some undiscernible reason, duels are not infrequent, which generally end in the death of one or both combatants. At other times, equally without apparent reason, civil war breaks out in a hive, which is sometimes attended with much slaughter.

Architecture.—Coming now to the construction of the cells and combs, there is no doubt that here we meet with

---

[1] *The Bee*, 1877, No. 1.
[2] Dr. Lindley Kemp, *Indications of Instinct*.

the most astonishing products of instinct that are presented in the animal kingdom. A great deal has been written on the practical exhibition of high mathematical principles which bees display in constructing their combs in the form that secures the utmost capacity for storage of honey with the smallest expenditure of building material. The shortest and clearest statement of the subject that I have met with is the following, which has been given by Dr. Reid :—

There are only three possible figures of the cells which can make them all equal and similar, without any useless interstices. These are the equilateral triangle, the square, and the regular hexagon. Mathematicians know that there is not a fourth way possible in which a plane may be cut into little spaces that shall be equal, similar, and regular, without useless spaces. Of the three figures, the hexagon is the most proper for convenience and strength. Bees, as if they knew this, make their cells regular hexagons.

Again, it has been demonstrated that, by making the bottoms of the cells to consist of three planes meeting in a point, there is a saving of material and labour in no way inconsiderable. The bees, as if acquainted with these principles of solid geometry, follow them most accurately. It is a curious mathematical problem, at what precise angle the three planes which compose the bottom of a cell ought to meet, in order to make the greatest possible saving, or the least expense of material and labour. This is one of the problems which belong to the higher parts of mathematics. It has accordingly been resolved by some mathematicians, particularly by the ingenious Maclaurin, by a fluctionary calculation, which is to be found in the Transactions of the Royal Society of London. He has determined precisely the angle required, and he found, by the most exact mensuration the subject would admit, that it is the very angle in which the three planes in the bottom of the cell of a honeycomb do actually meet.[1]

Marvellous as these facts undoubtedly are, they may now be regarded as having been satisfactorily explained. Long ago Buffon sought to account for the hexagonal form of the cells by an hypothesis of mutual pressure. Supposing the bees to have a tendency to build tubular

[1] Handcock on Instinct, p. 18.

cells, if a greater number of bees were to build in a given space than could admit of all the parallel tubes being completed, tubes with flat sides and sharp angles might result, and if the mutual pressure were exactly equal in all directions, these sides and angles would assume the form of hexagons. This hypothesis of Buffon was sustained by such physical analogies as the blowing of a crowd of soap-bubbles in a cup, the swelling of moistened peas in a confined space, &c. The hypothesis, however, as thus presented was clearly inadequate; for no reason is assigned why the mutual pressure, even if conceded to exist, should always be so exactly equal in all directions as to convert all the cylinders into perfect hexagons—even the analogy of the soap-bubbles and the moistened peas failing, as pointed out by Brougham and others, to sustain it, seeing that as a matter of fact bubbles and peas under circumstances of mutual pressure do not assume the form of hexagons, but, on the contrary, forms which are conspicuously irregular. Moreover, the hypothesis fails to account for the particular prismatic shape presented by the cell base. Therefore it is not surprising that this hypothesis should have gained but small acceptance. Kirby and Spence dispose of it thus:—'He (Buffon) gravely tells us that the boasted hexagonal cells of the bee are produced by the reciprocal pressure of the cylindrical bodies of these insects against each other!!'[1] The double note of admiration here may be taken to express the feelings with which this hypothesis of Buffon was regarded by all the more sober-minded naturalists. Yet it turns out to have been not very wide of the mark. As is often the case with the gropings of a great mind, the idea contains the true principle of the explanation, although it fails as an explanation from not being in a position to take sufficient cognizance of all the facts. Safer it is for lesser minds to restrain their notes of exclamation while considering the theories of a greater; however crude or absurd the latter may appear, the place of their birth renders it not impossible that some day they may prove to have been prophetic of truth revealed by fuller know-

[1] *Introd. Ent*, ii, p. 465.

ledge. Usually in such cases the final explanation is eventually reached by the working of a yet greater mind, and in this case the undivided credit of solving the problem is to be assigned to the genius of Darwin.

Mr. Waterhouse pointed out 'that the form of the cell stands in close relation to the presence of adjoining cells.' Starting from this fact, Mr. Darwin says,—

Let us look to the great principle of gradation, and see whether Nature does not reveal to us her method of work. At one end of a short series we have humble-bees, which use their old cocoons to hold honey, sometimes adding to them short tubes of wax, and likewise making separate and very irregular rounded cells of wax. At the other end of the series we have the cells of the hive-bee, placed in a double layer. . . . In the series between the extreme perfection of the cells of the hive-bee and the simplicity of those of the humble-bee we have the cells of the Mexican *Melipona domestica*, carefully described and figured by Pierre Huber. . . . It forms a nearly regular waxen comb of cylindrical cells, in which the young are hatched, and, in addition, some large cells of wax for holding honey. These latter cells are nearly spherical and of nearly equal sizes, and are aggregated into an irregular mass. But the important thing to notice is, that these cells are always made at that degree of nearness to each other that they would have intersected or broken into each other if the spheres had been completed; but this is never permitted, the bees building perfectly flat cells of wax between the spheres which thus tend to intersect. Hence each cell consists of an outer spherical portion; and of two, three, or more flat surfaces, according as the cell adjoins two, three, or more other cells. When one cell rests on three other cells, which, from the spheres being nearly of the same size, is very frequently and necessarily the case, the three flat surfaces are united into a pyramid; and this pyramid, as Huber has remarked, is manifestly a gross imitation of the three-sided pyramidal base of the cell of the hive-bee. . . .

Reflecting on this case, it occurred to me that if the Melipona had made its spheres at some given distance from each other, and had made them of equal sizes, and had arranged them symmetrically in a double layer, the resulting structure would have been as perfect as the comb of the hive-bee. Accordingly I wrote to Prof. Miller of Cambridge, and this geometer has kindly read over the following statement, drawn up from his information, and tells me that it is strictly correct.

This statement having fully borne out his theory, Mr. Darwin continues:—

Hence we may safely conclude that, if we could slightly modify the instincts already possessed by the Melipona, and in themselves not very wonderful, this bee would make a structure as wonderfully perfect as that of the hive-bee. We must suppose the Melipona to have the power of forming her cells truly spherical, and of equal sizes; and this would not be very surprising, seeing that she already does so to a certain extent, and seeing what perfectly cylindrical burrows many insects make in wood, apparently by turning round on a fixed point. We must suppose the Melipona to arrange her cells in level layers, as she already does her cylindrical cells; and we must further suppose —and this is the greatest difficulty— that she can somehow judge accurately at what distance to stand from her fellow-labourers when several are making their spheres; but she is already so far able to judge of distance that she always describes her spheres so as to intersect to a certain extent; and then she unites the points of intersection by perfectly flat surfaces. By such modifications of instinct, which in themselves are not very wonderful—hardly more wonderful than those which guide a bird to make its nest,—I believe that the hive-bee has acquired through natural selection her inimitable architectural powers.[1]

Mr. Darwin next tested this theory by the experiment of introducing into beehives plates of wax, and observing that the bees worked upon these plates just as the theory required. That is to say, they made their cells by excavating a number of little circular pits at equal distances from one another, so that by the time the pits had acquired the width of an ordinary cell, the sides of the pits intersected. As soon as this occurred the bees ceased to excavate, and instead began to build up flat walls of wax on the lines of intersection. Other experiments with very thin plates of vermilion-coloured wax showed that the bees all worked at about the same rate, and on opposite sides of the plates, so that the common bottoms of any two opposite pits were flat. These flat bottoms 'were situated, as far as the eye could judge, exactly along the planes of imaginary intersection between the basins on the opposite sides of the ridge of wax;' so that if the

[1] *Origin of Species*, 'Cell-making Instinct.'

plate of wax had been thick enough to admit of the opposite basins being deepened (and widened) into cells, the mutual intersection of *adjacent* as well as *opposite* bottoms would have given rise, as in the first experiment with the thick plate of wax, to the pyramidal bottoms. Experiments with the vermilion wax also showed, as Huber had previously stated, that a number of individual bees work by turns at the same cell; for by covering parts of growing cells with vermilion wax, Mr. Darwin—

Invariably found that the colour was most delicately diffused by the bees—as delicately as a painter could have done it with his brush—by atoms of the coloured wax having been taken from the spot on which it had been placed, and worked into the growing edges of the cells all round.

Such, omitting details, is the substance of Mr. Darwin's theory. In summary he concludes,—

The work of construction seems to be a sort of balance struck between many bees, all instinctively standing at the same relative distance from each other, all trying to sweep equal spheres, and then building up, or leaving ungnawed, the planes of intersection between these spheres.

This theory, while serving as a full and simple explanation of all the facts, has, as we have seen, been so fully substantiated by observation and experiment, that it deserves to be regarded as raised to the rank of a completed demonstration. It differs from the theory of Buffon in two important particulars: it embraces all the facts, and supplies a cause adequate to explain them. This cause is natural selection, which converts the random 'pressure' in Buffon's theory into a precisely regulated principle. Random pressure alone could never produce the beautifully symmetrical form of the hexagonal cell with the pyramidal bottom; but it could and must have produced the intersection of cylindrical cells among possibly many extinct species of bees, such as the Melipona. Whenever this intersection occurred in crowded nests, it must clearly have been of great benefit in securing economy of precious wax; for in every case where a flat wall of partition between two adjacent cells did duty instead of a double cylindrical wall of separate cells, there wax should have been saved. Thus we can see how natural selection would have worked towards the developing of an instinct to excavate cells near enough together to produce intersection; and once begun, there is no reason why this instinct should not have been perfected by the same agency, till we meet with its ideal perfection in the hive-bee. For as Mr. Darwin observes,—

With respect to the formation of wax, it is known that bees are often hard pressed to get sufficient nectar; and I am informed by Mr. Tegetmeier that it has been experimentally proved that from twelve to fifteen pounds of dry sugar are consumed by a hive of bees for the secretion of a pound of wax; so that a prodigious quantity of fluid nectar must be collected and consumed by the bees in a hive for the secretion of the wax necessary for the construction of their combs. Moreover, many bees have to remain idle for many days during the process of secretion. . . . Hence it would continually be more and more advantageous to our humble-bees if they were to make their cells more and more regular, nearer together, and aggregated into a mass, like the cells of Melipona; for in this case a large part of the bounding surface of each cell would serve to bound the adjoining cell, and much labour and wax would be saved. Again, from the same cause, it would be advantageous to the Melipona if she were to make her cells closer together, and more regular in every way than at present; for then, as we have seen, the spherical surfaces would wholly disappear and be replaced by plane surfaces; and the Melipona would make a comb as perfect as that of the hive-bee. Beyond this stage of perfection in architecture, natural selection could not lead; for the comb of the hive-bee, as far as we can see, is absolutely perfect in economising labour and wax.

The problem, then, as to the origin and perfection of the cell-making instinct appears thus to have been fully and finally solved. I shall now adduce a few facts to show that while the general instinct of building hexagonal cells has doubtless been acquired by natural selection in the way just explained, it is nevertheless an instinct not wholly of a blind or mechanical kind, but is constantly under the control of intelligent purpose. Thus Mr. Darwin observes,—

It was really curious to note in cases of difficulty, as when two pieces of comb met at an angle, how often the bees would pull down and rebuild in different ways the same cell, sometimes recurring to a shape which they had at first rejected.[1]

Again, Huber saw a bee building upon the wax which had already been put together by her comrades. But she did not arrange it properly, or in a way to continue the design of her predecessors, so that her building made an undesirable corner with theirs. 'Another bee perceived it, pulled down the bad work before our eyes, and gave it to the first in the requisite order, so that it might exactly follow the original direction.' Similarly, to quote Büchner,—

All the cells have not the same shape, as would be the case if the bees in building worked according to a perfectly instinctive and unchangeable plan. There are very manifold changes and irregularities. Almost in every comb irregular and unfinished cells are to be found, especially where the several divisions of a comb come together. The small architects do not begin their comb from a single centre, but begin building from many different points, so as to progress as rapidly as possible, and so that the greatest number may work simultaneously; they therefore build from above downwards, in the shape of flat truncated cones or hanging pyramids, and these several portions are afterwards united together during the winter building. At these lines of junction it is impossible to avoid irregular cells between the pressed together or unnaturally lengthened ones. The same is true more or less of the passage cells, which are made to unite the large cells of the so-called drone wax with the smaller ones of the working bees, and which are generally placed in two or three rows. The cells also which they usually build from the combs to the glass walls of their hives, in order to hold them up, show somewhat irregular forms. Finally, in places where special conditions of the situation do not otherwise permit, it may be observed that the bees, far from clinging obstinately to their plan, very well understand how to accommodate themselves to circumstancces not only in cell-building, but also in making their combs. F. Huber tried to mislead their instinct, or rather to put to the proof their reason and cleverness in every possible way, but they always emerged triumphant from the ordeal. For instance, he put bees in a hive

[1] *Origin of Species*, p. 225.

the floor and roof of which were made of glass, that is of a body which the bees use very unwillingly for the attachment of their combs, on account of its smoothness. Thus the possibility of building as usual from above downwards, and also from below upwards, was taken away from them; they had no point of support save the perpendicular walls of their dwelling. They thereupon built on one of these walls a regular stratum of cells, from which, building sideways, they tried to carry the comb to the opposite side of the hive. To prevent this Huber covered that side also with glass. But what way out of the difficulty was found by the clever insects? Instead of building further in the projected direction, they bent the comb round at the extreme point, and carried it at a right angle towards one of the inner sides of the hive which was not covered with glass, and there fastened it. The form and dimensions of the cells must necessarily have been altered thereby, and the arrangement of their work at the angle must have been quite different from the usual. They made the cells of the convex side so much broader than those of the concave that they had a diameter two or three times as great, and yet they managed to join them properly with the others. They also did not wait to bend the comb until they came to the glass itself, but recognised the difficulty beforehand,[1] which had been interposed by Huber while they were building with a view to overcome the first difficulty.

Special Habits.

The Mason-Bee.—This insect closes the roof of its larval cell with a kind of mortar, which sets as hard as stone. A little hole, closed only with soft mud, is, however, left in one part of the roof as a door of exit for the matured insect. It is said that when a mason-bee finds an old and deserted nest, it saves itself the trouble of making a new one—utilising the ready-made nest after having well cleaned it. In Algiers the mason-bees have been observed in this way to utilise empty snail-shells. According to Blanchard, some individuals avoid the labour of making their own nests or houses for their young, by possessing themselves of their neighbours' houses either by craft or by force. 'Does the mason-bee act like a machine,' says E. Menault, 'when it directs its work ac-

[1] *Mind in Animals,* pp. 252-3.

cording to circumstances, possesses itself of old nests, cleanses and improves them, and thereby shows that it can fully appreciate the immediate position? Can one believe that no kind of reflection is here necessary?'

The Tapestry-Bee.—The so-called tapestry-bee digs holes for her larvæ three or four inches deep in the earth, and lines the walls and floor of the chamber with petals of the poppy laid perfectly smooth. Several layers of petals are used, and when the eggs are introduced the chamber is closed by drawing all the leaves together at the top. Loose earth is then piled over the whole structure in order to conceal it. The so-called rose-bee (*Megachile centuncularis*) displays very similar habits.[1]

The Carpenter-Bee.—This was first observed and described by Réaumur.[2] It makes a long cylindrical tube in the wood of beams, palings, &c. This it divides into a number of successive chambers by partitions made of agglutinated saw-dust built across the tube at right angles to its axis. In each chamber there is deposited a single egg, together with a store of pollen for the nourishment of the future larva. The larvæ hatch out in succession and in the order of their age—*i.e.* the dates at which they were deposited. To provide for this, the bee bores a hole from the lower cell to the exterior, so that each larva, when ready to escape from its chamber, finds an open way from the tube. The larvæ have to cut their own way out through the walls of their respective chambers, and it is remarkable that they always cut through the wall that faces the tubular passage left by the parent; they never bore their way out in the opposite direction, which, were they to do so, would entail the destruction of all the other and immature larvæ.

The Carding-Bee.—This insect surrounds its nest with a layer of wax, and then with a thick covering of moss. For this purpose a number of bees co-operate, and in order to save time each bee does not find and carry its own moss, but, with a division of labour similar to that

---

[1] For a complete account of these habits see Bingley, *Animal Biography*, vol. iii., pp. 272-5.
[2] *Mém. sur les Insectes*, tom. vi., p. 39.

which we have already noticed in the case of certain ants, a row of bees is formed, and the bits of moss passed from one to another along the line. There is a long passage to the nest, through which the moss has to be passed, and it is said that at the mouth of the tunnel a guard is stationed to drive away ants or other intruders.

Wasps.—These usually construct their nests of wood-dust, which they scrape off the weather-worn surfaces of boards, palings, &c., and work into a kind of paper with their saliva. If they happen to find any real paper, they perceive that it so much resembles the product of their own manufacture that they utilise it forthwith. The wasps do not require any special cells or chambers for the storage of honey, as they do not lay up any supply for the winter. The cells which they construct are therefore used exclusively for the rearing of larvæ. In form these cells are sometimes cylindrical or globular, but more usually hexagonal, like those of the hive-bee. Although the mode of building is different from that employed by the bees, there can be little doubt that if it were as carefully investigated Mr. Darwin's theory of transition from the cylindrical to the hexagonal form would be found to apply here also, seeing that both forms so frequently occur in the same nest.

The Mason-Wasp.—The habits of this insect are described by Mr. Bates. It constructs its nest of clay. Each pellet that the insect brings it lays on the top of its nest-wall, and then spreads it out with its jaws, and treads it smooth with its feet. The nest, which is suspended on the branch of a tree, is then stocked with spiders and insects paralysed by stinging. The victims, not being wholly deprived of life, keep fresh until required as food of the developing larvæ.

The Butcher-Wasps.—These also paralyse their prey in a similar manner, and for a similar purpose. Fabre removed from a so-called sphex-wasp a killed grasshopper, which it was conveying to its nest and had momentarily laid down at the mouth of the burrow—as these insects always do on returning with prey, in order to see that nothing has intruded into the burrow during their absence. Fabre carried the dead or paralysed grasshopper to a considerable distance from the hole. On coming out the insect searched about until it found its prey. It then again carried it to the mouth of its burrow, and again laid it down while it once more went in to see that all was right at home. Again Fabre removed the grasshopper, and so on for forty times in succession—the sphex never omitting to go through its fixed routine of examining the interior of its burrow every time that it brought the prey to its mouth.

Mr. Mivart, in his 'Lessons from Nature,' points to the instinct of this animal in the stinging of the ganglion of its prey as one that cannot be explained on Mr. Darwin's theory concerning the origin of instincts. In my next work, which will have to deal with this theory, I shall consider Mr. Mivart's difficulty, and also the difficulty first pointed out by Mr. Darwin himself as to why neuter insects, separated as they appear to be from the possibility of communicating by heredity any instinctive acquirements of the individual to the species, should present any instincts at all.

General Intelligence.

Beginning with Sir John Lubbock's observations on this head, I shall first quote his statements with regard to way-finding :—

I have found, he says, that some bees are much more intelligent in this respect than others. A bee which I had fed several times, and which had flown about in the room, found its way out of the glass in a quarter of an hour, and when put in a second time came out at once. Another bee, when I closed the postern door, used to come round to the honey through an open window.

Bees seem to me much less clever in finding things than I had expected. One day (April 14, 1872), when a number of them were very busy on some barberries, I put a saucer with some honey between two bunches of flowers; these were repeatedly visited, and were so close that there was hardly room for the saucer betweeen them, yet from 9.30 to 3.30 not a single bee took any notice of the honey. At 3.30 I put some honey on one of the bunches of flowers, and it was eagerly sucked by the bees; two kept continually returning till past five in the evening.

One day when I came home in the afternoon I found that at least a hundred bees had got into my room through the postern and were on the window, yet not one was attracted by an open jar of honey which stood in a shady corner about 3 feet 6 inches from the window.

One day (29th April, 1872) I placed a saucer of honey close to some forget-me-nots, on which bees were numerous and busy; yet from 10 A.M. till 6 only one bee went to the honey.

I put some honey in a hollow in the garden wall opposite the hives at 10.30 (this wall is about five feet high and four feet from the hives); yet the bees did not find it during the whole day.

On the 30th March, 1873, a fine sunshiny day, when the bees were very active, I placed a glass containing honey at 9 in the morning on the wall in front of the hives; but not a single bee went to the honey the whole day. On April 20 I tried the same experiment, with the same result.

September 19.—At 9.30 I placed some honey in a glass about four feet from and just in front of the hive; but during the whole day not a bee observed it.

As it then occurred to me that it might be suggested that there was something about this honey which rendered it unattractive to the bees, on a following day I placed it again on the top of the wall for three hours, during which not a single bee came, and then moved it close to the alighting-board of the hive. It remained unnoticed for a quarter of an hour, when two bees observed it; and others soon followed in considerable numbers. . . . On the whole, wasps seem to me more clever in finding their way than bees. I tried wasps with the glass mentioned on p. 124 [*i.e.* the bell-jar], but they had no difficulty in finding their way out.

We shall now conclude this *résumé* of Sir John Lubbock's observations by quoting two other passages bearing on the general intelligence of bees and wasps:—

The following fact struck me as rather remarkable. The wasp already mentioned at the foot of p. 135 one day smeared her wings with syrup, so that she could not fly. When this happened to a bee, it was only necessary to carry her to the alighting-board; when she was soon cleaned by her comrades. But I did not know where this wasp's nest was, and therefore could not pursue a similar course with her. At first, then, I was afraid that she was doomed. I thought, however, that I would wash her, fully expecting, indeed, to terrify her so much that she would not return again. I therefore caught her, put her in a bottle half full of water, and shook her up well till the honey was washed off. I then transferred her to a dry bottle and put her in the sun. When she was dry I let her out, and she at once flew to her nest. . To my surprise, in thirteen minutes she returned, as if nothing had happened, and continued her visits to the honey all the afternoon.

This experiment interested me so much that I repeated it with another marked wasp, this time, however, keeping the wasp in the water till she was quite motionless and insensible. When taken out of the water she soon recovered; I fed her; she went quietly away to her nest as usual, and returned after the usual absence. The next morning this wasp was the first to visit the honey.

I was not able to watch any of the above-mentioned wasps for more than a few days; but I kept a specimen of *Polistes Gallica* for no less than nine months.

This is the wasp which has already been alluded to under the heading 'Memory;' but it is evident that the capacity which the insect displayed of becoming tamed implies no small degree of general intelligence; its hereditary instincts were conspicuously modified by the individual experiences incidental to its domestication.

The remaining passages that deserve quotation are the following:—

It is sometimes said of bees that those of one hive all know one another, and immediately recognise and attack any intruder from another hive. At first sight this certainly implies a great deal of intelligence. It is, however, possible that the bees of particular hives have a particular smell. Thus Langshaft, in his interesting 'Treatise on the Honey-Bee,' says: 'Members of different colonies appear to recognise their hive companions by the sense of smell; and I believe that if colonies are sprinkled with scented syrup, they may generally be safely mixed. Moreover, a bee returning to its own hive with a load of treasure is a very different creature from a hungry marauder; and it is said that a bee, if laden with honey, is allowed to enter any hive with impunity.' Mr. Langshaft continues, 'There is an air of roguery about a thieving bee which, to the expert, is as characteristic as are the motions of a pickpocket to a skilful policeman. Its sneaking look, and nervous, guilty agitation, once seen, can never be mistaken.' It is, at any rate, natural that a bee which enters a wrong hive by accident should be much surprised and alarmed, and would thus probably betray herself.

On the whole, then, I do not attach much importance to their recognition of one another as an indication of intelligence.

Since their extreme eagerness for honey may be attributed rather to their anxiety for the common weal than to their desire for personal gratification, it cannot fairly be imputed as greediness; still the following scene, one which most of us have witnessed, is incompatible surely with much intelligence. The sad fate of their unfortunate companions does not in the least deter others who approach the tempting lure from madly alighting on the bodies of the dying and dead, to share the same miserable end. No one can understand the extent of their infatuation until he has seen a confectioner's shop assailed by myriads of hungry bees. I have seen thousands strained out from the syrup in which they had perished; thousands more alighting even upon the boiling sweets, the floor covered and windows darkened with bees, some crawling, others flying, and others still, so completely besmeared as to be able neither to crawl nor fly, not one in ten able to carry home its ill-gotten spoils, and yet the air filled with new hosts of thoughtless comers.

Passing on now to the statements of other observers, Huber first noticed the remarkable fact that when beehives are attacked by the death's-head moth the bees close the entrance of their hive with wax and propolis to keep out the marauder. The barricade, which is built immediately behind the gateway, completely stops it up —only a small hole being left large enough to admit a bee, and therefore of course too small to admit the moth. Huber specially states that it was not until the beehives had been *repeatedly* attacked and robbed by the death's-head moth, that the bees closed the entrance of their hive with wax and propolis. *Pure* instinct would have induced the bees to provide against the first attack. Huber also observed that a wall built in 1804 against the death's-head hawk-moth was destroyed in 1805. In the latter year there were no death's-head moths, nor were any seen during the following. But in the autumn of 1807 a large number again appeared, and the bees at once protected themselves against their enemies. The bulwark was destroyed again in 1808.

Again, Huber (*loc. cit.*, tom. ii., p. 280) gives a case of apparent exercise of reason, or power of inference from a particular case to other and general cases. A piece of comb fell down and was fixed in its new position by wax. The bees then strengthened the attachments of all the other combs, clearly because they inferred that they too might be in danger of falling. This is a very remarkable case, and leads Huber to exclaim, 'I admit that I was unable to avoid a feeling of astonishment in the presence of a fact from which the purest reason seemed to shine out.'

A closely similar, and therefore corroborative case of an even more remarkable kind is thus narrated in Watson's 'Reasoning Power of Animals' (p. 448):—

Dr. Brown, in his book on the bee, gives another illustration of the reasoning power of bees, observed by a friend of his. A centre comb in a hive, being overburdened with honey, had parted from its fastenings, and was pressing against another comb, so as to prevent the passage of the bees between them. This accident excited great bustle in the colony, and as soon as their proceedings could be observed, it was found that they had constructed two horizontal beams between the two combs, and had removed enough of the honey and wax above them to admit the passage of a bee, while the detached comb had been secured by another beam, and fastened to the window with spare wax. But what was most remarkable was, that, when the comb was thus fixed, they removed the horizontal beams first constructed, as being of no further use. The whole occupation took about ten days.

Again, Mr. Darwin's MS. quotes from Sir B. Brodie's 'Psychological Inquiries' (1854, p. 88) the following case, which is analogous to the above, except that the supports required had to be made in a vertical instead of in a horizontal direction:—

On one occasion, when a large portion of the honeycomb had been broken off, they pursued another course. The fragment had somehow become fixed in the middle of the hive, and the bees immediately began to erect a new structure of comb on the floor, so placed as to form a pillar supporting the fragment, and preventing its further descent. They then filled up the space above, joining the comb which had become detached to that from which it had been separated, and they concluded their labours by removing the newly constructed comb below, thus proving that they had intended it to answer a merely temporary purpose.

Similarly, Dr. Dzierzon, an experienced keeper of bees, and the observer who first discovered the fact of their parthenogenesis, makes the general remark,—

The cleverness of the bees in repairing perfectly injuries to their cells and combs, in supporting on pillars pieces of their building accidentally knocked down by a hasty push, in fastening them with rivets, and bringing everything again into proper unity, making hanging bridges, chains, and ladders, compels our astonishment.

Lastly, as still further corroboration of such facts, I shall quote the following from Jesse's ' Gleanings: '[1]—

Bees show great ingenuity in obviating the inconvenience they experience from the slipperiness of glass, and certainly beyond what we can conceive that mere instinct would enable them to do. I am in the habit of putting small glass globes on the top of my straw hives, for the purpose of having them filled with honey; and I have invariably found that before the bees commence the construction of combs, they place a great number of spots of wax at regular distances from each other, which serve as so many footstools on the slippery glass, each bee resting on one of these with its middle pair of legs, while the fore claws were hooked with the hind ones of the bee next above him; thus forming a ladder, by means of which the workers were enabled to reach the top, and begin to make their combs there.

Herr Kleine, in his pamphlet on Italian Bees and Bee-keeping (Berlin, 1855), says that on substituting during the absence of the bees a hive filled with empty comb for their own hive, the returning bees exhibit the utmost perplexity. As the substituted hive stands in the exact spot previously occupied by their own hive, the returning bees fly into it without observing the change. But finding only empty combs inside, 'they stop, do not know

[1] Vol. i., pp. 22-3 (3rd ed.).

where they are, come out of the hole again without depositing their loads, fly off, look most carefully round the stand to assure themselves that they have made no mistake, and go in once more when convinced that they are at the right place. The same thing is repeated over and over again, until the bees at last bow to the incomprehensible and unavoidable, lay down their loads, and set to work at those tasks made necessary by the new circumstances of the hive. But as all the newly arriving bees behave in similar fashion, the disturbance lasts till late in the evening, and the uncertainty and anxiety of the bees is so great that the bee-master cannot contemplate it without deep sympathy.' Under such circumstances the bees take quickly to a substituted queen; 'for the feeling of the first comers that they have no right to the new dwelling, having, as they suppose, made some inexplicable mistake which they cannot remedy, prevents them from feeling any hostility to the new queen which they find; they probably consider themselves as merely on sufferance, and feel that they should be grateful that no action is taken against them for their illegal entry, as generally happens in bee-experience.' Hence the writer adopts this device when he desires to exchange or substitute queens.

Büchner, after alluding to this case, supplements it with the following:—

The wind threw down from the stand of a bee-master—a friend of the author's, whose name will soon become known— a straw beehive, the inmates of which were surprised in full work, and no small disorder in the interior was the result. The owner repaired the hive, put the loose comb back in its place, and replaced it in such a manner that the wind could not again catch it, hoping that the accident would have no further results. But when he examined the hive a few days later, he found that the bees had left their old home in the lurch, and had tried to enter other hives, clearly because they could no longer trust the weather, and feared that the terrible accident might again befall them.

. Dr. Erasmus Darwin, in his 'Zoonomia,' asserts that bees, when transported to Barbadoes, where there is no winter, cease to lay up honey. In contradiction to this statement, however, Kirby and Spence say, 'It is known to every naturalist acquainted with the fact, that many different species of bees store up honey in the hottest climates, and that there is no authentic instance on record of the hive-bees altering in any age or climate their peculiar operations.'

On the other hand, more recent observation has shown that Dr. Darwin's statement is probably correct. For, according to a note in *Nature*,[1] European bees, when transported to Australia, retain their industrious habits only for the first two or three years. After that time they gradually cease to collect honey till they become wholly idle. In a subsequent number of the same periodical (p. 411) a correspondent writes that the same fact is observable with bees transported to California, but is obviated by abstracting honey as the bees collect it.

There seems to be no doubt that bees and wasps are able to distinguish between persons, and even to recognise those whom they are accustomed to see, and to regard as friends. Bee-masters who attend much to their bees, so as to give the insects a good chance of knowing them, are generally of the opinion that the insects do know them, as shown by the comparatively sparing use of their stings. Again, many instances might be quoted, such as that given by Guerinzius,[2] who allowed a species of wasp native to Natal to build in the doorposts of his house, and who observed that although he often interfered with the nest, he was only once stung, and this by a young wasp; while no Caffre could venture to approach the door, much less to pass through it.[3] This power of distinguishing between persons indicates a higher order of intelligence than we might have expected to meet with among insects; and, according to Bingley, bees will not only learn to distinguish persons, but even lend themselves to tuition by those whom they know. For he says,

---

[1] Vol. xvii., p. 373.
[2] See Brehm, *Thierleben*, ix., p. 252.
[3] An exactly similar case is recorded by Stodmann in his *Travels in Surinam*, ii., p. 286.

'Mr. Wildman, whose remarks on the management of bees are well known, possessed a secret by which he could at any time cause a hive of bees to swarm upon his head, shoulders, or body, in a most surprising manner. He has been seen to drink a glass of wine with the bees all over his head and face more than an inch deep; several fell into the glass, but did not sting him. He could even act the part of a general with them, by marshalling them in battle array on a large table. Then he divided them into regiments, battalions, and companies, according to military discipline, waiting only for his word of command. The moment he uttered the word *march!* they began to march in a very regular manner in rank and file, like soldiers. To these, his Lilliputians, he also taught so much politeness that they never attempted to sting any of the numerous company which, at different times, resorted to admire this singular spectacle."

Huber's observation, since amply confirmed, of bees biting holes through the base of corollas in order to get at the honey which the length of the corollas prevent them from reaching in the ordinary way, also seems to indicate a rational adjustment to unusual circumstances. For the bees do not resort to this expedient until they find from trial that they cannot reach the nectar from above; but having once ascertained this, they forthwith proceed to pierce the bottoms of all the flowers of the same species. From an interesting account by Mr. Francis Darwin [1] (unfortunately too long to quote) it appears that, even when the nectar may be reached from above, bees may still resort to the expedient of biting through corollas in order to save time.

In connection with biting holes in corollas I may quote an observation communicated to me by a correspondent, Sir J. Clarke Jervoise. Speaking of a humble-bee, he says: 'I watched him into the flower of a foxglove, and, when out of sight, I closed the lips of the flower with my finger and thumb. He did not hesitate a moment, but cut his way out at the further end as if he had been served the same trick before. I never did it.'

[1] *Nature*, ix., p. 189.

Bees are highly particular in the matter of keeping their hives pure, and their sanitary arrangements often exhibit intelligence of a high order.

The following is quoted from Büchner (*loc. cit.*, p. 248):—

Impure air within the hive is that which the bees must above all things fear and avoid, for with the pressure together of so many individuals in a comparatively small space, it would not only be directly harmful to individual bees, but would produce among them dangerous diseases. They therefore also never void their excrements within, but always outside the hive. While this is very easy to do in summer, it is, on the contrary, very difficult in the winter, when the bees sit close together and generally motionless in the upper part of the hive, and when, from impure air and foul evaporations, as well as from bad and insufficient food, dysentery-like diseases break out among them, and often carry off the whole community in a brief space of time. In such cases they utilise the first fine day to relieve themselves, and in the spring they take a long general cleansing flight. But they also know how to take advantage of special circumstances so as to perform the process of purification in the way least harmful to the hive. Herr Heinrich Lehr, of Darmstadt, a bee-keeping friend of the author, has sent the following communication:—During an epidemic of dysentery in winter, from which most of his hives suffered (as the bees were no longer able to retain their excrements), one hive suffered less than the others. Exact investigation showed that this hive was soiled all over at the back with the excrement of the bees, and that the inmates had here made a kind of drain. On this spot a little opening had been made by the falling off of the covering clay, which led directly to the upper part of the hive, where the bees were accustomed to sit together during the winter. This excellent opportunity, whereby they could reach in the shortest way an otherwise difficult object, and one rendered complicated by circumstances, did not escape them.

It sometimes happens that mice, slugs, &c., enter a beehive. They are then killed and covered with a coating of propolis. Réaumur says[1] that he once saw a snail enter a hive in this way. The hard shell was an effective protection against the stings of the bees, so the insects smeared round the edges of the shell with wax and

---

[1] See *Kirby and Spence*, vol. ii, p. 229.

resin, fastening down the animal to the wall of the hive, so that it died of starvation or want of air. If the encasing of an animal (such as a mouse) with propolis is not sufficient to prevent its putrefaction, the bees gnaw away all the putrescible parts of the carcass and carry them out of the hive, leaving only the skeleton behind. The dead bodies of their companions are also carried out of the hive and deposited at a distance. There is no question about this fact (which it will be remembered is analogous to that already mentioned in the case of ants); according to Büchner, however, bees not only remove their dead, but also, occasionally at least, bury them. But as he gives very inadequate evidence in support of this assertion, we may safely set it aside as insufficiently proven.

Büchner, however, gives an admirable summary, and makes some judicious remarks on the well-known and highly remarkable habit which bees practise for the obvious purpose of ventilating their hives. As this account gives all the facts in a brief compass, I cannot do better than quote it:—

Very interesting, and closely connected with this characteristic of cleanliness, is the conduct of the so-called ventilating-bees, which have to take care that in summer or hot weather the air necessary for respiration of the bees in the interior of the hive is renewed, and the too high temperature cooled down. The latter precaution is necessary, not only on account of the bees working within the hive, to whom, as already said, a temperature risen beyond a certain point would be intolerable, but also to guard against the melting or softening of the wax. The bees charged with the care of the ventilation divide themselves into rows and stages in regular order through all parts of the hive, and by swift fanning of their wings send little currents of air in such fashion that a powerful stream or change of air passes through all parts of the hive. Other bees stand at the mouth of the hive, which fan in the same way and considerably accelerate the wind from within. The current of air thus caused is so strong that little bits of paper hung in front of the mouth are rapidly moved, and that, according to F. Huber, a lighted match is extinguished. The wind can be distinctly felt if the hand be held in front.

The motion of the wings of the ventilating bees is so rapid that it is scarcely perceptible, and Huber saw some bees working their wings in this way for five-and-twenty minutes. When they are tired they are relieved by others. According to Jesse, the bees in very hot weather, in spite of all their efforts, are unable to sufficiently lower the temperature, and prevent the melting of some of the wax; they then get into a condition of great excitement, and it is dangerous to approach them. In such a case they also try to mend matters by a number leaving the hive and settling in large masses on its surface, so as to protect it as much as possible from the scorching rays of the sun.

Although the described plan of ventilation is remarkable enough in itself, it is yet more remarkable in that it is clearly only the result of bee-keeping, and is evoked by this misfortune. For there could be no need of such ventilation for bees in a state of nature, whose dwellings in hollow trees and clefts of rocks leave nothing to be desired as to roominess and airiness, while in the narrow artificial hive this need at once comes out strongly. In fact, the fanning of the bees almost entirely ceased when Huber brought them into large hives five feet high, in which there was plenty of air. It follows, therefore, that the fanning and ventilating can have absolutely nothing to do with an inborn tendency or instinct, but have been gradually evoked by necessity, thought, and experience.

As the following observation on the cautious sagacity of wasps is, so far as I am aware, new, and as it certainly does not admit of mal-observation, I introduce it on the authority of a correspondent, the Rev. Mr. J. W. Mossman, who writes from Tarrington Rectory, Wragby. He found an apple in his orchard which had fallen from a tree in apparently good condition; but on taking it up observed that it was little more than a shell filled with wasps. Giving the apple a shake, he saw a wasp slowly emerging from a single small aperture in the rind:—

This aperture was sufficient, and only just sufficient, to admit of the ingress or egress of a single wasp. The circumstance which struck me as very remarkable was this—that the wasp did not make its way through the aperture with its head first, as I should have expected, but with its tail, darting out its sting to its utmost extent, and brandishing it furiously. In this manner it came out of the apple backwards. Then, finding itself in the open air upon the outer surface of the apple, it turned round, and without any attempt to molest me, flew off in the usual way. The moment this first wasp had emerged, the sting and tail of another was seen protruding. This, too, I watched with much interest, and exactly the same process was repeated as in the case of the first. I held the apple in my hand until some ten or a dozen wasps had made their exit in the same identical manner in each individual case. I then threw down the apple, inside of which, however, there were still apparently a good many wasps.

It seemed to me at the time, and I have always felt since, that the wasps coming out of the apple backwards, brandishing their stings as a defensive weapon against possible enemies, whom of course they were not able to see, was an evidence of what would be called thought and reflection in the case of human beings. It seems to me that these wasps must have reflected that if they came out of the narrow aperture in the apple, which was their only possible means of ready egress, in the usual manner, head first, they might be taken at a disadvantage by a possible enemy, and destroyed in detail. They, therefore, with great prudence and foresight, came out of the apple backwards, protecting themselves by means of their chief offensive and defensive weapons, their stings, which, according to their normal method of locomotion, would have been useless to them as long as they were making their exit.

With regard to the tactics displayed by hunting wasps I may quote the following cases :—

Mr. Seth Green, writing to the *New York World* of May 14, says that one morning when he was watching a spider's nest, a wasp alighted within an inch or two of the nest, on the side opposite the opening. Creeping noiselessly around towards the entrance of the nest the wasp stopped a little short of it, and for a moment remained perfectly quiet ; then reaching out one of his antennæ he wriggled it before the opening and withdrew it. This overture had the desired effect, for the boss of the nest, as large a spider as one ordinarily sees, came out to see what was wrong and to set it to rights. No sooner had the spider emerged to that point at which he was at the worst disadvantage than the wasp, with a quick movement, thrust his sting into the body of his foe, killing him easily and almost instantly. The experiment was repeated on the part of the wasp, and when there was no response from the inside he became satisfied, probably, that he held the fort. At all events, he proceeded to enter the nest and slaughter the young spiders, which were afterwards lugged off one at a time.

Mr. Henry Cecil writes as follows (*Nature*, vol. xviii., p. 311):—

I was sitting one summer's afternoon at an open window (my bedroom) looking into a garden, when I was surprised to observe a large and rare species of spider run across the window-sill in a crouching attitude. It struck me the spider was evidently alarmed, or it would not have so fearlessly approached me. It hastened to conceal itself under the projecting ledge of the window-sill inside the room, and had hardly done so when a very fine large hunting wasp buzzed in at the open window and flew about the room, evidently in search of something. Finding nothing, the wasp returned to the open window and settled on the window-sill, running backwards and forwards as a dog does when looking or searching for a lost scent. It soon alighted on the track of the poor spider, and in a moment it discovered its hiding-place, darted down on it, and no doubt inflicted a wound with its sting. The spider rushed off again, and this time took refuge under the bed, trying to conceal itself under the framework or planks which supported the mattress. The same scene occurred here; the wasp now appeared to follow the spider by sight, but ran backwards and forwards in large circles like a hound. The moment the trail of the spider was found the wasp followed all the turns it had made till it came on it again. The poor spider was chased from hiding-place to hiding-place, out of the bedroom, across a passage, and into the middle of another large room, where it finally succumbed to the repeated stings inflicted by the wasp. Rolling itself up into a ball the wasp then took possession of its prey, and after ascertaining it could make no resistance, tucked it up under its *very long hind legs*, just as a hawk or eagle carries off its quarry, when I interposed and secured both for my collection.

Mr. Belt, in his work already frequently quoted, gives the following account of a struggle which not unfrequently occurs between wasps and ants for the sweet secretion of 'frog-hoppers:'—

Similarly as, on the savannahs, I had observed a wasp attending the honey-glands of the bull's-horn acacia along with the ants; so at Santo Domingo another wasp, belonging to quite a different genus (*Nectarina*), attended some of the clusters of frog-hoppers, and for the possession of others a constant skirmishing was going on. The wasp stroked the young hoppers, and sipped up the honey when it was exuded, just like the ants.

When an ant came up to a cluster of leaf-hoppers attended by a wasp, the latter would not attempt to grapple with its rival on the leaf, but would fly off and hover over the ant; then when its little foe was well exposed, it would dart at it and strike it to the ground. The action was so quick that I could not determine whether it struck with its fore-feet or its jaws; but I think it was with the feet. I often saw a wasp trying to clear a leaf from ants that were already in full possession of a cluster of leaf-hoppers. It would sometimes have to strike three or four times at an ant before it made it quit its hold and fall. At other times one ant after the other would be struck off with great celerity and ease, and I fancied that some wasps were much cleverer than others. In those cases where it succeeded in clearing the leaf, it was never left long in peace; for fresh relays of ants were continually arriving, and generally tired the wasp out. It would never wait for an ant to get near it, doubtless knowing well that if its little rival once fastened on its leg, it would be a difficult matter to get rid of it again. If a wasp first obtained possession, it was able to keep it; for the first ants that came up were only pioneers, and by knocking these off, it prevented them from returning and scenting the trail to communicate the intelligence to others.

Dr. Erasmus Darwin records an observation 'Zoonomia,' i., p. 183) which, from having since been so widely quoted, deserves to be called classical. He saw a wasp upon the ground endeavouring to remove a large fly which was too heavy for it to carry off. The wasp cut off the head and abdomen, and flew away with the thorax alone. The wind, however, catching the wings of this portion made it still too unwieldy for the wasp to guide. It therefore again alighted, and nipped off first one wing and then the other, when it was able to fly off with its booty without further difficulty.

This observation has since been amply confirmed. I shall quote some of the confirmatory cases.

Mr. R. S. Newall, F.R.S., in *Nature*, vol. xxi., p. 494, says:—

Many years ago I was examining an apple tree, when a wasp alighted on a leaf which formed a caterpillar's nest neatly rolled up. The wasp examined both ends, and finding them closed, it soon clipped a hole in the leaf at one end of the nest about one-eighth of an inch in diameter. It then went to the other end and made a noise which frightened the caterpillar, which came rushing out of the hole. It was immediately seized by the wasp, who finding it too large to carry off at once, cut it in two and went off with his game. I waited a little and saw the wasp come back for the other half, with which it also flew away.

Again, Büchner (*loc. cit.*, p. 297) gives the following account in the words of his informant, Herr H. Löwenfels, who himself witnessed the incident :—

> I here found a robber-wasp busied in lifting from the ground a large fly which it had apparently killed. It succeeded indeed in its attempt, but had scarcely raised its prey a few inches above the ground when the wind caught the wings of the dead fly, and they began to act like a sail. The wasp was clearly unable to resist this action, and was blown a little distance in the direction of the wind, whereupon it let itself fall to the ground with its prize. It now made no more attempts to fly, but with eager industry pulled off with its teeth the fly's wings which hindered it in its object. When this was quite done it seized the fly, which was heavier than itself, and flew off with it untroubled on its journey through the air at a height of about five feet.

Büchner also records the two following remarkable observations, which from being so similar corroborate one another. The first is received from Herr Albert Schlüter, who writing from Texas says that he there saw a cicada pursued by a large hornet, which threw itself upon its prey and seemed to sting it to death :—

> The murderer walked over its prey, which was considerably larger than itself, grasped its body with its feet, spread out its wings, and tried to fly away with it. Its strength was not sufficient, and after many efforts it gave up the attempt. Half a minute went by; sitting astride on the corpse and motionless —only the wings occasionally jerking—it seems to reflect, and indeed not in vain. A mulberry tree stood close by, really only a trunk—for the top had been broken off, clearly by the last flood—of about ten or twelve feet high. The hornet saw this trunk, dragged its prey toilsomely to the foot of it, and then up to the top. Arrived thereat, it rested for a moment, grasped its victim firmly, and flew off with it to the prairies. That which it was unable to raise off the ground it could now carry easily once high in the air.

The other instance is as follows:—

Th. Meenan ('Proc. of the Acad. of Nat.,' Philadelphia, Jan. 22, 1878) observed a very similar case with *Vespa maculata*. He saw one of these wasps try in vain to raise from the ground a grasshopper it had killed. When all its efforts proved to be in vain, it pulled its prey to a maple tree, about thirty feet off, mounted it with its prize, and flew away from it. 'This,' adds the writer, 'was more than instinct. It was reflection and judgment, and the judgment was proved to be correct.'

Depriving bees of their antennæ has the effect of producing an even more marked bewilderment than results from this operation in the case of ants. A queen thus mutilated by Huber ran about in confusion, dropping her eggs at random, and appeared unable to take with precision the food that was offered her. She showed no resentment to a similarly mutilated stranger queen that was introduced: the workers also heeded not the mutilated stranger; but when an unmutilated stranger was introduced they fell upon her. When the mutilated queen was allowed to escape, none of the workers followed.

CHAPTER V.

TERMITES.

The habits of the Termites, or so-called White Ants, have not been so closely studied as they deserve. Our chief knowledge concerning them is derived from the observations of Jobson, in his 'History of Gambia;' Bastian, in 'The Nations of Eastern Asia;' Forsteal, Lespès, König, Sparman, Hugen, Quatrefages, Fritz Müller, and most of all, Smeathman, in 'Philosophical Transactions,' vol. lxxi. In Africa these insects raise their hills to a height of between ten and twenty feet, and construct them of earth, stones, pieces of wood, &c., glued together by a sticky saliva. The hills are in the form of a cone, and so strong that it is said the buffaloes are in the habit of using them as watch-towers on which to post sentries, and that they will even support the weight of an elephant. The growth of these gigantic mounds is gradual, increasing with the increase of the population. From the mound in all directions there radiate subterranean tunnels, which may be as much as a foot in width, and which serve as roadways. Besides these tunnels there are a number of other subterranean tubes, which serve the purpose of drainage to carry off the floods of water to which the nest is exposed during tropical showers. Büchner calculates that a pyramid built by man on a scale proportional to his size would only equal one of these nests if it attained to the height of 3,000 feet. The following is this author's description of the internal structure :—

These internal arrangements are so various and so complicated that pages of description might be written thereupon. There are myriads of rooms, cells, nurseries, provision chambers, guard-rooms, passages, corridors, vaults, bridges, subterranean streets and canals, tunnels, arched ways, steps, smooth inclines, domes, &c., &c., all arranged on a definite, coherent, and well-considered plan. In the middle of the building, sheltered as far as possible from outside dangers, lies the stately royal dwelling, resembling an arched oven, in which the royal pair reside, or rather are imprisoned; for the entrances and outlets are so small, that although the workers on service can pass easily in and out, the queen cannot; for during the egg-laying her body swells out to an enormous size, two or three thousand times the size and weight of an ordinary worker. The queen, therefore, never leaves her dwelling, and dies therein. Round the palace, which is at first small, but is later enlarged in proportion as the queen increases in size until it is at least a yard long and half a yard high, lie the nurseries, or cells for the eggs and larvæ; next these the servants' rooms, or cells for the workers which wait on the queen; then special chambers for the soldiers on guard, and, between these, numerous store-rooms, filled with gums, resins, dried plant-juices, meal, seeds, fruits, worked-up wood, &c. According to Bettziech-Beta, there is always in the midst of the nest a large common room, which is used either for popular assemblies or as the meeting and starting point of the countless passages and chambers of the nest. Others are of the opinion that this space serves for purposes of ventilation.

Above and below the royal cell are the rooms of the workers and soldiers which are specially charged with the care and defence of the royal pair. They communicate with each other, as well as with the nursery-cells and store-rooms, by means of galleries and passages which, as already said, open into the common room in the middle under the dome. This room is surrounded by high, boldly projected arched ways, which lose themselves further out in the walls of the countless rooms and galleries. Many roofs outside and in protect this room and the surrounding chambers from rain, which, as already said, is drained away by countless subterranean canals, made of clay and of a diameter of ten or twelve centimetres. There are also, under the layer of clay covering the whole building, broad spirally winding passages running from below to the highest points, which communicate with the passages of the interior, and apparently, as they mainly consist of smooth inclines, serve for carrying provisions to the higher parts of the nest.[1]

[1] *Loc. cit.*, p. 189.

The termites, like many species of true ants, are divided into two distinct castes, the workers and the soldiers. If a breach is made in the walls of the dome the soldiers rush out to meet the enemy, and fight desperately with any enemy that they may find. Here, again, I cannot do better than quote Büchner's epitome of facts :—

If the assailant withdraws beyond their reach and inflicts no further injury, they retire within their dwelling in the course of half an hour, as though they had come to the conclusion that the enemy who had done the mischief had fled. Scarcely have the soldiers disappeared when crowds of workers appear in the breach, each with a quantity of ready-made mortar in its mouth. As soon as they arrive they stick this mortar round the open place, and direct the whole operation with such swiftness and facility that in spite of their great number they never hinder each other, nor are obliged to stop. During this spectacle of apparent restlessness and confusion the observer is agreeably surprised to see arising a regular wall, filling up the gap. During the time that the workers are thus busied the soldiers remain within the nest, with the exception of a few, which walk about apparently idly, never touching the mortar, among the hundreds and thousands of workers. Nevertheless one of them stands on guard close to the wall which is being built. It turns gently each way in turn, lifting its head at intervals of one or two minutes to strike the building with its heavy mandibles, making the before-mentioned crackling noise. This signal is immediately answered by a loud rustling from the interior of the nest and from all the subterranean passages and holes. There is no doubt that this noise arises from the workers, for as often as the sign is given they work with increased energy and speed. A renewal of the attack instantaneously changes the scene. 'At the first stroke,' says Smeathman, 'the workers run into the many tunnels and passages which run through the building, and this happens so quickly that they seem regularly to vanish. In a few seconds they are all gone, and in their stead appear the soldiers once more, as numerous and as pugnacious as before. If they find no enemy, they turn back slowly into the interior of the hill, and immediately the mortar-laden workers again appear, and among them a few soldiers, which behave just as on the first occasion. So one can have the pleasure of seeing them work and fight in turn, as often as one chooses; and it will be found each time that one set never fight, and the other never work, however great the need may be.'[1]

Similar facts have been observed by Fritz Müller of the South American species.

The Termites, being like the Ecitons blind, like them make all their expeditions under the protection of covered ways. These are underground tunnels in all cases where circumstances permit, but on arriving at a rock or other impenetrable obstruction, they build a tubular passage upon the surface. According to Büchner,—

They can even carry their viaducts through the air, and that in such bold arches that it is difficult to understand how they were projected. In order to reach a sack of meal which was well protected below, they broke through the roof of the room in which it was, and built a straight tube from the breach they had made down to the sack. As soon as they tried to carry off their booty to a safe place, they became convinced that it was impossible to pull it up the straight road. In order to meet this difficulty, they adopted the principle of the smooth incline, the use of which we have already seen in the interior of their nests, and built close to the first tube a second, which wound spirally within, like the famous clock tower of Venice. It was now an easy task to carry their booty up this road and so away. . . . Either from the desire to remain undiscovered, or from their liking for darkness, they have the remarkable habit of destroying and gnawing everything from within outwards, and of leaving the outside shell standing, so that from the outside appearance the dangerous state of the inside is not perceptible. If, for instance, they have destroyed a table or other piece of household furniture, in which they always manage from the ground upwards to hit exactly the places on which the feet of the article rest, the table looks perfectly uninjured outside, and people are quite astonished when it breaks down under the slightest pressure. The whole inside is eaten away, and only the thinnest shell is left standing. If fruits are lying on the table, they also are eaten out from the exact spot on which they rest on the surface of the table.

In similar fashion things consisting wholly of wood, such as wooden ships, trees, &c., are destroyed by them so that they finally break in without any one having noticed the mischief. Yet it is said that they go so prudently to work in their de-

[1] *Ibid.*, p. 119.

struction that the main beams, the sudden breakage of which would threaten the whole building and themselves therewith, are either spared, or else so fastened together again with a cement made out of clay and earth that their strength is greater than ever!(?) Hagen also states that they never cut right through the corks which stop up stored bottles of wine, but leave a very thin layer, which is sufficient to prevent the outflow of the wine and the consequent destruction of the workers. The same author relates that in order to reach a box of wax lights they made a covered road from the ground up to the second story of a house.[1]

It is needless to give a special description of any of the other habits of these insects, such as their swarming, breeding, &c., for they all more or less closely resemble the analogous habits of ants and bees. It is very remarkable that insects of two distinct orders should both manifest such closely similar social habits of such high complexity, and it rather surprises me that more has not been made of this point by writers opposed to the principles of evolution. Of course if the point were raised, the argument in answer would require to be, either that the similar instincts were derived from common and very remote progenitors (in which case the fact would form by far the most remarkable instance of the permanency of instincts among changing species), or more probably, that similar causes operating in the two orders have produced similar effects—complex and otherwise unique though these effects undoubtedly are.

In connection with the theory of evolution I may conclude this chapter with the following quotation from Smeathman, as it shows how natural relation may develop for the benefit of the species instincts which are detrimental to the individual. Speaking of the soldiers he says :—

I was always amused at the pugnacity displayed when, in making a hole in the earthy cemented archway of their covered roads, a host of these little fellows mounted the breach to cover the retreat of the workers. The edges of the rupture bristled with their armed heads as the courageous warriors ranged

[1] *Geistesleben der Thiere*, pp. 134 and 199-200.

themselves in compact line around them. They attacked fiercely any intruding object, and as fast as their front ranks were destroyed, others filled up their places. When the jaws closed in the flesh, they suffered themselves to be torn in pieces rather than loosen their hold. It might be said that this instinct is rather a cause of their ruin than a protection when a colony is attacked by the well-known enemy of termites, the ant-bear; but it is the soldiers only which attach themselves to the long worm-like tongue of this animal, and the workers, on whom the prosperity of the young brood immediately depends, are left for the most part unharmed. I always found, on thrusting my finger into a mixed crowd of termites, that the soldiers only fastened upon it. Thus the fighting caste do in the end serve to protect the species by sacrificing themselves to its good [1]

[1] Phil. Trans., *loc. cit*

CHAPTER VI.

SPIDERS AND SCORPIONS.

Emotions.

THE emotional life of spiders, so far as we can observe it as expressed in their actions, seems to be divided between sexual passion (including maternal affection) and the sterner feelings incidental to their fiercely predatory habits. But the emotions, although apparently few and simple in character, are exceedingly strong in force. In many species the male spider in conducting his courtship has to incur an amount of personal danger at the hands (and jaws) of his terrific spouse, which might well daunt the courage of a Leander. Ridiculously small and weak in build, the males of these species can only conduct the rites of marriage with their enormous and voracious brides by a process of active manœuvring, which if unsuccessful is certain to cost them their lives. Yet their sexual emotions are so strong that, as proved by the continuance of the species, no amount of personal risk is sufficient to deter them from giving these emotions full play. There is no other case in the animal kingdom where courtship is attended with any approach to the gravity of danger that is here observable. Among many animals the males have to meet a certain amount of inconvenience from the coquetry or disinclination of the females; but here the coquetry and disinclination has passed into the hungry determination of a ferocious giantess. The case, therefore, because unique, is of interest from an evolutionary point of view. We can see a direct advantage to species from the danger incurred by males on account of mutual jealousy; for this, giving rise to what Mr. Darwin has called 'the law of battle,' must obviously be a constant source of the creation and the maintenance of specific proficiency; the law of battle determines that only the strongest and most courageous males shall breed. But the benefit to species is not so obvious where the danger of courtship arises from the side of the female. Still, that there must be some benefit is obvious, seeing that the whole structure of the male, if we take that of the female as the original type, has been greatly modified with reference to this danger: had the latter been wholly useless, either it would not have been allowed to arise, or the species must have become extinct. The only suggestion I can make to meet this aberrant case is that the courage and determination required of the male, besides being no doubt of use to him in other relations in life, may be of benefit to the species by instilling these qualities into the psychology both of his male and female descendants.

The courage and rapacity of spiders as a class are too well and generally known to require special illustration. One instance, however, may be quoted to show the strength of their maternal emotions. Bonnet threw a spider with her bag of eggs into the pit of an ant-lion. The latter seized the eggs and tore them away from the spider; but although Bonnet forced her out of the pit, she returned, and chose to be dragged in and buried alive rather than leave her charge.

The only other point that occurs to me with reference to the emotions of spiders is the somewhat remarkable one concerning their apparent fondness of music. The testimony is so varied and abundant on this matter that we can scarcely doubt the truth of the facts. These simply are that spiders—or at any rate some species or individuals —approach a sounding musical instrument, 'especially when the music is tender and not too loud.' They usually approach as near as possible, often letting themselves down from the ceiling of the room by a line of web, and remain suspended above the instrument. Should the music become loud, they often again retreat. Professor C. Reclain, during a concert at Leipsic, saw a spider descend in this way from one of the chandeliers while a violin solo was being played; but as soon as the orchestra began to sound it quickly ran back again.[1] Similar observations have been published by Rabigot, Simonius, von Hartmann, and others.

A highly probable explanation of these facts has recently been given by Mr. C. V. Boys, which relieves us of the necessity of imputing to animals so low in the scale any rudiment of æsthetic emotion as aroused by musical tones. As the observation is an interesting one, I shall quote it *in extenso*:—

Having made some observations on the garden spider which are I believe new, I send a short account of them, in the hope that they may be of interest to the readers of *Nature*.

Last autumn, while watching some spiders spinning their beautiful geometrical webs, it occurred to me to try what effect a tuning-fork would have upon them. On sounding an A fork, and lightly touching with it any leaf or other support of the web, or any portion of the web itself, I found that the spider, if at the centre of the web, rapidly slues round so as to face the direction of the fork, feeling with its fore-feet along which radial thread the vibration travels. Having become satisfied on this point, it next darts along that thread till it reaches either the fork itself or a junction of two or more threads, the right one of which it instantly determines as before. If the fork is not removed when the spider has arrived it seems to have the same charm as any fly; for the spider seizes it, embraces it, and runs about on the legs of the fork as often as it is made to sound, never seeming to learn by experience that other things may buzz besides its natural food.

If the spider is not at the centre of the web at the time that the fork is applied, it cannot tell which way to go until it has been to the centre to ascertain which radial thread is vibrating, unless of course it should happen to be on that particular thread, or on a stretched supporting thread in contact with the fork.

If, when a spider has been enticed to the edge of the web the fork is withdrawn, and then gradually brought near, the spider is aware of its presence and of its direction, and reaches out as far as possible in the direction of the fork; but if a sounding fork is gradually brought near a spider that has not been disturbed, but which is waiting as usual in the middle of

[1] *Body and Mind,* p. 275.

the web, then, instead of reaching out towards the fork, the spider instantly drops—at the end of a thread, of course. If under these conditions the fork is made to touch any part of the web, the spider is aware of the fact, and climbs the thread and reaches the fork with marvellous rapidity. The spider never leaves the centre of the web without a thread along which to travel back. If after enticing a spider out we cut this thread with a pair of scissors, the spider seems to be unable to get back without doing considerable damage to the web, generally gumming together the sticky parallel threads in groups of three and four.

By means of a tuning-fork a spider may be made to eat what it would otherwise avoid. I took a fly that had been drowned in paraffin and put it into a spider's web, and then attracted the spider by touching the fly with a fork. When the spider had come to the conclusion that it was not suitable food, and was leaving it, I touched the fly again. This had the same effect as before, and as often as the spider began to leave the fly I again touched it, and by this means compelled the spider to eat a large portion of the fly.

The few house-spiders that I have found do not seem to appreciate the tuning-fork, but retreat into their hiding-places as when frightened; yet the supposed fondness of spiders for music must surely have some connection with these observations; and when they come out to listen, is it not that they cannot tell which way to proceed?

The few observations that I have made are necessarily imperfect, but I send them, as they afford a method which might lead a naturalist to notice habits otherwise difficult to observe, and so to arrive at conclusions which I in my ignorance of natural history must leave to others.[1]

General Habits.

Coming now to general habits, our attention is claimed by the only general habit that is of interest—namely, that of web-building. The instinct of constructing nets for the capture of prey occurs in no other class of animals, while in spiders it not only attains to an extraordinary degree of perfection (so that, in the opinion of some geometers, the instinct is not less wonderful in this respect than is that displayed by the hive-bee in the con-

[1] *Nature*, xxiii., pp. 149-50.

struction of its cells), but also ramifies into a number of diverse directions. Thus we have, in different species, wide open networks spread between the branches of bushes, &c., closely woven textures in the corners of buildings, earth tubes lined with silk, the strong muslin-like snare of the Mygale, which, as first noticed by Madame Merian,[1] and since confirmed by Bates,[2] is able to retain a struggling humming-bird while this most beautiful animal in creation is being devoured by the most repulsive; and many other varieties might be mentioned. It may at first sight appear somewhat remarkable that this instinct of spreading snares should on the one hand occur only in one class of the animal kingdom, while on the other hand, in the class where it does occur, it should attain such extreme perfection, and run into so much variety. But we must here remember that the development of the instinct obviously depends upon the presence of a web-secreting apparatus, which is a comparatively rare anatomical feature. In caterpillars, which are not predaceous, the web is used only for the purposes of protection and locomotion; and it is easy to see that the spreading of snares would here be of no use to the animals. But in spiders, of course, the case is otherwise. Once granting the power of forming a web, and it is evident that there is much potential service to which this power may be put with reference to the voracious habits of the animal; and therefore it is not to be wondered that both the anatomical structures and their correlated instincts should attain to extreme perfection in sundry lines of development. The origin of the web-building structure was probably due to the use of the web for purposes of locomotion or of cocoon-spinning, as we see it still so used in the same way that it is used by caterpillars for descending from heights, and in the case of the gossamer spider for travelling immense distances through the air. As the anatomical structures in question differ very greatly in the case of spiders and in that of caterpillars, we may wonder why analogous if not homolo-

[1] *Naturalist on the Amazon*, p. 83.
[2] For many other confirmations see Sir E. Tennent, *Nat. Hist. Ceylon*, pp. 468–69.

gous structures should never have been developed in the case of any other animal having predaceous habits—especially, perhaps, in that of the imago form of predaceous insects. It is easy to see how, if there were any original tendency to secrete a viscid substance in the neighbourhood of the anus, this might be utilised in descending from low elevations (as certain kinds of slugs use their viscid slime as threads whereby to let themselves down from low branches to the ground); and so we can understand how natural selection might thus have the material supplied out of which to develop such highly specialised organs as the spinnerets of a spider. But if we are inclined to wonder why this should not have happened among other animals, we must remember that any expectation that it should rests on negative grounds; we have no reason to suppose that in any other case the initial tendency to secrete a viscid substance was present. One inference, however, in the case of spiders seems perfectly valid. As this comparatively rare faculty of web-spinning occurs so generally throughout the class, it must have had its earliest origin very far back in the history of that class, though probably not so far back as to include the common progenitors of the spiders and the scorpions, seeing that the latter do not spin webs.

I shall now give a few details on the manner in which spiders' webs are made. Without going into the anatomy of the subject further than to observe that a spider's 'thread' is a composite structure made up of a number of finer threads, which leave their respective spinneret-holes in an almost fluid condition, and immediately harden by exposure to the air, I shall begin at once to describe the method of construction.

The so-called 'geometric spider' constructs her web by first laying down the radiating and unadhesive rays, and then, beginning from the centre, spins a spiral line of unadhesive web, like that of the rays which it intersects. This line, in being woven through the radii in a spiral from centre to circumference, serves as a scaffolding for the spider to walk over, and also keeps the rays properly stretched. She next spins another spiral line, but this time from the circumference to near the centre, and formed of web, covered with a viscid secretion to retain prey. Lastly, she constructs her lair to hide and watch for prey, at some distance from the web but connected with it by means of a line of communication or telegraph, the vibrations of which inform her of the struggling of an insect in the net.[1]

According to Thompson,—

The web of the garden spider—the most ingenious and perfect contrivance that can be imagined—is usually fixed in a perpendicular or somewhat oblique direction in an opening between the leaves of some plant or shrub; and as it is obvious that round its whole extent lines will be required to which those ends of radii that are farthest from the centre can be attached, the construction of those exterior lines is the spider's first operation. It seems careless about the shape of the area they are to enclose, well aware that it can as readily inscribe a circle in a triangle as a square; and in this respect it is guided by the distance or proximity of the points to which it can attach them. It spares no pains, however, to strengthen and keep them in a proper degree of tension. With the former view it composes each line of five or six or even of more threads glued together; and with the latter it fixes to them from different points a numerous and intricate apparatus of smaller threads; and having thus completed the foundation of its snare, it proceeds to fill up the outline. Attaching a thread to one of the main lines, it walks along it, guiding it with one of its hind legs, that it may not touch in any part and be prematurely glued, and crosses over to the opposite side, where, by applying its spinners, it firmly fixes it. To the middle of this diagonal thread, which is to form the centre of its net, it fixes a second, which in like manner it conveys and fastens to another part of the lines including the area. The work now proceeds rapidly. During the preliminary operations it sometimes rests, as though its plan required meditation; but no sooner are the marginal lines of the net firmly stretched, and two or three radii spun from its centre, than it continues its labour so quickly and unremittingly that the eye can scarcely follow its progress. The radii, to the number of about twenty, giving the net the appearance of a wheel, are speedily finished. It then proceeds to the centre, quickly turns itself round, pulls each thread with its feet to ascertain its strength, breaking any one that seems defective, and

[1] Kirby, vol. ii., p. 298.

replacing it by another. Next it glues, immediately round the centre, five or six small concentric circles, distant about half a line from each other, and then four or five larger ones, each separated by the space of half an inch or more. These last serve as a sort of temporary scaffolding to walk over, and to keep the radii properly stretched while it glues to them the concentric circles that are to remain, which it now proceeds to construct. Placing itself at the circumference, and fastening its thread to the end of one of the radii, it walks up that one, towards the centre, to such a distance as to draw the thread from its body of a sufficient length to meet the next. Then stepping across and conducting the thread with one of its hind legs, it glues it with its spinners to the point in the adjoining radius to which it is to be fixed. This process it repeats until it has filled up nearly the whole space from the circumference to the centre with concentric circles, distant from each other about two lines. It always, however, leaves a vacant interval around the smallest first spun circles that are nearest to the centre, and bites away the small cotton-like tuft that united all the radii, which being held now together by the circular threads have thus probably their elasticity increased; and in the circular opening, resulting from this procedure, it takes its station and watches for its prey, or occasionally retires to a little apartment formed under some leaf, which it also uses as a slaughterhouse.[1]

According to Büchner,—

The long main threads, with the help of which the spider begins and attaches its web, are always the thickest and strongest; while the others, forming the web itself, are considerably weaker. Injuries to the web at any spot the spider very quickly repairs, but without keeping to the original plan, and without taking more trouble than is absolutely necessary. Most spiders' webs, therefore, if closely looked into, are found to be somewhat irregular. When a storm threatens, the spider, which is very economical with its valuable spinning material, spins no web, for it knows that the storm will tear it in pieces and waste its pains, and it also does not mend a web which has been torn. If it is seen spinning or mending, on the other hand, fine weather may be generally reckoned on. . . . The emerged young at first spin a very irregular web, and only gradually learn to make a larger and finer one, so that here, as everywhere else, practice and experience play a great part. . .

[1] Thompson, *Passions of Animals*, p. 145.

The position must also offer favourable opposite points for the attachment of the web itself. People have often puzzled their brains, wondering how spiders, without being able to fly, had managed first to stretch their web through the air between two opposite points. But the little creature succeeds in accomplishing this difficult task in the most various and ingenious ways. It either, when the distance is not too great, throws a moist viscid pellet, joined to a thread, which will stick where it touches; or hangs itself by a thread in the air and lets itself be driven by the wind to the spot; or crawls there, letting out a thread as it goes, and then pulls it taut when arrived at the desired place; or floats a number of threads in the air and waits till the wind has thrown them here or there. The main or radial threads which fasten the web possess such a high degree of elasticity, that they tighten themselves between two distant points to which the spider has crawled, without it being necessary for the latter to pull them towards itself. When the little artist has once got a single thread at its disposition, it strengthens this until it is sufficiently strong for it to run backwards and forwards thereupon, and to spin therefrom the web.[1]

Special Habits.

Water-spider.—The water-spider (*Argyroneta aquatica*), as is well known, displays the curious instinct of building her nest below the surface of water, and constructing it on the principle of a diving-bell. The animal usually selects still waters for this purpose, and makes her nest in the form of an oval hollow, lined with web, and held secure by a number of threads passing in various directions and fastened to the surrounding plants. In this oval bell, which is open below, she watches for prey, and, according to Kirby,[2] passes the winter after having closed the opening. The air needful for respiration the spider carries from the surface of the water. To do this she swims upon her back in order to entangle an air-bubble upon the hairy surface of her abdomen. With this bubble she descends, 'like a globe of quicksilver,' to the opening of her nest, where she liberates it and returns for more.

[1] *Loc. cit.*, p. 316 *et seq.*
[2] *Hist. Habits and Inst. of Animals*, vol. ii.. p. 296.

The Vagrant or Wolf Spider.—This insect catches its prey by stealthily stalking it until within distance near enough to admit of a sudden dart being successful in effecting capture. Some species, before making the final dart (*e.g. Salticus scenicus*), fix a line of web upon the surface over which they are creeping, so that whether their station is vertical or horizontal with reference to the prey, they can leap fearlessly, the thread in any case preventing their fall. Dr. H. F. Hutchinson says that he has seen this spider crawling over a looking-glass stalking its own reflection.[1]

The following is quoted from Büchner:—

Less idyllic than the water-spider is our native hunting-spider (*Dolomedes fimbriata*), which belongs to those species which spin no web, but hunt their victims like animals of prey. As the *Argyroneta* is the discoverer of the diving-bell, so may this be regarded as the discoverer or first builder of a floating raft. It is not content with hunting insects on land, but follows them on the water, on the surface of which it runs about with ease. It, however, needs a place to rest on, and makes it by rolling together dry leaves and such like bodies, binding them into a firm whole with its silken threads. On this raft-like vessel it floats at the mercy of wind and waves; and if an unlucky water-insect comes for an instant to the surface of the water to breathe, the spider darts at it with lightning speed, and carries it back to its raft to devour at its ease. Thus everywhere in nature are battle, craft, and ingenuity, all following the merciless law of egoism, in order to maintain their own lives and to destroy those of others!

Trap-door Spiders.—These display the curious instinct of providing their nests with trap-doors. The nest consists of a tube excavated in the earth to the depth of half a foot or more. In all save one species the tube is unbranched; it is always lined with silk, which is continuous with the lining of the trap-door or doors, of which it forms the hinge. In the species which constructs a branching tube, the branch is always single, more or less straight, takes origin at a point situated a few inches from the orifice of the main tube, is directed upwards at an acute

[1] *Nature*, vol. xx., p. 581.   [2] *Loc. cit.*, p. 323.

angle with that tube, and terminates blindly just below the surface of the soil. At its point of junction with or departure from the main tube it is provided with a trap-door resembling that which closes the orifice of the main tube, and of such a size and arrangement that when closed against the opening of the branch tube it just fills that opening ; while when turned outwards, so as to uncork this opening, it just fills the diameter of the main tube : the latter, therefore, is in this species provided with two trap-doors, one at the surface of the soil, and the other at the fork of the branched tube.

Each species of trap-door spider is very constant in building a particular kind of trap-door ; but among the different species there are four several kinds of trap-doors to be distinguished. 1st. The single-door cork nest, wherein the trap-door is a thick structure, and fits into the tube like a cork into a bottle. 2nd. The single-door wafer nest, wherein the trap-door is as thin as a piece of paper. 3rd. The double-door unbranched nest, wherein there is a second trap-door situated a few inches below the first one. And 4th, the double-door branched nest already described. In all cases the trap-doors open outwards, and when the nest is placed, as it usually is, on a sloping bank, the trap-door opens upwards ; hence there is no fear of its gaping, for gravity is on the side of holding it shut.

The object of the trap-door is to conceal the nest, and for this purpose it is always made so closely to resemble the general surface of the ground on which it occurs, that even a practised eye finds it difficult to detect the structure when closed. In order to make the resemblance to the surrounding objects as perfect as possible, the spider either constructs the surface of its door of a portion of leaf, or weaves moss, grass, &c., into the texture. Moggridge says,[1]—

Thus, for example, in one case where I had cut out a little clod of mossy earth, about two inches thick and three square on the surface, containing the top of the tube and the moss-covered cork door of *N. cœmentaria*, I found, on revisiting the

---

[1] *Harvesting Ants and Trap-door Spiders*, p. 120.

place six days later, that a new door had been made, and that the spider had mounted up to fetch moss from the undisturbed bank above, planting it in the earth which formed the crown of the door. Here the moss actually called the eye to the trap, which lay in the little plain of brown earth made by my digging.

If an enemy should detect the trap-door and endeavour to open it, the spider frequently seizes hold of its internal surface, and, applying her legs to the walls of the tube, forcibly holds the trap-door shut. In the double trap-door species it is surmised that the second trap-door serves as an inner barrier of defence, behind which the spider retires when obliged to abandon the first one. In the branched tube species (which, so far as at present known, only occurs in the south of Europe) it is surmised that the spider, when it finds that an enemy is about to gain entrance at the first trap-door, runs into the branch tube and draws up behind it the second trap-door. The surface of this trap-door, being overlaid with silk like the walls of the tube, is then invisible; so that the enemy no doubt passes down the main tube to find it empty, without observing the lateral branch in which the spider is concealed behind the closed door.

As showing that these animals are to no small extent able to adapt their dwellings to unusual circumstances, I shall here quote the following from Moggridge (*loc. cit.*, p. 122):—

Certain nests which were furnished with two doors of the cork type were observed by Mr. S. S. Saunders in the Ionian Islands. The door at the surface of these nests was normal in position and structure, but the lower one was placed at the very bottom of the nest, and inverted, so that, though apparently intended to open downwards, it was permanently closed by the surrounding earth. The presence of a carefully constructed door in a situation which forbade the possibility of its ever being opened seemed, indeed, something difficult to account for. However, it occurred to Mr. Saunders that, as these nests were found in the cultivated ground round the roots of olive trees, they may occasionally have got turned topsy-turvy when the soil was broken up. The spider then, finding her door buried below in the ground and the bottom of the tube at the surface, would have either to seek new quarters or to adapt the nest to its altered position, and make an opening and door at the exposed end. In order to try whether one of these spiders would do this, Mr. Saunders placed a nest, with its occupant inside, upside down in a flower-pot. After the lapse of ten days a new door was made, exactly as he had conjectured it would be, and the nest presented two doors like those which he had found at first.

The most remarkable fact connected with these animals, if we regard their peculiar instinct from the standpoint of the descent theory, is the wide range of their geographical distribution. In all quarters of the globe species of trap-door spiders are found occurring in more or less localised areas; and as it is improbable that so peculiar an instinct should have arisen independently in more than one line of descent, we can only conclude that the wide dispersion of the species presenting it has been subsequent to the origin and perfecting of the instinct. This conclusion of course necessitates the supposition that the instinct must be one of enormous antiquity; and in this connection it is worthy of remark that we seem to have independent evidence to show that such is the case. It is a principle of evolution that the earlier any structure or instinct appears in the development of the race, the sooner will it appear in the development of the individual; and read by the light of this principle we should conclude, quite apart from all considerations as to the wide geographical distribution of trap-door spiders, that their instincts—as, indeed, is the case with the characteristic instincts of many other species of spiders—must be of immense age. Thus, again to quote Moggridge,—

It seems to be the rule with spiders generally that the offspring should leave the nest and construct dwellings for themselves when very young.

Mr. Blackwall, speaking of British spiders, says:—'Complicated as the processes are by which these symmetrical nets are produced, nevertheless young spiders, acting under the influence of instinctive impulse, display, even in their first attempts to fabricate them, as consummate skill as the most experienced individuals.'

Again, Mr. F. Pollock[1] relates of the young of *Epeira aurelia*, which he observed in Madeira, that when seven weeks old they made a web the size of a penny, and that these nets have the same beautiful symmetry as those of the full-grown spider.

And, speaking of trap-door spiders, Moggridge says,—

I cannot help thinking that these very small nests, built as they are by minute spiders probably not very long hatched from the egg, must rank among the most marvellous structures of this kind with which we are acquainted. That so young and weak a creature should be able to excavate a tube in the earth many times its own length, and know how to make a perfect miniature of the nest of its parents, seems to be a fact which has scarcely a parallel in nature.[2]

Regarding the steps whereby the instinct of building trap-doors probably arose, Büchner quotes Moggridge thus:—

To show, lastly, how various are the transitional forms and gradations so important in deciding upon the gradual origin of the forms of nests, Moggridge also alludes to the similar buildings made by other genera of spiders. *Lycosa Narbonensis*, a spider of Southern France much resembling the Apuleian tarantula, and belonging to the family of the wolf spiders, makes cylindrical holes in the earth, about one inch wide and three or four inches deep, in a perpendicular direction; when they have attained this depth they run further horizontally, and end in a three-cornered room, from one to two inches broad, the floor of which is covered with the remnants of dead insects. The whole nest is lined within with a thick silken material, and has at its opening—closed by no door—an aboveground chimney-shaped extension, made of leaves, needles, moss, wood, &c., woven together with spider threads. These chimneys show various differences in their manner of building, and are intended chiefly, according to Moggridge, to prevent the sand blown about by the violent sea-winds from penetrating into the nests. During winter the opening is wholly and continuously woven over, and it is very well possible, or probable, that the process of reopening such a warm covering in the spring,

---

[1] 'The History and Habits of *Epeira aurelia*,' in *Annals and Mag. of Nat. Hist.* for June 1865.
[2] *Harvesting Ants and Trap-door Spiders*, p. 126. This admirable work, with its appendix, contains a very full account of the whole economy of the interesting animals with which it is concerned.

after this opening was three-quarters completed, and was large enough to let the spider pass out, may have long ago awaked in the brain of some species of spider the idea of making a permanent and moveable door. But from this to the practical construction of so perfect a door as we have learned to know, and even to the building of the exceedingly complicated nest of the *N. Manderstjernæ*, through all the gradations which we already know, and which doubtless exist in far greater number, is no great or impossible step.

General Intelligence.

Coming now to the general intelligence of spiders, I think there can be no reasonable doubt, from the force of concurrent testimony, that they are able to distinguish between persons, and approach those whom they have found to be friendly, while shunning strangers. This power of discrimination, it will be remembered, also occurs among bees and wasps, and therefore its presence in spiders is not antecedently improbable. I myself know a lady who has 'tam d' spiders to recognise her, so that they come out to be fed when she enters the room where they are kept; and stories of the taming of spiders by prisoners are abundant. The following anecdote recorded by Büchner is in this connection worth quoting:—

Dr. Moschkau, of Gohlis, near Leipsic, writes as follows to the author, on August 28, 1876:—'In Oderwitz (?), where I lived in 1873 and 1874, I noticed one day in a half-dark corner of the anteroom a tolerably respectable spider's web, in which a well-fed cross-spider had made its home, and sat at the nest-opening early and late, watching for some flying or creeping food. I was accidentally several times a witness of the craft with which it caught its victim and rendered it harmless, and it soon became a regular duty to carry it flies several times during a day, which I laid down before its door with a pair of pincers. At first this feeding seemed to arouse small confidence, the pincers perhaps being in fault, for it let many of the flies escape again, or only seized them when it knew that they were within reach of its abode. After a while, however, the spider came each time and took the flies out of the pincers and spun them over. The latter business was sometimes done so superficially, when I gave flies very quickly one after the other, that some of the already ensnared flies found time and opportunity to escape. This game was carried on by me for some weeks, as it seemed to me curious. But one day when the spider seemed very ravenous, and regularly flew at each fly offered to it, I began teasing it. As soon as it had got hold of the fly I pulled it back again with the pincers. It took this exceedingly ill. The first time, as I finally left the fly with it, it managed to forgive me, but when I later took a fly right away, our friendship was destroyed for ever. On the following day it treated my offered flies with contempt, and would not move, and on the third day it had disappeared.[1]

Jesse relates the following anecdote, which seems to display on the part of a spider somewhat remote adaptation of means to novel circumstances. He confined a spider with her eggs under a glass upon a marble mantelpiece. Having surrounded the eggs with web,—

She next proceeded to fix one of her threads to the upper part of the glass which confined her, and carried it to the further end of the piece of grass, and in a short time had succeeded in raising it up and fixing it perpendicularly, working her threads from the sides of the glass to the top and sides of the piece of grass. Her motive in doing this was obvious. She not only rendered the object of her care more secure than it would have been had it remained flat on the marble, but she was probably aware that the cold from the marble would chill her eggs, and prevent their arriving at maturity: she therefore raised them from it in the manner I have described.[2]

Mr. Belt gives the following account of the intelligence which certain species of South American spiders display in escaping from the terrible hosts of the Eciton ants:—

Many of the spiders would escape by hanging suspended by a thread of silk from the branches, safe from the foes that swarmed both above and below.

I noticed that spiders generally were most intelligent in escaping, and did not, like the cockroaches and other insects, take shelter in the first hiding-place they found, only to be driven out again, or perhaps caught by the advancing army of ants. I have often seen large spiders making off many yards in advance, and apparently determined to put a good distance

[1] *Loc. cit.*, p. 319.
[2] *Gleanings*, vol. i., p. 103.

between themselves and the foe. I once saw one of the false spiders, or harvest-men (*Phalangidæ*), standing in the midst of an army of ants, and with the greatest circumspection and coolness lifting, one after the other, its long legs, which supported its body above their reach. Sometimes as many as five out of its eight legs would be lifted at once, and whenever an ant approached one of those on which it stood, there was always a clear space within reach to put down another, so as to be able to hold up the threatened one out of danger.[1]

Mr. L. A. Morgan, writing to 'Nature' (Jan. 22, 1880), gives an account of a spider conveying a large insect from the part of the web where it was caught to the 'larder,' by the following means. The spider first went two or three times backwards and forwards between the head of the insect and the main strand of the web. After this he went about cutting all the threads around the insect till the latter hung by the head strands alone. The spider then fixed a thread to the tail end, and by this dragged the carcass as far on its way to the larder as the head strands would permit. As soon as these were taut, he made the tail rope fast, went back to the head rope and cut it; then he attached himself to the head and pulled the body towards the larder, until the tail rope was taut. In this way, by alternately cutting the head and tail ropes and dragging the insect bit by bit, he conveyed it safely to the larder.

But the practical acquaintance with mechanical principles which this observation displays is perhaps not so remarkable as that which is sometimes shown by spiders when they find that a widely spread web is not tightly enough stretched, and as a consequence is to an inconvenient extent swayed about by the wind. Under such circumstances these animals have been observed to suspend to their webs small stones or other heavy objects, the weight of which serves to steady the whole system. Gleditsch saw a spider so circumstanced let itself down to the ground by means of a thread, seize a small stone, remount, and fasten the stone to the lower part of its web, at a height sufficient to enable animals and men to walk

[1] *Naturalist in Nicaragua*, p. 19.

beneath it. After alluding to this case, Büchner observes (*loc. cit.*, p. 318),—

But a similar observation was made by Professor E. H. Weber, the famous anatomist and physiologist, and was published many years ago in Müller's Journal. A spider had stretched its web between two posts standing opposite each other, and had fastened it to a plant below for the third point. But as the attachment below was often broken by the garden work, by passers-by, and in other ways, the little animal extricated itself from the difficulty by spinning its web round a little stone, and fastened this to the lower part of its web, swinging freely, and so to draw the web down by its weight instead of fastening it in this direction by a connecting thread. Carus ('Vergl. Psycho.,' 1866, p. 76) also made a similar observation. But the most interesting observation on this head is related by J. G. Wood ('Glimpses into Petland'), and repeated by Watson (*loc. cit.*, p. 455). One of my friends, says Wood, was accustomed to grant shelter to a number of garden spiders under a large verandah, and to watch their habits. One day a sharp storm broke out, and the wind raged so furiously through the garden that the spiders suffered damage from it, although sheltered by the verandah. The mainyards of one of these webs, as the sailors would call them, were broken, so that the web was blown hither and thither, like a slack sail in a storm. The spider made no fresh threads, but tried to help itself in another way. It let itself down to the ground by a thread, and crawled to a place where lay some splintered pieces of a wooden fence thrown down by the storm. It fastened a thread to one of the bits of wood, turned back with it, and hung it with a strong thread to the lower part of its nest, about five feet from the ground. The performance was a wonderful one, for the weight of the wood sufficed to keep the nest tolerably firm, while it was yet light enough to yield to the wind, and so prevent further injury. The piece of wood was about two and a half inches long, and as thick as a goose-quill. On the following day a careless servant knocked her head against the wood, and it fell down. But in the course of a few hours the spider had found it and brought it back to its place. When the storm ceased, the spider mended her web, broke the supporting thread in two, and let the wood fall to the ground!

If so well-observed a fact requires any further confirmation, I may adduce the following account, which is of the more value as corroborative evidence from the writer not appearing to be aware that the fact had been observed before. This writer is Dr. John Topham, whom the late Dr. Sharpey, F.R.S., assured me is a competent observer, and who publishes the account in 'Nature' (xi. 18):—

A spider constructed its web in an angle of my garden, the sides of which were attached by long threads to shrubs at the height of nearly three feet from the gravel path beneath. Being much exposed to the wind, the equinoctial gales of this autumn destroyed the web several times.

The ingenious spider now adopted the contrivance here represented. It secured a conical fragment of gravel with its larger end upwards by two cords, one attached to each of its opposite sides, to the apex of its wedge-shaped web, and left it suspended as a moveable weight to be opposed to the effect of such gusts of air as had destroyed the webs previously occupying the same situation.

The spider must have descended to the gravel path for this special object, and having attached threads to a stone suited to its purpose, must have afterwards raised this by fixing itself upon the web, and pulling the weight up to a height of more than two feet from the ground, where it hung suspended by elastic cords. The excellence of the contrivance is too evident to require further comment.

An almost precisely analogous case, with a sketch, is published by another observer in 'Land and Water,' Dec. 12, 1877.

Scorpions.

Before quitting the Arachnida I must allude to some recent correspondence on the alleged tendency of the scorpion to commit suicide when surrounded by fire. This alleged tendency has long been recognised in popular fables, and has been used by Byron as a poetical metaphor in certain well-known lines. But until the publication of the correspondence to which I allude, no one supposed the tendency in question to have any existence in fact. This correspondence took place in 'Nature' (vol. xi.), and as the subject is an interesting one, I shall reproduce the more important contributions to it *in extenso*. It was opened by Mr. W. G. Biddie as follows:—

I shall feel obliged if you will record in 'Nature' a fact with reference to the common black scorpion of Southern India, which was observed by me some years ago in Madras.

One morning a servant brought to me a large specimen of this scorpion, which, having stayed out too long in its nocturnal rambles, had apparently got bewildered at daybreak, and been unable to find its way home. To keep it safe the creature was at once put into a glazed entomological case. Having a few leisure minutes in the course of the forenoon I thought I would see how my prisoner was getting on, and to have a better view of it the case was placed in a window in the rays of the hot sun. The light and heat seemed to irritate it very much, and this recalled to my mind a story which I had read somewhere that a scorpion, on being surrounded with fire, had committed suicide. I hesitated about subjecting my pet to such a terrible ordeal, but taking a common botanical lens, I focussed the rays of the sun on its back. The moment this was done it began to run hurriedly about the case, hissing and spitting in a very fierce way. This experiment was repeated some four or five times with like results, but on trying it once again, the scorpion turned up its tail and plunged the sting, quick as lightning, into its own back. The infliction of the wound was followed by a sudden escape of fluid, and a friend standing by me called out, 'See, it has stung itself: it is dead;' and sure enough in less than half a minute life was quite extinct. I have written this brief note to show (1) that animals may commit suicide; (2) that the poison of certain animals may be destructive to themselves.

The following corroborative evidence on the subject was then supplied by Dr. Allen Thomson, F.R.S. ('Nature,' vol. xx., p. 577):—

Doubts having been expressed at various times, even by learned naturalists, as to the reality of the suicide or self-destruction of the scorpion by means of its own poison, and these doubts having been again stated in 'Nature,' vol. xx., p. 553, by Mr. B. F. Hutchinson, of Peshawur, as the result of his own observations, I think it may be useful to give an articulate account of the phenomenon as it has been related to me by an eye-witness, which removes all possible doubt as to its occurrence under certain circumstances.

While residing many years ago, during the summer months, at the baths of Sulla in Italy, in a somewhat damp locality, my informant together with the rest of the family was much annoyed by the frequent intrusion of small black scorpions into the house, and their being secreted among the bedclothes, in shoes, and other articles of dress. It thus became necessary to be constantly on the watch for these troublesome creatures, and to take means for their removal and destruction. Having been informed by the natives of the place that the scorpion would destroy itself if exposed to a sudden light, my informant and her friends soon became adepts in catching the scorpions and disposing of them in the manner suggested. This consisted in confining the animal under an inverted drinking-glass or tumbler, below which a card was inserted when the capture was made, and then, waiting till dark, suddenly bringing the light of a candle near to the glass in which the animal was confined. No sooner was this done than the scorpion invariably showed signs of great excitement, running round and round the interior of the tumbler with reckless velocity for a number of times. This state having lasted for a minute or more, the animal suddenly became quiet, and turning its tail on the hinder part of its body over its back, brought its recurved sting down upon the middle of the head, and piercing it forcibly, in a few seconds became quite motionless, and in fact quite dead. This observation was repeated very frequently; in truth, it was adopted as the best plan of getting rid of the animals. The young people were in the habit of handling the scorpions with impunity immediately after they were so killed, and of preserving many of them as curiosities.

In this narrative the following circumstances are worthy of attention:—

(1) The effect of light in producing the excitement amounting to despair, which causes the animal to commit self-destruction;

(2) The suddenness of the operation of the poison, which is probably inserted by the puncture of the head into the upper cerebral ganglion; and

(3) The completeness of the fatal symptoms at once induced.

I am aware that the phenomena now described have been observed by others, and they appear to have been familiarly known to the inhabitants of the district in which the animals are found. Sufficient confirmation of the facts is also to be found in the narratives of 'G. Biddie' and 'M. L.' contained in 'Nature,' vol. ix., pp. 29–47, and it will be observed that the circumstances leading the animal to self-destruction in these instances were somewhat similar to those narrated by my informant. It is abundantly clear, therefore, that the view taken by Mr. Hutchinson, viz., that the 'popular idea regarding scorpionic suicide is a delusion based on an impossibility,' is wholly untenable; indeed, the recurved direction of the sting, which he refers to as creating the impossibility of the animal destroying itself, actually facilitates the operation of inflicting the wound. I suppose Mr. Hutchinson, arguing from the analogy of bees or wasps, imagined that the sting would be bent forwards upon the body, whereas the wound of the scorpion is invariably inflicted by a recurvation of the tail over the back of the animal.

It will be perceived that these observations were not made by Dr. Allen Thomson himself, and that there are certain inherent discrepancies in the account which he has published—such, for instance, as the reason given for trying and repeating the experiment, the method being clearly a cumbersome one to employ if the only object were that of 'disposing of' the animals. Nevertheless, as Dr. Thomson is a high authority, and as I learn from him that he is satisfied regarding the capability and veracity of his informant, I have not felt justified in suppressing his evidence. Still I think that so remarkable a fact unquestionably demands further corroboration before we should be justified in accepting it unreservedly. For if it is a fact, it stands as a unique case of an instinct detrimental alike to the individual and to the species.

CHAPTER VII.

REMAINING ARTICULATA.

THE Hymenoptera being so much the most intelligent order, not merely of insects, but of Invertebrata, and the Arachnida having been now considered, very little space need be occupied with the remaining classes of the Articulata.

Coleoptera.

Sir John Lubbock, in his first paper on Bees and Wasps, quotes the following case from Kirby and Spence, with the remarks which I append:—

The first of these anecdotes refers to a beetle (*Ateuchus pilularius*) which, having made for the reception of its eggs a pellet of dung too heavy for it to move, repaired to an adjoining heap, and soon returned with three of his companions. 'All four now applied their united strength to the pellet, and at length succeeded in pushing it out; which being done, the three assistant beetles left the spot and returned to their own quarters.' This observation rests on the authority of an anonymous German artist; and though we are assured that he was a 'man of strict veracity,' I am not aware that any similar fact has been recorded by any other observer.

Catesby, however, says:—

I have attentively admired their industry, and their mutual assisting of each other in rolling these globular balls from the place where they made them, to that of their interment, which is usually a distance of some yards, more or less. This they perform back foremost, by raising their hind parts and pushing away the ball with their hind feet. Two or three of them are sometimes engaged in trundling one ball, which from meeting with impediments, on account of the unevenness of the ground, is sometimes deserted by them. It is, however, attempted by others with success, unless it happen to roll into some deep hollow or ditch, where they are accustomed to leave it; but they continue their work by rolling off the next ball that comes in their way. None of them seem to know their own balls, but an equal care for the whole appears to affect all the community. They form these pellets while the dung remains moist, and leave them to harden in the sun before they attempt to roll them. In their rolling of them from place to place, both they and the balls may frequently be seen tumbling about over the little eminences that are in their way. They are not, however, easily discouraged, and by repeating their attempts usually surmount the difficulties.[1]

Büchner speaks of the fact that dung-beetles co-operate in their work as one that is well established, but gives no authorities or references.[2] A friend of my own, however, informs me that she has witnessed the fact; and in view of analogous observations made on other species of Coleoptera, I see no reason to doubt this one. Some of these observations I may here append.

Herr Gollitz writes to Büchner thus:—

Last summer, in the month of July, I was one day in my field, and found there a mound of fresh earth like a molehill, on which a striped black and red beetle, with long legs, and about the size of a hornet, was busy taking away the earth from a hole that led like a pit into the mound, and levelling the place. After I had watched this beetle for some time, I noticed a second beetle of the same kind, which brought a little lump of earth from the interior to the opening of the hole, and then disappeared again in the mound; every four or five minutes a pellet came out of the hole, and was carried away by the first-named beetle. After I had watched these proceedings for about half an hour, the beetle which had been working underground came out and ran to its comrade. Both put their heads together, and clearly held a conversation, for immediately afterwards they changed work. The one which had been working outside went into the mound, the other took the outside labour, and all went on vigorously. I watched the affair still for a little longer, and went away with the notion that these insects could understand each other just like men. Klingelhöffer, of Darmstadt (in Brehm, *loc. cit.*, ix., p. 86), says:—A golden running beetle came to a cockchafer lying on its back in the garden, intending to eat it, but was unable to master it; it ran to the next bush, and

[1] Quoted by Bingley, *Animal Biography*, vol. iii., p. 118.
[2] *Loc. cit.*, p. 344.

returned with a friend, whereupon the two overpowered the cockchafer, and pulled it off to their hiding-place.

Similarly, there is no doubt that the burying beetles (*Nicrophorus*) co-operate.

Several of them unite together to bury under the ground, as food and shelter for their young, some dead animal, such as a mouse, a toad, a mole, a bird, &c. The burial is performed because the corpse, if left above ground, would either dry up, or grow rotten, or be eaten by other animals. In all these cases the young would perish, whereas the dead body lying in the earth and withdrawn from the outer air lasts very well. The burying beetles go to work in a very well-considered fashion, for they scrape away the earth lying under the body, so that it sinks of itself deeper and deeper. When it is deep enough down, it is covered over from above. If the situation is stony, the beetles with united forces and great efforts drag the corpse to some place more suitable for burying. They work so diligently that a mouse, for instance, is buried within three hours. But they often work on for days, so as to bury the body as deeply as possible. From large carcasses, such as those of horses, sheep, &c., they only bury pieces as large as they can manage.[1]

Lastly, Clarville gives a case of a burying beetle which wanted to carry away a dead mouse, but, finding it too heavy for its unaided strength, went off, like the beetles previously mentioned, and brought four others to its assistance.[2]

A friend of Gleditsch fastened a dead toad, which he desired to dry, upon the top of an upright stick. The burying beetles were attracted by the smell, and finding that they could not reach the toad, they undermined the stick, so causing it to fall with the toad, which was then buried safe out of harm's way.[3]

A converse exemplification of beetle-intelligence is given by G. Berkeley.[4] He saw a beetle carrying a dead spider up a heath plant, and hanging it upon a twig of the heath in so secure a position, that when the insect had left it Mr. Berkeley found that a sharp shake of the heather would not bring the dead spider down. As the burying

---

[1] Büchner, *loc. cit.*, p. 344.
[2] Quoted in Strauss, *Insects*, s. 389.
[3] Kirby and Spence, *loc. cit.*, pp. 321-2.
[4] *Life and Recollections*, vol. ii., p. 356.

beetle preserves its treasure by hiding it out of sight below ground, so this beetle no doubt secured the same end but by other means; 'seeing,' as Mr. Berkeley observes, 'that if it did not hang up its prey, it might fall into the hands of other hunters, it took all possible pains to find out the best store-room for it.'

The above instances of beetle-intelligence lead me to credit the following, which has been communicated to me by Dr. Garraway, of Faversham. On a bank of moss in the Black Forest he saw a beetle alight with a caterpillar which it was carrying, and proceed to excavate a cylindrical hole in the peat, about an inch and a half deep, into which, when completed, it dropped the caterpillar, and then flew away through the pines. 'I was struck,' says my correspondent, 'with the creature's folly in leaving the whole uncovered, as every curious wayfaring insect would doubtless be tempted to enter therein. However, in about a minute the beetle returned, this time carrying a small pebble, of which there were none in the immediate vicinity, and having carefully fitted this into the aperture, fled away into space.'

Earwig.

I must devote a short division of this chapter to the earwig. M. Geer describes a regular process of incubation as practised by the mother insect. He placed one with her eggs in a box, and scattered the eggs on the floor of the latter. The earwig, however, carried them one by one into a certain part of the box, and then remained constantly sitting upon the heap without ever quitting it for a moment. When the eggs were hatched, the young earwigs kept close to their mother, following her about everywhere, and often running under her abdomen, just as chickens run under a hen.[1]

A young lady, who objects to her name being published, informs me that her two younger sisters (children) are in the habit of feeding every morning with sugar an earwig, which they call 'Tom,' and which crawls up a certain curtain regularly every day at the same hour, with the apparent expectation of getting its breakfast. This re-

---
[1] Quoted by Bingley, *loc. cit.*, vol. iii., pp. 150–51.

sembles analogous instances which have been mentioned in the case of spiders.

Dipterous Insects.

The gad-fly, whose eggs are hatched out in the intestines of the horse, exhibits a singular refinement of instinct in depositing them upon those parts of the horse which the animal is most likely to lick. For, according to Bingley and other writers, 'the inside of the knee is the part on which these flies principally deposit their eggs; and next to this they fix them upon the sides, and the back part of the shoulder; but almost always in places liable to be licked by the tongue.' The female fly deposits her eggs while on the wing, or at least scarcely appears to settle when she extends her ovipositor to touch the horse. She lays only a single egg at a time—flying away a short distance after having deposited one in order to prepare another, and so on.

The following anecdote, which I quote from Jesse, seems to indicate no small degree of intelligence on the part of the common house-fly—intelligence, for instance, the same both in kind and degree as that which was displayed by Sir John Lubbock's pet wasp already mentioned:

Slingsby, the celebrated opera dancer, resided in the large house in Cross-deep, Twickenham, next to Sir Wathen Waller's, looking down the river. He was fond of the study of natural history, and particularly of insects, and he once tried to tame some house-flies, and preserve them in a state of activity through the winter. For this purpose, quite at the latter end of autumn, and when they were becoming almost helpless, he selected four from off his breakfast-table, put them upon a large handful of cotton, and placed it in one corner of the window nearest the fireplace. Not long afterwards the weather became so cold that all flies disappeared except these four, which constantly left their bed of cotton at his breakfast-time, came and fed at the table, and then returned to their home. This continued for a short time, when three of them became lifeless in their shelter, and only one came down. This one Slingsby had trained to feed upon his thumb-nail, by placing on it some moist sugar mixed with a little butter. Although there had been at intervals several days of sharp frost, the fly never missed taking his daily meal in this way till after Christmas, when, his kind preserver having invited a friend to dine and sleep at his house, the fly, the next morning, perched upon the thumb of the visitor, who, being ignorant that it was a pet of his host's, clapped his hand upon it, and thus put an end to Mr. Slingsby's experiment.[1]

Crustacea.

There is no doubt that these are an intelligent group of animals, although I have been able to collect but wonderfully little information upon the subject. Mr. Moseley, F.R.S., in his very interesting work, 'Notes by a Naturalist on the *Challenger*,' says (p. 70):—

In the tropics one becomes accustomed to watch the habits of various species of crabs, which there live so commonly an aërial life. The more I have seen of them, the more have I been astonished at their sagacity.

And again (pp. 48-9):—

A rock crab (*Grapsus stringosus*) was very abundant, running about all over the rocks, and making off into clefts on one's approach. I was astonished at the keen and long sight of this crab. I noticed some made off at full speed to their hiding-places at the instant that my head showed above a rock fifty yards distant. . . .

At Still Bay, on the sandy beach of which a heavy surf was breaking, I encountered a sand crab (*Œcypoda ippeus*), which was walking about, and got between it and its hole in the dry sand above the beach. The crab was a large one, at least three inches in breadth of its carapace. . . . With its curious column-like eyes erect, the crab bolted down towards the surf as the only escape, and as it saw a great wave rushing up the shelving shore, dug itself tight into the sand, and held on to prevent the undertide from carrying it into the sea. As soon as the wave had retreated, it made off full speed for the shore. I gave chase, and whenever a wave approached, the crab repeated the manœuvre. I once touched it with my hand whilst it was buried and blinded by the sandy water, but the surf compelled me to retreat, and I could not snatch hold of it for fear of its powerful claws. At last I chased it, hard pressed, into the surf in a hurry, and being unable to get proper hold in time, it was washed into the sea. The crab evidently dreaded going into the sea. . . . They soon die when kept a short time beneath the water.

[1] *Gleanings*, vol. ii., pp. 165-6.

The land crabs of the West Indies and North America descend from their mountain home in May and June, to deposit their spawn in the sea. They travel in such swarms that the roads and woods are covered with them. They migrate in a straight line, and rather than allow themselves to be deflected from it, 'they scale the houses, and surmount every other obstacle that lies in their way' (Kirby). They travel chiefly by night, and when they arrive at the sea-shore they 'bathe three or four different times,' and then 'commit their eggs to the waves.' They return to the mountains by the same route, but only the most vigorous survive the double journey.

Prof. Alex. Agassiz details some interesting observations on the behaviour of young hermit crabs reared by himself 'from very young stages,' when first presented with shells of mollusks. 'A number of shells, some of them empty, others with the animal living, were placed in a glass dish with the young crabs. Scarcely had the shells reached the bottom before the crabs made a rush for the shells, turned them round and round, invariably at the mouth, and soon a couple of the crabs decided to venture in, which they did with remarkable alacrity.' The crabs which obtained for their share the shells still inhabited by living mollusks, 'remained riding round upon the mouth of their future dwelling, and, on the death of the mollusk, which generally occurred soon after in captivity, commenced at once to tear out the animal, and having eaten him, proceeded to take its place within the shell.'[1]

There is a species of small crustacean (*Podocerus capillatus*) described by Mr. Bates, which builds a nest to contain its eggs. The nest is in the form of a hollow cone, built upon seaweed, and composed of fine thread-like material closely interlaced. 'These nests,' says Mr. Bates, 'are evidently used as a place of refuge and security, in which the parent protects and keeps her brood of young until they are old enough to be independent of the mother's care.'

Dr. Erasmus Darwin tells us, on the authority of a friend on whose competency as an observer he relied, that the common crab during the moulting season stations as

[1] *American Journ. Sc. and Art*, vol. x., Oct. 1875.

sentinel an unmoulted or hard-shelled individual, to prevent marine enemies from injuring moulted individuals in their unprotected state. While thus mounting guard the hard-shelled crab is much more courageous than at other times, when he has only his own safety to consider. But these observations require to be corroborated.

In 'Nature' (xv., p. 415) there is a notice of a lobster (*Homarus marinus*) in the Rothesay Aquarium which attacked a flounder that was confined in the same tank with him, and having devoured a portion of his victim, buried the rest beneath a heap of shingle, on which he 'mounted guard.' 'Five times within two hours was the fish unearthed, and as often did the lobster shovel the gravel over it with his huge claws, each time ascending the pile and turning his bold defensive front to his companions.'

The following is quoted from Mr. Darwin's 'Descent of Man' (pp. 270-1):—

A trustworthy naturalist, Mr. Gardner, whilst watching a shore-crab (*Gelasimus*) making its burrow, threw some shells towards the hole. One rolled in, and three other shells remained within a few inches of the mouth. In about five minutes the crab brought out the shell which had fallen in, and carried it away to the distance of a foot; it then saw the three other shells lying near, and evidently thinking that they might likewise roll in, carried them to the spot where it had laid the first. It would, I think, be difficult to distinguish this act from one performed by man by the aid of reason.

Mr. Darwin also alludes to the curious instinctive habits of the large shore-crab (*Birgus latro*), which feeds on fallen cocoa-nuts ' by tearing off the husk fibre by fibre; and it always begins at that end where the three eye-like depressions are situated. It then breaks through one of these eyes by hammering with its heavy front pincers, and turning round, extracts the albuminous core with its narrow posterior pincers.'

Remarkable cases occur of commensalism between certain crabs and sea-anemones, and they betoken much intelligence. Thus Professor Möbius says in his 'Beiträge zur Meeresfauna der Insel Mauritius' (1880) that there are two crabs belonging to different genera which have the habit of firmly grasping a sea-anemone in each claw and carrying them about, presumably to secure some benefit to themselves. The more familiar case of the species of anemone which lives on the shells tenanted by hermit crabs is of special interest to us on account of a remarkable observation published by Mr. Gosse, F.R.S. (*Zoologist*, June, 1859). He found that on his detaching the anemone (*Adamsia*) from the shell, the hermit crab always took it up in its claws and held it against the shell 'for the space of ten minutes at a time, until fairly attached by a good strong base.' It was said by the late Dr. Robert Ball that when the common *Sagartia parasitica* is attached to a stone and a hermit crab is placed in its vicinity, the anemone will leave the stone and attach itself to the hermit's shell (*Critic*, March 24, 1860).

Intelligence of Larvæ of Certain Insects.

I shall now allude to some of the more interesting facts touching the psychology of insects when in their immature or larval state. This is an interesting topic from the point of view which we occupy as evolutionists, because a caterpillar is really a locomotive and self-feeding embryo, whose entire mental constitution is destined to undergo a metamorphosis no less complete and profound than that which is also destined to take place in its corporeal structure. Yet although the caterpillar has an embryo psychology, its instincts and even intelligence often seem to be higher or more elaborated than is the case with the imago form. Where such is the case the explanation of course must be that it is of more importance to the species that the larval form should be in a certain measure intelligent than that the imago form should be so. Every larva is a potential imago, or breeding individual; therefore its life is of no less value to the species during its larval than during its adult existence; and if certain instincts or grades of intelligence are of more use to it during the former than during the latter period, of course natural selection would determine the unusual event which we seem here in some cases to see—namely, that the

INTELLIGENCE OF LARVÆ—ANT-LION.

embryo should stand on a higher level of psychological development than the adult.

I may most fitly begin under this heading with the remarkable instincts of the so-called 'ant-lion,' which is the larva of a neuropterous insect, the common *Myrmeleon* (*M. formicarium*). I quote the following account of its habits from Thompson's 'Passions of Animals' (p. 258):—

The devices of the ant-lion are still more extraordinary if possible. He forms, with astonishing labour and perseverance, a pit in the shape of a funnel, in a dry sandy soil, under some old wall or other spot protected from the wind. His pit being finished, he buries himself among the sand at the bottom, leaving only his horns visible, and thus waits patiently for his prey. When an ant or any other small insect happens to walk on the edge of the hollow, it forces down some of the particles of sand, which gives the ant-lion notice of its presence. He immediately throws up the sand which covers his head to overwhelm the ant, and with its returning force brings it to the bottom. This he continues to do till the insect is overcome and falls between his horns. Every endeavour to escape, when once the incautious ant has stepped within the verge of the pit, is vain, for in all its attempts to climb the side the deceptive sand slips from under its feet, and every struggle precipitates it still lower. When within reach its enemy plunges the points of its jaws into its body, and having sucked out all its juices, throws out the empty skin to some distance.

According to Bingley, if the ant-lion, while excavating its pitfall,—

Comes to a stone of some moderate size, it does not desert the work on this account, but goes on, intending to remove that impediment the last. When the pit is finished, it crawls backward up the side of the place where the stone is; and, getting its tail under it, takes great pains and time to get it on a true poise, and then begins to crawl backward with it up the edge to the top of the pit, to get it out of the way. It is a common thing to see an ant-lion labouring in this manner at a stone four times as big as its own body; and as it can only move backwards, and the poise is difficult to keep, especially up a slope of such crumbling matter as sand, which moulders away from under its feet, and necessarily alters the position of its body, the stone very frequently rolls down, when near the verge, quite to the bottom. In this case the animal attacks it again in the same way, and is often not discouraged by five or six miscarriages, but continues its struggle so long that it at length gets over the verge of the place. When it has done this, it does not leave it there, lest it should roll in again; but is always at the pains of pushing it further on, till it has removed it to a necessary distance from the edge of the pit.[1]

Passing on now to the intelligence of caterpillars, Mr. G. B. Buckton, F.R.S., writing from Haslemere, says:—

Many caterpillars of *Pieris rapæ* have, during this autumn, fed below my windows. On searching for suitable positions for passing into chrysalides, some eight or ten individuals, in their direct march upwards, encountered the plate-glass panes of my windows; on these they appeared to be unable to stand. Accordingly in every case they made silken ladders, some of them five feet long, each ladder being formed of a single continuous thread, woven in elegant loops from side to side. . . . The reasoning, however, seems to be but narrow, for one ladder was constructed parallel to the window-frame for nearly three feet, on which secure footing could be had by simply diverting the track two inches.[2]

In this case it appears clear that we have to do with instinct, and not with reason. No doubt it is the congenital habit of these caterpillars to overcome impediments in this way; but the instinct is one of sufficient interest to be here stated.

The following is quoted from Kirby and Spence:—

A caterpillar described by Bonnet, which, from being confined in a box, was unable to obtain a supply of the bark with which its ordinary instinct directs it to make its cocoon, substituted pieces of paper that were given to it, tied them together with silk, and constructed a very passable cocoon with them. In another instance the same naturalist having opened several cocoons of a moth (*Noctura verbasci*), which are composed of a mixture of grains of earth and silk, just after being finished, the larvæ did not repair the injury *in the same manner*. Some employed both earth and silk; others contented themselves with spinning a silken veil before the opening.[3]

The same authorities state, as result of their own observation, that the—

---

[1] *Animal Biography*, vol. iii., pp. 244-5.
[2] *Nature*, vii., p. 49.
[3] *Intr. to Ent.*, ii., p. 475.

INTELLIGENCE OF LARVÆ—CATERPILLARS.

Common cabbage caterpillar, which, when building web under stone or wooden surfaces, previously covers a space with a web to form a base for supporting its dependent pupa, when building a web beneath a muslin surface dispenses with this base altogether: it perceives that the woven texture of the muslin forms facilities for attaching the threads of the cocoon securely enough to support the weight of the cocoon without the necessity of making the usual square inch or so of basal support.[1]

The instincts of the larva of the *Tinea* moth are thus described by Réaumur:—

It feeds upon the elm, using the leaves both as food and clothing. To do this it only eats the parenchyma of the leaf, preserving the upper and under epidermal membranes, between which it then insinuates itself as it progressively devours the parenchyma. It, however, carefully avoids separating these membranes where they unite at the extreme edge of the leaf, which is designed to form 'one of the seams of its coat.' The cavity when thus excavated between the two epidermal membranes is then lined with silk, made cylindrical in shape, cut off at the two ends and all along the side remote from the 'seam,' and then the two epidermal membranes sewn together along the side where they have had to be cut in order to separate them from the tree. The larva now has a coat exactly fitting its body, and open at each end. By the one opening it feeds, and by the other discharges its excrement, 'having on one side a nicely jointed seam—that which is commonly applied to its back—composed of the natural marginal junction of the membranes of the leaf.'

Réaumur cut off the edge of a newly finished coat, so as to expose the body of the larva at that point. The animal did not set about making a new coat *ab initio*, as we might expect that it would on the popular supposition that a train of instinctive actions is always as mechanical as the running down of a set of cog-wheels, and that wherever a novel element is introduced the machinery must be thrown out of gear, so that it cannot meet a new emergency of however simple a character, and must therefore re-start the whole process over again from the beginning. In this case the larva sewed up the rent; and not only so, but 'the scissors having cut off one of the projections intended to enter into the construction of

[1] *Ibid.*, p. 175.

the triangular end of the case, it entirely changed the original plan, and made that end the head which had been first designed for the tail.'

Another remarkable case of the variation of instinct in the Lepidoptera is stated by Bonnet. There are usually, he says, two generations of the Angoumois moth: the first appear in early summer, and lay their eggs upon the ears of wheat in the fields; the second appear later in the summer, or in the autumn, and these lay their eggs upon wheat in the granaries; from these eggs there comes the first generation of next year's moths. This is a highly remarkable case—supposing the facts to be as Bonnet states; for it seems that the early summer moths, although born in the granaries, immediately fly to the unreaped fields to lay their eggs in the standing corn, while the autumn moths never attempt to leave the granaries, but lay their eggs upon the stored wheat.[1]

Westwood says that—

A species of Tasmanian caterpillar (*Noctua Ewingii*) swarms over the land in enormous companies, which regularly begin to march at four o'clock in the morning, and as regularly halt at midday. *Liparis chrysorrhaea*, a kind of caterpillar, spins for the winter a common web, in which several hundred individuals find a common shelter.[2]

According to Kirby and Spence,—

The larva of the ichneumon, while feeding upon its caterpillar host, spares the walls of the intestines until it is time for it to escape, when, the life of the caterpillar being no longer necessary to its development, it perforates these walls.[3]

The larvæ *Theda isocrates* live in a group of seven or eight in the fruit of pomegranate. In consequence of their excavations within the fruit, the latter is apt to fall; and to prevent its doing so the larvæ throw out a thread of attachment wherewith to secure the fruit to the branch, so that if the stalk withers, this thread serves to suspend the fruit.[4]

The caterpillar of the Bombyx moth, which is a native of France, exhibits very wonderful instincts. The larva is gregarious in its habits, each society (family) consisting of perhaps

---

[1] *Œuvres*, ix., p. 370.   [2] *Trans. Ent. Soc.*, vol. ii.
[3] *Introd. Ent.*, Letter xi.
[4] Westwood, *Trans. Ent Soc.*, vol. ii., p. 1.

600 or 800 individuals. When young they have no fixed habitation, but encamp sometimes in one place, and sometimes in another, under the shelter of their web; but when they have attained two-thirds of their growth, they weave for themselves a common tent. About sunset the regiment leaves its quarters. ... At their head is a chief, by whose movements their procession is regulated. When he stops all stop, and proceed when he proceeds; three or four of his immediate followers succeed in the same line, the head of the second touching the tail of the first; then comes an equal series of pairs, next of threes, and so on, as far as fifteen or twenty. The whole procession moves regularly on with an even pace, each file treading in the steps of those that precede it. If the leader, arriving at a particular point, pursues a different direction, all march to that point before they turn.[1]

The following additional facts concerning these remarkable habits may be quoted. I take them from the account published by Mr. Davis in 'Loudoun's Magazine of Natural History:'—

The caterpillars, he observed, were Bombyces, and were seen crossing a road in single file, each so close to its predecessor that the line was quite continuous, 'moving like a living cord.' The number of caterpillars was 154, and the length of the line 27 feet. When Mr. Davis removed one from the line the caterpillar immediately in front suddenly stood still, then the next, and next, and so on to the leader. Similarly, those behind the point of interruption successively halted. After a pause of a few moments, the first caterpillar behind the break in the line endeavoured to fill up the vacant space, and so recover contact or communication, which after a time it succeeded in doing, when the information that the line was again closed was passed forward in some way from caterpillar to caterpillar till it reached the leader, when the whole line was again put in motion. The individual which had been abstracted remained rolled up and motionless; but on being placed near the moving column it immediately unrolled, and made every attempt to get readmitted into the procession. After many endeavours it succeeded, the one below falling into the rear of the interloper. On repeating the experiment by removing a caterpillar fifty from the head of the procession, Mr. Davis found that it took just thirty seconds by his watch for information of the fact to reach the leader. All the same results followed as in the previous

---

[1] Kirby and Spence, *Entomology*, Letter xvi.

case. It was observable that the animals were guided neither by sight nor smell while endeavouring to close up the interrupted line; for the caterpillar next behind the interruption, on whom the duty of closing up devolved, 'turned right and left, and often in a wrong direction, when within half an inch of the one immediately before him; when he at last touched the object of his search, the fact was communicated again by signal; and in thirty seconds the whole line was in rapid march.' This gentleman adds that the object of the march was the search for new pasture. The caterpillars feed on the Eucalyptus, and when they have completely stripped one tree of its leaves, they all congregate on the trunk, and proceed as described to another tree.

De Villiers[1] gives an account of his observations on the manner in which these caterpillars (*Cnethocampii pitzocampa*) are able to pass information, which does not quite agree with the above observation of Mr. Davis. For he says that, in a train of 600 caterpillars, interference by him in any part of the train was communicated through the whole series instantaneously—all the 600 caterpillars stopping immediately and with one consent like a single organism.

According to Kirby and Spence there is a kind of caterpillar (*Pieris cratægi*) which lives in little colonies of ten or twelve in common chambers lined with silk. In one part they make of the same material a little bag or pocket, which is used by the community or household as a water-closet. When full of excrement the caterpillars empty it by turning out the pellets with their feet.[2]

Only two other instances of noteworthy intelligence as exhibited by larvæ have fallen within my reading. One of these is mentioned by Réaumur, who says that the larvæ of *Hemerobius chrysops* chase aphides, and having killed them, clothe themselves in their skins; and the other case is the very remarkable one mentioned in his newly published work by W. MacLachlan, F.R.S., of caddis-worms adjusting the specific gravity of their tubes to suit that of the water in which they live, by attaching heavy or light material to them according as they require sinking or flotation.

[1] *Trans. Ent. Soc. France*, vol. i., p. 201.
[2] *Introduction to Entomology*, Letter xxvi.

CHAPTER VIII.

FISH.

ALTHOUGH we here pass into the sub-kingdom of animals the intelligence of which immeasurably surpasses that of the other sub-kingdoms, it is remarkable that these lowest representatives of the higher group are psychologically inferior to some of the higher members of the lower groups. Neither in its instincts nor in general intelligence can any fish be compared with an ant or a bee—a fact which shows how slightly a psychological classification of animals depends upon zoological affinity, or even morphological organisation. For although a highly competent authority, namely Van Baer, has said that a bee is as highly organised an animal as a fish, though on a different type,[1] no one would be found to assert that an ant or a bee is so much more highly organised than a fish as its higher intelligence would require, supposing degrees of intelligence to stand in necessary relation to degree of organic development. And this consideration is not materially altered if, instead of regarding the whole organism, we look to the nervous system alone. There is no doubt that the cerebral hemispheres of a fish, although small as compared with these organs in the higher Vertebrata, are, bulk for bulk, enormous as compared with the œsophageal ganglia or 'brain' of an insect; while the disproportion becomes still greater if the cerebral hemispheres of a fish are compared with their supposed analogues in the brain of an ant, viz., the pedunculated and convoluted lobes which surmount the cephalic ganglion. But here the relative smallness of the ant as a whole must be taken into con-

[1] *Phil. Frags.*, translated by Huxley, *Taylor's Mag.*, 1853, p. 196.

sideration, and also the fact that its brain is relatively much more massive as well as more highly organised than that which occurs in any other order of invertebrated animals, except, perhaps, the octopus and his allies. Therefore, although the brain of a fish is formed upon a type which by increase of size and complexity is destined in function far to eclipse all other types of nerve-centre, we have to observe that in its lowest stage of evolution as presented to science in the fishes, this type is functionally inferior to the invertebrate type, where this reaches its highest stage of evolution in the Hymenoptera.

Emotions.

Fish display emotions of fear, pugnacity; social, sexual, and parental feelings; anger, jealousy, play, and curiosity. So far the class of emotions is the same as that with which we have met in ants, and corresponds with that which is distinctive of the psychology of a child about four months old. I have not, however, any evidence of sympathy, which would be required to make the list of emotions identical; but sympathy may nevertheless be present.

Fear and pugnacity are too apparent in fish to require special proof. The social or gregarious feelings are strongly shown by the numberless species which swim in shoals, the sexual feelings are proved by courtships, and the parental by those species which build nests and guard their young. Schneider saw several species of fish at the Naples Aquarium protecting their eggs. In one case the male mounted guard over a rock where the eggs were deposited, and swam with open mouth against intruders. The following accounts of the nidification of certain species of fish show that the parental instincts are not unlike those which obtain in birds, and are comparable in point of strength with the same instincts as they occur in ants, bees, and spiders.

Agassiz remarks [1] that while examining the marine products of the Sargasso Sea, Mr. Mansfield picked up and brought to him a round mass of sargassum, about the size of the two

[1] Silliman's *American Journal*, Feb. 1872.

fists placed together. The whole consisted, to all appearance, of nothing but gulf-weed, the branches and leaves of which were, however, evidently knit together, and not merely balled into a roundish mass. The elastic threads which held the gulf-weed together were beaded at intervals, sometimes two or three beads being close together, or a branch of them hanging from the cluster of threads. This nest was full of eggs scattered throughout the mass, and not placed together in a cavity. It was evidently the work of the *Chironectes*. This rocking fish-cradle is carried along as an undying arbour, affording at the same time protection and afterwards food for its living freight. It is suggested that the fish must have used their peculiar pectoral fins when constructing this elaborate nest.

The well-known tinker or ten-spined stickleback (*Gasterosteus pungitius*) is one of our indigenous fish which constructs a nest. On May 1, 1864, a male [1] was placed in a well-established aquarium of moderate size, to which, after three days, two ripe females were added. Their presence at once roused him into activity, and he soon began to build a nest of bits of dirt and dead fibre, and of growing confervoid filaments, upon a jutting point of rock among some interlacing branches of *Myriophyllum spicatum*—all the time, however, frequently interrupting his labours to pay his addresses to the females. This was done in most vigorous fashion, he swimming, by a series of little jerks, near and about the female, even pushing against her with open mouth, but usually not biting. After a little coquetting she responds and follows him, swimming just above him as he leads the way to the nest. When there, the male commences to flirt—he seems unaware of its situation, will not swim to the right spot, and the female, after a few ineffectual attempts to find the proper passage into it, turns tail to swim away, but is then viciously pursued by the male. When he first courts the female, if she, not being ready, does not soon respond, he seems quickly to lose his temper, and, attacking her with great apparent fury, drives her to seek shelter in some crevice or dark corner. The coquetting of the male near the nest, which seems due to the fact that he really has not quite finished it, at length terminates by his pushing his head well into the entrance of the nest, while the female closely follows him, placing herself above him, and apparently much excited. As he withdraws she passes into the nest, and pushes quite through it, after a very brief delay, during which she deposits her ova. The male now fertilises the eggs, and drives the female

[1] Ransom, *Ann. and Mag. Nat. Hist*, 1865, xvi., p. 449.

away to a safe distance; then, after patting down the nest, he proceeds in search of another female. The nest is built and the ova deposited in about twenty-four hours. The male continued to watch it day and night, and during the light hours he also continually added to the nest.

The marine fifteen-spined stickleback (*Gasterosteus spinachia*) affords another instance of nest-constructing fishes. The places selected for their nests are usually harbours, or some sheltered spots to where pure sea water reaches. The fish either find growing, or even collect some of the softer kinds of green or red seaweed, and join them with so much of the coralline tufts (*Janiæ*) growing on the rock as will serve the purpose of affording firmness to the structure, and constitute a pear-shaped mass five or six inches long, and about as stout as a man's fist. A thread, which is elastic and resembles silk, is employed for the purpose of binding the materials together: under a magnifier it appears to consist of several strands connected by a gluey substance, which hardens by exposure to the water.[1]

M. Carbonnier, who has studied the habits of the Chinese butterfly-fish (*Macropodus*) in his private aquarium in Paris, where he had some in confinement, observed that the male constructs a nest of froth of considerable size, 15 to 18 centimetres horizontal diameter, and 10 to 12 high. He prepares the bubbles in the air (which he sucks in and then expels), strengthening them with mucous matter from his mouth, and brings them into the nest. Sometimes the buccal secretion will fail him, whereupon he goes to the bottom in search of confervæ, which he sucks and bites for a little in order to stimulate the act of secretion. The nest prepared, the female is induced to enter. Not less curious is the way in which the male brings the eggs from the bottom into the nest. He appears unable to carry them up in his mouth; instead of this, he first swallows an abundant supply of air, then descending, he places himself beneath the eggs, and suddenly, by a violent contraction of the muscles in the interior of his mouth and pharynx, he exhales the air which he had accumulated by the gills. This air, finely divided by the lamellæ and fringes of the gills, escapes in the form of two jets of veritable gaseous powder, which envelopes the eggs and raises them to the surface. In this manœuvre the *Macropodus* entirely disappeared in a kind of air-mist, and when this had dissipated he reappeared with a

[1] Quoted from Francis Day, F.L.S., 'Instincts and Emotions of Fish,' *Journ. Linn. Soc.*, vol. xv., pp. 36–7, where see for other cases of nest-building among fish.

multitude of air-bubbles like little pearls clinging all over his body.[1]

Again, in detailing Mr. Baker's observations on the three-spined stickleback, published in the Philosophical Transactions, this author says:—

It has been remarked that after the deposition of the eggs the nest was opened more to the action of the water, and the vibratory motion of the body of the male fish, hovering over its surface, caused a current of water to be propelled across the surface of the ova, which action was repeated almost continuously. After about ten days the nest was destroyed and the materials removed; and now were seen the minute fry fluttering upwards here and there, by a movement half swimming, half leaping, and then falling rapidly again upon or between the clear pebbles of the shingle bottom. This arose from their having the remainder of the yelk still attached to their body, which, acting as a weight, caused them to sink the moment the swimming effort had ceased. Around, across, and in every direction the male fish, as the guardian, continually moved. Now his labours became more arduous, and his vigilance was taxed to the utmost extreme, for the other fish (two tench and a gold carp), some twenty times larger than himself, as soon as they perceived the young fry in motion, continuously used their utmost endeavours to snap them up. The courage of the little stickleback was now put to its severest test; but, nothing daunted, he drove them all off, seizing their fins and striking with all his strength at their heads and at their eyes. His care of the young brood when encumbered with the yelk was very extraordinary; and as this was gradually absorbed and they gained strength, their attempts to swim carried them to a greater distance from the parent fish; his vigilance, however, seemed everywhere, and if they rose by the action of their fins above a certain height from the shingle bottom, or flitted beyond a given distance from the nest, they were immediately seized in his mouth, brought back, and gently puffed or jetted into their place again. The same care of the young, bringing them back to their nest up till about the sixth day after hatching, has been remarked by Dr. Ransom in the ten-spined stickleback (*G. pungitius*).[1]

The well-known habit of the lophobranchiate fish, of

[1] *Ibid.*

incubating their eggs in their pouches, also displays highly elaborated parental feeling.[1] M. Risso says that when the young of the pipe-fish are hatched out, the parents show them marked attachment, and that the pouch then serves them as a place of shelter or retreat from danger.[2]

M. Carbonnier has recorded how the male of the curiously grotesque telescope-fish, a variety of *Carassius auratus* (Linn.), acts as accoucheur to the female. Three males pursued one female which was heavy with spawn, and rolled her like a ball upon the ground for a distance of several metres, and continued this process without rest or relaxation for two days, until the exhausted female, who had been unable to recover her equilibrium for a moment, had at last evacuated all her ova.[3]

That adult fish are capable of feeling affection for one another would seem to be well established : thus Jesse relates how he once captured a female pike (*Esox lucius*) during the breeding season, and that nothing could drive away the male from the spot at which he had perceived his partner slowly disappear, and whom he had followed to the edge of the water.

Mr. Arderon[4] gave an account of how he tamed a dace, which would lie close to the glass watching its master; and subsequently how he kept two ruffs (*Acerina cernua*) in an aquarium, where they became very much attached to one another. He gave one away, when the other became so miserable that it would not eat, and this continued for nearly three weeks. Fearing his remaining fish might die, he sent for its former companion, and on the two meeting they became quite happy again. Jesse gives a similar account of two gold carp.[5]

Anger is strikingly shown by many fish, and notoriously by sticklebacks when their territory is invaded by a neighbour. These animals display a strange instinct of appropriating to themselves a certain part of the tank in which they may be confined, and furiously attacking any other stickleback which may presume to cross the imaginary frontier. Under such circumstances of provocation I have seen the whole animal change colour, and, darting at

---

[1] Kaup, *Catal. Lopho. Fish in Brit. Mus.* 1856, p. i.
[2] Yarrell, *Brit. Fishes*, 2nd ed. ii. p. 436.
[3] *Compt. Rend.*, Nov. 4, 1872, p. 1127.
[4] *Phil. Trans. Royal Society*, 1747.
[5] F. Day, *loc. cit.*

the trespasser, show rage and fury in every movement. Of course, here, as elsewhere, it is impossible to be sure how far apparent expression of an emotion is due to the presence of that mental state which we recognise as the emotion in ourselves; but still the best guide we have to follow is that of apparent expression.

Following this principle, we are also entitled to attribute to fish the emotions conducive to play; for nothing can well be more expressive of sportive glee than many of their movements. As for jealousy, the fights of many male fish for the possession of females constitutes evidence of emotion which would be called by this name in the higher animals. Schneider, in his recent work already often quoted, says that he has observed a male fish (*Labrus*) show jealousy only towards other individual males of his own species—chasing these away from the neighbourhood of his female, but not objecting to the approach of fish of other species.

Curiosity is shown by the readiness, or even eagerness, with which fish will approach to examine any unfamiliar object. So much is this the case that fishermen, like hunters, sometimes trade upon this faculty:—

> And the fisher, with his lamp
> And spear, about the low rocks damp
> Crept, and struck the fish which came
> To worship the delusive flame.[1]

Stephenson, the engineer, on sinking lighted lanterns in the water, also found that fish were attracted to them.[2]

Special Habits.

As curious instances of special instincts in fish we may notice the well-known habit of the angler (*Lophius piscator*), which conceals itself in mud and seaweed, while waving in the water certain filaments with which it is provided above its snout. Other fish, attracted by these moving objects, approach, and are thereupon seized by the

---

[1] Shelley, *Lines written in the Bay of Lerici*.
[2] See Smiles, *Lives of Engineers*, vol. iii., p. 69.

angler. We must also allude to the *Chelmon rostratus*, which shoots its prey by means of a drop of water projected from the mouth with considerable force and unerring aim. The mark thus shot at is always some small object, such as a fly, at rest above the surface of the water, so that when suddenly hit it falls into the water.[1] This remarkable instinct can only, I think, have originated as a primordially intentional adjustment, and as such shows a high degree of intelligence on the part of these fishes' ancestors. Moreover, the wonderful co-ordination of sight and muscular movements required to judge the distance, to make due allowance for refraction, and to aim correctly, shows that the existing representatives are not unworthy of their ancestors.

Several species of fish in different parts of the world have the habit of quitting pools which are about to dry up, and taking excursions across country in search of more abundant water. Eels have this habit, and perform their migrations by night. Dr. Hancock, in the 'Zoological Journal,' gives an account of a species of *Doras*, the individuals of which are about a foot in length, and travel by night in large shoals, or ' droves,' when thus searching for water. A strong serrated arm constitutes the first ray of the pectoral fin; and, using this as a kind of foot, the animal pushes itself forward by means of its tail, thus moving nearly as fast as a man can walk. Another migrating fish (*Hydrargzra*) was found by thousands in the fresh waters of Carolina by Bosc. It travels by leaps, and, according to Bosc, always directs itself towards the nearest water, although he purposely placed them so that they could not see it.

But perhaps the strangest among this class of habits is that of the climbing perch (*Perca scandens*), first discovered by Daldorff in Tranquebar; for this animal not only creeps over land, but even climbs the fan palm in search of certain Crustacea which form its food. In climbing it uses its open gill-covers as hands wherewith to suspend itself, while it deflects its tail laterally upwards so as to bring to bear upon the bark certain little spines with

[1] See 'On the Jaculator-Fish,' by Schlosser, *Phil. Trans.* 1761.

which its anal fin is provided; it then pushes itself upwards by straightening the tail, while it closes the gill-covers not to prevent progress, and so on. Sir E. Tennent, however, without disputing the evidence that these fish do climb trees, says,—

The probability is, as suggested by Buchanan, that the ascent which was witnessed by Daldorff was accidental, and ought not to be regarded as the habit of the animal.[1]

A great number of species of fish perform migrations. In relation to intelligence, the most interesting of these is the migration of salmon, which annually leave the sea to spawn in rivers, though there is some doubt whether the same individuals spawn every year. There is no doubt, however, that the same individuals frequently, though not invariably, revisit the same rivers for their successive spawnings. This fact may be due either to the remembrance of locality, similar to that which is unquestionably manifested by birds, or to the salmon not swimming far along the coast during other seasons of the year, and therefore in the spawning season when seeking a river happening to hit upon the same one. The latter hypothesis is one which Mr. Herbert Spencer tells me he is inclined to adopt, and, being a salmon-fisher, he has paid attention to the subject. He informs me of an observation by a friend of his own, who saw a salmon, when about to spawn, swimming along the coast-line, and all round a boathouse, apparently seeking any stream that it might first encounter.

The distances up rivers to which salmon will swim in the spawning season is no less surprising than the energy with which they perform the feat, and the determination with which they overcome all obstacles. They reach Bohemia by the Elbe, Switzerland by the Rhine, and, which is much more wonderful, the Cordilleras of America by the Maragnon.

They employ only three months in ascending to the sources of the Maragnon (a journey of 3,000 miles), the current of which is remarkably rapid, which is at the rate of nearly forty

[1] *Natural History of Ceylon.* p. 351.

miles a day; in a smooth stream or lake their progress would increase in a fourfold ratio. Their tail is a very powerful organ, and its muscles have wonderful energy; by placing it in their mouths they make of it a very elastic spring, for letting it go with violence they raise themselves in the air to the height of from twelve to fifteen feet, and so clear the cataract that impedes their course: if they fail in their first attempt, they continue their efforts till they have accomplished it.[1]

General Intelligence.

With reference to the general intelligence of fish, allusion may first be made to their marked increase of wariness in waters which are much fished. This shows no small degree of intelligence, for the caution is proved to be the result of observation by the fact that young trout under such circumstances are less wary than old ones. Moreover, many fish will abandon old haunts when much disturbed. Again, according to Kirby, the carp thrusts itself into the mud in order that the net may pass over it, or, if the bottom be stony, makes great leaps to clear it.

At the Andaman Islands fish are captured by the convicts by means of weirs fixed across the openings of creeks. After existing a week or so, it is observed that captures invariably cease; and it is believed that such is due to barnacles, &c., clustering on to the wood of which they are composed. It does not seem improbable that the fish have learned to avoid a locality out of terror at those which enter but do not again return.[2]

Lacepède[3] relates that some fish, which had been kept for many years in a basin of the Tuileries, would come when called by their names. Probably it was the sound of the voice and not the articulate words to which they responded; for Lacepède also relates that in many parts of Germany trout, carp, and tench were summoned to their food by the sound of a bell; and the same thing has been recorded of various fish in various localities, notably by Sir Joseph Banks, who used to collect his fish by sounding a bell.[4]

[1] Kirby, *Hist. Habits and Instincts of Animals*, vol i. p. 119.
[2] F. Day, *loc. cit.*   [3] *Hist. des Poiss., Introd.*, cxxx.
[4] For sundry other similar cases see Mr. Day's excellent paper already quoted.

In 'Nature' (vol. xi., p. 48) Mr. Mitchell gives the following instance of intelligence on the part of a small perch. Having one day disturbed its nest full of young fry, Mr. Mitchell next day went to look for the nest; 'but we searched in vain for the fish and her young. At length, a few yards further up stream, we discovered the parent guarding her fry with jealous care in a cavity scooped out of the coarse sand. . . . This is the first and only instance that has come under my notice of a fish watching over her young, and conveying them, when threatened with danger, to some other place.'

In 'Nature' (December 19, 1878) there is also published a communication which was made by Mr. J. Faraday to the Manchester Anglers' Association, concerning a skate which he observed in the aquarium of that town:—

A morsel of food thrown into the tank fell directly in an angle formed by the glass front and the bottom. The skate, a large example, made several vain attempts to seize the food, owing to its mouth being on the underside of its head and the food being close to the glass. He lay quite still for a while as though thinking, then suddenly raised himself into a slanting posture, the head inclined upwards, and the under surface of the body towards the food, when he waved his broad expanse of fins, thus creating an upward current or wave in the water, which lifted the food from its position and carried it straight to his mouth.

It will be observed, however, that this observation is practically worthless, from the observer having neglected to repeat the conditions in order to show that the movements of the fish were not, in their adaptation to these circumstances, purely accidental. Therefore I should not have alluded to this observation, had I not found that it has been quoted by several writers as a remarkable display of intelligence on the part of the fish.

I must not take leave of this class without making some allusion to the alleged habits of the so-called 'pilot-fish,' and also to those of 'thresher' and 'sword-fish.' I class these widely different habits together because they are alike in being dubious; different observers give different accounts, and therefore, until more information is forthcoming, we must suspend our judgment with regard to the habits in question. The following describes what these habits are believed by many observers to be.

Captain Richards, R.N., says that he saw a blue shark following a bait which was thrown out to him from the ship. The shark, which was attended by four pilot-fish, repeatedly approached the bait; but every time he did so one of the latter rushed in and prevented him. After a time the shark swam away; but when he had gone a considerable distance, he turned back again, swam quickly after the vessel, and before the pilot-fish could overtake him, seized the bait and was caught. While hoisting him on board, one of the pilots was seen to cling to his side until above water, when it dropped off. All the pilots then swam about for a time, as if searching for their friend, 'with every apparent mark of anxiety and distress.'[1] Colonel Smith fully corroborates this observation; but Mr. Geoffrey, on the other hand, saw a pilot-fish take great pains to bring a shark to the bait.[2] Probably the truth is that the pilot-fish attend the shark in order to obtain the crumbs that fall from his feasts, and that the cases in which they appear to prevent his taking the bait are without any psychological significance.

With regard to the alleged co-operation of the threshing and sword-fish in the destruction of whales, all that can be said is that the statements, although antecedently improbable, are sufficient in number not to be ignored. Mr. Day appears to accept the evidence as adequate, and gives the following cases:—

Captain Arn, in a voyage to Memel in the Baltic, gives the following interesting narrative:—One morning during a calm, when near the Hebrides, all hands were called up at 2 A.M. to witness a battle between several of the fish called threshers or fox-sharks (*Alopecias vulpes*), and some sword-fish on one side, and an enormous whale on the other. It was in the middle of the summer; and the weather being clear, and the fish close to the vessel, we had a fine opportunity of witnessing the contest. As soon as the whale's back appeared above the water, the

---

[1] Cnv., *Anim. Kingd.* x. p. 636.
[2] F. Day, *loc. cit.*

threshers springing several yards into the air descended with great violence upon the object of their rancour, and inflicted upon him the most severe slaps with their long tails, the sounds of which resembled the reports of muskets fired at a distance. The sword-fish in their turn attacked the distressed whale, stabbing from below: and thus beset on all sides and wounded, when the poor creature appeared, the water around him was dyed with blood. In this manner they continued tormenting and wounding him for many hours, until we lost sight of him; and I have no doubt they in the end completed his destruction.

The master of a fishing-boat has recently observed that the thresher-shark serves out the whales, the sea sometimes being all blood. One whale, attacked by these fish, once took refuge under his vessel, where it lay an hour and a half without moving a fin. He also remarked having seen the threshers jump out of the water as high as the mast-head and down upon the whale, while the sword fish was wounding him from beneath, the two sorts of fish evidently acting in concert.

CHAPTER IX.

BATRACHIANS AND REPTILES.

On the intelligence of frogs and toads very little has to be said. Frogs seem to have definite ideas of locality; for several of my correspondents inform me that they have known cases in which these animals, after having been removed for a distance of 200 or 300 yards from their habitual haunts, returned to them again and again. This, however, may I think perhaps be due to these haunts having a moistness which the animals are able to perceive at a great distance. But be this as it may, certainly the distance at which frogs are able to perceive moisture is surprising. Thus, for instance, Warden gives a case in which a pond containing a number of frogs dried up, and the frogs thereupon made straight for the nearest water, although this was at a distance of eight kilometres.[1]

A curious special instinct is met with in the toad *Bufo obstetricans*, from which it derives its name; for the male here performs the function of an accoucheur to the female, by severing from her body the gelatinous cord by which the ova are attached.

Another special instinct or habit manifested by toads is described by M. Duchemin in a paper before the Academy of Sciences at Paris.[2] The habit consists in the killing of carp by squatting on the head of the fish and forcing the fore-feet into its eyes. Probably this habit arises from sexual excitement on the part of the toads.

I have one case, communicated to me by a correspondent, of a frog which learnt to know her voice, and to come when called. As fish will sometimes do the same

---

[1] *Account of the United States*, vol. ii., p. 9.
[2] April 11, 1870.

thing, the account is sufficiently credible for me to quote:—

I used to open the gate in the railings round the pond, and call out 'Tommy' (the name I had given it), and the frog would jump out from the bushes, dive into the water, and swim across to me—get on my hand sometimes. When I called 'Tommy,' it would nearly always come, whatever the time of day, though it was only fed after breakfast; but it seemed quite tame.

A very similar case is recorded by Mr. Pennent [1] of a toad which was domesticated for thirty-six years, and knew all his friends.

There is no doubt that frogs are able to appreciate coming changes of weather, and to adapt their movements in anticipation of them; but these facts show delicate sensibility rather than remarkable intelligence.

The following observation of Edward, the Scottish naturalist, however, shows considerable powers of observation on the part of frogs. After describing the great noise made by a number of frogs on a moonlight night, he says:—

Presently, when the whole of the vocalists had reached their highest notes, they became hushed in an instant. I was amazed at this, and began to wonder at the sudden termination of the concert. But, looking about, I observed a brown owl drop down, with the silence of death, on to the top of a low dyke close by the orchestra.[2]

Reptiles.

Like the other cold-blooded Vertebrata, the reptiles are characterised by a sluggishness and low development of mental power which is to some extent proverbial. Nevertheless, that some members of the class present vivid emotions is not to be questioned. Thus, to quote from Thompson:—

The common guana (*Lacerta iguana*) is naturally extremely gentle and harmless. Its appearance, however, is much against

---

[1] See Bingley, *Animal Biography*, vol. ii., p. 406.
[2] Smiles, *Life of Edwards*, p. 124.

it, especially when agitated by fear or anger. Its eyes then seem on fire; it hisses like a serpent, swells out the pouch under its throat, lashes about its long tail, erects the scales on its back, and extending its wide jaws, holds its head, covered over with tubercles, in a menacing attitude. The male, during the spring of the year, exhibits great attachment towards the female. Throwing aside his usual gentleness of character, he defends her even with fury, attacking with undaunted courage every animal that seems inclined to injure her; and at this time, though his bite is by no means poisonous, he fastens so firmly, that it is necessary either to kill him or to beat him with great violence on the nose, in order to make him quit his hold.[1]

Several species of snake incubate their eggs and show parental affection for their young when they are hatched out; but neither in these nor in any other of their emotions do the reptiles appear to rise much above the level of fish. The case, however, which I shall afterwards quote, of the tame snakes kept by Mr. and Mrs. Mann, seems to show a somewhat higher degree of emotional development than could be pointed to as occurring in any lower Vertebrata. Moreover, according to Pliny, so much affection subsists between the male and female asp, that when the one is killed the other seeks to avenge its death; and this statement is so far confirmed—or rather, its origin explained—by Sir Emerson Tennent that he says when a cobra is killed, its mate is often found on the same spot a day or two afterwards.

Passing on to the general intelligence of reptiles, we shall find that this also, although low as compared with the intelligence of birds and mammals, is conspicuously higher than that of fish or batrachians.

Taking first the case of special instincts, Mr. W. F. Barrett, in a letter to Mr. Darwin, bearing the date May 6, 1873, and contained among the MSS. already alluded to, gives an account of cutting open with a penknife the egg of an alligator just about to hatch. The young animal, although blind, 'instantly laid hold of the finger, and attempted to bite.' Similarly, Dr. Davy, in his 'Account of Ceylon,' gives an interesting observation of his own on a young crocodile, which he cut out of the egg,

[1] *Passions of Animals*, p. 229.

and which, as soon as it escaped, started off in a direct line for a neighbouring stream. Dr. Davy placed his stick before it to try to make the little animal deviate from its course ; but it stoutly resisted the opposition, and raised itself into a posture of offence, just as an older animal would have done.

Humboldt made exactly the same observation with regard to young turtles, and he remarks that as the young normally quit the egg at night, they cannot see the water which they seek, and must therefore be guided to it by discerning the direction in which the air is most humid. He adds that experiments were made which consisted in putting the newly hatched animals into bags, carrying them to some distance from the shore, and liberating them with their tails turned towards the water. It was invariably found that the young animals immediately faced round, and took without hesitation the shortest way to the water.

Scarcely less remarkable than the instincts of the young turtles are those of the old ones. Their watchful timidity at the time of laying their eggs is thus described by Bates:—

Great precautions are obliged to be taken to avoid disturbing the sensitive turtles, who, previous to crawling ashore to lay, assemble in great shoals off the sand-bank. The men during this time take care not to show themselves, and warn off any fisherman who wishes to pass near the place. Their fires are made in a deep hollow near the borders of the forest, so that the smoke may not be visible. The passage of a boat through the shallow waters where the animals are congregated, or the sight of a man or a fire on the sand-bank, would prevent the turtles from leaving the water that night to lay their eggs; and if the causes of alarm were repeated once or twice they would forsake the praia for some other quieter place. . . . I rose from my hammock by daylight, shivering with cold—a praia, on account of the great radiation of heat in the night from the sand, being towards the dawn the coldest place that can be found in this climate. Cardozo and the men were already up watching the turtles. The sentinels had erected for this purpose a stage about fifty feet high, on a tall tree near their station, the ascent to which was by a roughly made ladder of woody lianas. They are enabled, by observing the turtles from this watch-tower, to ascertain the date of successive deposits of eggs, and thus guide the commandante in fixing the time for the general invitation to the Ega people. The turtles lay their eggs by night, leaving the water, when nothing disturbs them, in vast crowds, and crawling to the central and highest part of the praia. These places are, of course, the last to go under water when, in unusually wet seasons, the river rises before the eggs are hatched by the heat of the sand. One could almost believe, from this, that the animals used forethought in choosing a place; but it is simply one of those many instances in animals where unconscious habit has the same result as conscious prevision. The hours between midnight and dawn are the busiest. The turtles excavate with their broad webbed paws deep holes in the fine sand: the first comer, in each case, making a pit about three feet deep, laying its eggs (about 120 in number) and covering them with sand; the next making its deposit at the top of that of its predecessor, and so on until every pit is full. The whole body of turtles frequenting a praia does not finish laying in less than fourteen or fifteen days, even when there is no interruption. When all have done, the area (called by the Brazilians *taboleiro*) over which they have excavated is distinguishable from the rest of the praia only by signs of the sand having been a little disturbed.[1]

The same naturalist says of the alligator,

These little incidents show the timidity and cowardice (? prudence and caution) of the alligator. He never attacks man when his intended victim is on his guard; but he is cunning enough to know when this may be done with impunity. Of this we had proof a few days afterwards, &c.[2]

Of the alligator, Jesse writes:[3]—

But a most singular instance of attachment between two animals, whose natures and habits were most opposite, was related to me by a person on whose veracity I can place the greatest reliance. He had resided for nine years in the American States, where he superintended the execution of some extensive works for the American Government. One of these works consisted in the erection of a beacon in a swamp in one of the rivers,

---

[1] *Naturalist on the Amazon*, pp. 285-6.
[2] *Ibid.* The astonishing facts relating to the migration of turtles in the laying season will be treated under the general heading 'Migration' in my forthcoming work.
[3] *Gleanings* vol. i., pp. 163-4.

where he caught a young alligator. This animal he made so perfectly tame that it followed him about the house like a dog, scrambling up the stairs after him, and showing much affection and docility. Its great favourite, however, was a cat, and the friendship was mutual. When the cat was reposing herself before the fire (this was at New York), the alligator would lay himself down, place his head upon the cat, and in this attitude go to sleep. If the cat was absent the alligator was restless; but he always appeared happy when the cat was near him. The only instance in which he showed any ferocity was in attacking a fox, which was tied up in the yard. Probably, however, the fox resented some playful advances which the other had made, and thus called forth the anger of the alligator. In attacking the fox he did not make use of his mouth, but beat him with so much severity with his tail, that, had not the chain which confined the fox broken, he would probably have killed him. The alligator was fed on raw flesh, and sometimes with milk, for which he showed a great fondness. In cold weather he was shut up in a box, with wool in it; but, having been forgotten one frosty night, he was found dead in the morning. This is not, I believe, a solitary instance of amphibia becoming tame, and showing a fondness for those who have been kind to them. Blumenbach mentions that crocodiles have been tamed; and two instances have occurred under my own observation of toads knowing their benefactors, and coming to meet them with considerable alacrity.

With regard to the higher intelligence of reptiles, I may quote the following instances.

Three or four different correspondents tell me of cases which they have themselves observed, of snakes and tortoises unmistakably distinguishing persons. In one of these cases the tortoise would come to the call of the favoured person, and when it came would manifest its affection by tapping the boot of this person with its mouth; 'but it would not answer anyone else.' A separation of some weeks did not affect the memory of this tortoise for his friend.[1]

[1] The tortoise which has gained such immortal celebrity by having fallen under the observation of the author of the *Natural History of Selborne*, likewise distinguished persons in this way. For 'whenever the good old lady came in sight, who had waited on it for more than thirty years, it always hobbled with awkward alacrity towards its benefactress, whilst to strangers it was altogether inattentive.'

The following interesting observation on the intelligence of snakes shows, not only that these animals are well able to distinguish persons, and that they remember their friends for a period of at least six weeks, but also that they possess an intensity of amiable emotion scarcely to be expected in this class. Clearly the snakes in question were not only perfectly tame, but entertained a remarkable affection for those who tended and petted them. The facts were communicated to me by Mr. Walter Severn, the well-known artist, who was a friend of Mr. and Mrs. Mann, the gentleman and his wife to whom the snakes belonged. Mr. and Mrs. Mann having got into trouble with their neighbours on account of the fear and dislike which their pets occasioned, legal proceedings were instituted, and so the matter came before the public. Mr. Severn then wrote a letter to the *Times*, in order to show that the animals were harmless. From this letter the following is an extract:—

I happen to know the gentleman and lady against whom a complaint has been made because of the snakes they keep, and I should like to give a short account of my first visit to them.

Mr. M., after we had talked for a little time, asked if I had any fear of snakes; and after a timid 'No, not very,' from me, he produced out of a cupboard a large boa-constrictor, a python, and several small snakes, which at once made themselves at home on the writing-table among pens, ink, and books. I was at first a good deal startled, especially when the two large snakes coiled round and round my friend, and began to notice me with their bright eyes and forked tongues; but soon finding how tame they were, I ceased to feel frightened. After a short time Mr. M. expressed a wish to call Mrs. M., and left me with the boa deposited on an arm-chair. I felt a little queer when the animal began gradually to come near, but the entrance of my host and hostess, followed by two charming little children, put me at my ease again. After the first interchange of civilities, she and the children went at once to the boa, and, calling it by the most endearing names, allowed it to twine itself most gracefully round about them. I sat talking for a long time, lost in wonder at the picture before me. Two beautiful little girls with their charming mother sat before me with a boa-constrictor (as thick round as a small tree) twining playfully round the lady's waist and neck, and forming a kind of turban round her head, expecting to be petted and made much of like a kitten. The children over and over again took its head in their hands and kissed its mouth, pushing aside its forked tongue in doing so. The animal seemed much pleased, but kept turning its head continually towards me with a curious gaze, until I allowed it to nestle its head for a moment up my sleeve. Nothing could be prettier than to see this splendid serpent coiled all round Mrs. M. while she moved about the room, and when she stood to pour out our coffee. He seemed to adjust his weight so nicely, and every coil with its beautiful marking was relieved by the black velvet dress of the lady. It was long before I could make up my mind to end the visit, and I returned soon after with a friend (a distinguished M.P.[1]), to see my snake-taming acquaintance again. . . .

These (the snakes) seemed very obedient, and remained in their cupboard when told to do so.

About a year ago Mr. and Mrs. M. were away for six weeks, and left the boa in charge of a keeper at the Zoo. The poor reptile moped, slept, and refused to be comforted, but when his master and mistress appeared he sprang upon them with delight, coiling himself round them, and showing every symptom of intense delight.[2]

The end of this python was remarkable and pathetic. Mr. Severn tells me that some years after he had published the above letter Mr. Mann was seized with an apoplectic fit. His wife, being the only other person in the house at the time, ran out to fetch a doctor. She was absent about ten minutes, and on returning found that the serpent during her absence had crawled upstairs from the room below into that where her husband was lying, and was stretched beside him dead. Such being the fact, we are left to speculate whether the double seizure of the man and the snake was a mere coincidence, or whether the sight of its stricken master, acting on the emotions of a possibly not healthy animal, precipitated its death. Looking to the extreme suddenness of the latter, as well as to the fact of the animal having pined so greatly for his friends while it was confined at the Zoological Gardens, I think the probability rather points to the death of the

---

[1] This gentleman was Lord Arthur Russell.
[2] The *Times*, July 25, 1872.

animal having been accelerated by emotional shock. But of course the question is an open one.

So much for the power of reptiles to establish such definite and complete associations as are required for the recognition of persons—associations, however, to which, as we have seen, frogs, and even insects may attain. As for other associations, a correspondent writes to me:—

I believe tortoises are able to establish a definite association between particular colours on a flat surface and food. Only the day before reading your article on animal intelligence I noticed the endeavours of a small tortoise to eat the *yellow* flowers of an inlaid writing-table, and I have often remarked the same recognition with regard to red.

Lord Monboddo relates the following anecdote of a serpent:—

I am well informed of a tame serpent in the East Indies, which belonged to the late Dr. Vigot, and was kept by him in the suburbs of Madras. This serpent was taken by the French, when they invested Madras in the late war, and was carried to Pondicherry in a close carriage. But from thence he found his way back again to his old quarters, which it seems he liked better, though Madras is distant from Pondicherry about one hundred miles. This information, he adds, I have from a lady who then was in India, and had seen the serpent often before his journey and after his return.

Considering the enormous distances over which turtles are able to find their way in the season of migration, this display of the homing faculty to so great a degree in a serpent is not to be regarded as incredible.

Mr. E. L. Layard, in his 'Rambles in Ceylon' says of the cobra:[1]—

I once watched one which had thrust its head through a narrow aperture and swallowed one (*i.e.* a toad). With this encumbrance he could not withdraw himself. Finding this, he reluctantly disgorged the precious morsel, which began to move off. This was too much for snake philosophy to bear, and the toad was again seized; and again, after violent efforts to escape, was the snake compelled to part with it. This time, however, a lesson had been learnt, and the toad was seized by one leg, withdrawn, and then swallowed in triumph.

[1] See *Annas. and Mag. of Nat. Hist.*, 2nd series, vol. ix., p. 333.

Mr. E. C. Buck, B.C.S., says in 'Nature' (vol. viii., p. 303):—

I have witnessed exactly a similar plan pursued by a large number of Ganges crocodiles, which had been lying or swimming about all day in front of my tent, at the mouth of a small stream which led from some large inland lakes to the Ganges. Towards dusk, at the same moment every one of them left the bank on which they were lying, or the deep water in which they were swimming, and formed a line across the stream, which was about twenty yards wide. They had to form a double line, as there was not room for all in a single line. They then swam slowly up the shallow stream, driving the fish before them, and I saw two or three fish caught before they disappeared.

An account of reptile psychology would be incomplete without some reference to the alleged facts of snakes charming other animals by 'fascination,' and being themselves charmed by the arts of music, &c. The testimony on both subjects is conflicting, and especially with regard to the fascination of other animals by snakes. Thus:—

Mr. Pennant says that this snake (rattle-snake) will frequently lie at the bottom of a tree on which a squirrel is seated. He fixes his eyes on the animal, and from that moment it cannot escape; it begins a doleful outcry, which is so well known that a passer-by, on hearing it, immediately knows that a snake is present. The squirrel runs up the tree a little way, comes down again, then goes up, and afterwards comes still lower. The snake continues at the bottom of the tree with its eyes fixed on the squirrel, and his attention is so entirely taken up, that a person accidentally approaching may make a considerable noise without so much as the snake turning about. The squirrel comes lower, and at last leaps down to the snake, whose mouth is already distended for its reception. Le Vaillant confirms this fascinating terror by a scene he witnessed. He saw on the branch of a tree a species of shrike, trembling as if in convulsions, and at the distance of nearly four feet, on another branch, a large snake that was lying with outstretched neck and fiery eyes, gazing steadily at the poor animal. The agony of the bird was so great that it was deprived of the power of moving away; and when one of the party killed the snake, it (*i.e.* the bird) was found dead upon the spot—and that entirely from fear; for, on examination, it appeared not to have received the slightest wound. The same traveller adds that a short time afterwards he observed a small mouse in similar agonising convulsions, about two yards from a snake, whose eyes were intently fixed upon it; and on frightening away the reptile, and taking up the mouse, it expired in his hand.[1]

Many other observations, more or less similar, might be quoted; but, on the other hand, Sir Joseph Fayrer tells me that 'fascination is only fright;' and this appears to be the opinion of all persons who have had the opportunity of looking into the subject in a scientific manner. The truth probably is that small animals are occasionally much alarmed by the sight of a snake looking at them, and as a consequence of this more easily fall a prey. In some cases, it is likely enough, strong terror so unnerves the animal as to make it behave in the manner which the witnesses describe; in making half-palsied efforts to escape, it may actually fall or draw nearer to the object of its dread. Perhaps, therefore, Dr. Barton, of Philadelphia, is a little too severe on previous observers when he says that—

The report of this fascinating property has had its rise in nothing more than the fears and cries of birds and other animals in the protection of their nests. . . . The result of not a little attention has taught me that there is but one wonder in the business—the wonder that the story should ever have been believed by any man of understanding and observation.

But, be this as it may, it is certainly remarkable, as Sir J. Fayrer in his letter to me observes, 'how little fear some animals show until the moment that they are seized and struck.'

As for snake-charming, the facts seem to be that cobras and other serpents are attracted by the sound of a pipe to creep out of their hiding-places, when they are captured and tamed. It is certain that the fangs are not always drawn, and also that from the first moment of capture, before there has been time for any process of training, a real snake-charmer is able to make the reptile 'dance.' Thus, for instance, Sir E. Tennent publishes the following letter from Mr. Reyne. After describing all his

[1] Thompson, *Passions of Animals*, p. 118; see also Bingley, *Animal Biography*, vol. ii., pp. 447-8.

precautions to ensure that the snake-charmer had no tamed snakes concealed about his person, Mr. Reyne proceeds to tell how he made the man accompany him to the jungle, where, attracted by the music of a pipe which the man played, a large cobra came from an ant-hill which Mr. Reyne knew it to occupy:—

On seeing the man it tried to escape, but he caught it by the tail and kept swinging it round until we reached the bungalow. He then made it dance, but before long it bit him above the knee. He immediately bandaged the leg above the bite and applied a snake-stone to the wound to extract the poison. He was in great pain for a few minutes, but after that it gradually went away, the stone falling off just before he was relieved.[1]

Thus the only remarkable thing about the charming of a freshly caught snake seems to be that the charmer is able to make the animal 'dance'—for the fact of the snake approaching the unfamiliar sound of music is not in itself any more remarkable than a fish approaching the unfamiliar sight of a lantern. It does not, however, appear that this dancing is anything more than some series of gestures or movements which may be merely the expressions, more or less natural, of uneasiness or alarm. Anything else that charmed snakes may do is probably the result of training; for there is no doubt that cobras admit of being tamed, and even domesticated. Thus, for instance, Major Skinner, writing to Sir E. Tennent, says:—

In one family near Negombo, cobras are kept as protectors, in the place of dogs, by a wealthy man who has always large sums of money in his house. But this is not a solitary case of the kind. . . . The snakes glide about the house, a terror to the thieves, but never attempting to harm the inmates.[2]

Thus, on the whole, we may accept Dr. Davey's opinion —who had good opportunities for observation—that the snake-charmers control the cobras by working upon the well-known timidity and reluctance of these animals to use their fangs till they become virtually tame.

---

[1] *Natural History of Ceylon*, p. 814.
[2] Tennent, *loc. cit.*, p. 299.

CHAPTER X.

BIRDS.

ADEQUATELY to treat of the intelligence of birds a separate volume would be required; here it must be enough to deal with this class as I shall afterwards deal with the Mammalia—namely, by giving an outline sketch of the more prominent features of their psychology.

Memory.

The memory of birds is well developed. Thus, although we are much in the dark on the whole subject of migration —so much so that I reserve its discussion with all the problems that this presents for a separate chapter in my next work—we may at least conclude that the return of the same pair of swallows every year to the same nest must be due to the animals remembering the precise locality of their nests. Again, Buckland gives an account of a pigeon which remembered the voice of its mistress after an absence of eighteen months;[1] but I have not been able to

---

[1] *Curiosities*, &c., p. 126. Wilson also, in his *American Ornithology*, gives the following sufficiently credible account of the memory of a crow:— 'A gentleman who resided on the Delaware, a few miles below Easton, had raised [reared] a crow, with whose tricks and society he used frequently to amuse himself. This crow lived long in the family, but at length disappeared, having, as was then supposed, been shot by some vagrant gunner, or destroyed by accident. About eleven months after this, as the gentleman one morning, in company with several others, was standing on the river shore, a number of crows happened to pass by; one of them left the flock, and flying directly towards the company, alighted on the gentleman's shoulder, and began to gabble away with great volubility, as one long-absent friend naturally enough does on meeting another. On recovering from his surprise the gentleman instantly recognised his old acquaintance, and endeavoured, by several civil but sly manœuvres, to lay hold of him; but the crow, not altogether relishing quite so much familiarity, having now had a taste of the sweets of

meet with satisfactory evidence of the memory of a bird enduring for a longer time than this.

As it is a matter of interest in comparative psychology to trace as far as possible into detail the similarities of a mental faculty as it occurs in different groups of animals, and as the faculty of memory first admits of detailed study in the class which we are now considering, I shall here devote a paragraph to the facts concerning the exhibition of memory by birds where its mechanism best admits of being analysed; I refer to the learning of articulate phrases and tunes by talking and musical birds. The best observations in this connection with which I am acquainted are those of Dr. Samuel Wilks, F.R.S., and therefore I shall quote *in extenso* the portion of his paper which refers to the memory of parrots: other portions of this paper I shall have occasion to quote in my next work:—

When my parrot first came into my possession, several years ago, it was quite unlettered, and I therefore had an opportunity of observing the mode in which it acquired the accomplishment of speech. I was very much struck with its manner of learning, and the causes for its speaking on special occasions. The first seemed to resemble very much the method of children in learning their lessons, and the second to be due to some association or suggestion—the usual provocative for set speeches at all periods of human life. A parrot is well known to imitate sounds in a most perfect manner, even to the tone of the voice, besides having a compass which no human being can approach, ranging from the gravest to the most acute note. My bird, though possessing a good vocabulary of words and sentences, can only retain them for a few months unless kept constantly in practice by the suggestive recurrence of some circumstance which causes their continual utterance. If forgotten, however, they are soon revived in the memory by again repeating them a few times, and much more speedily than any new sentence can be acquired. In beginning to teach the parrot a sentence, it has to be repeated many times, the bird all the while listening most attentively by turning the opening of the ear as close as possible to the speaker. After a few hours it is heard attempting

liberty, cautiously eluded all his attempts; and suddenly glancing his eye on his distant companions, mounted in the air after them, soon overtook and mingled with them, and was never afterwards seen to return.'

to say the phrase, or, I should say, trying to learn it. It evidently has the phrase somewhere in store, for eventually this is uttered perfectly, but at first the attempts are very poor and ludicrous. If the sentence be composed of a few words, the first two or three are said over and over again, and then another and another word added, until the sentence is complete, the pronunciation at first being very imperfect, and then becoming gradually more complete, until the task is accomplished. Thus hour after hour will the bird be indefatigably working at the sentence, and not until some days have elapsed will it be perfect. The mode of acquiring it seems to me exactly what I have observed in a child learning a French phrase; two or three words are constantly repeated, and then others added, until the whole is known, the pronunciation becoming more perfect as the repetition goes on. I found also on whistling a popular air to my parrot that she picked it up in the same way, taking note by note until the whole twenty-five notes were complete. Then the mode of forgetting, or the way in which phrases and airs pass from its recollection, may be worth remarking. The last words or notes are first forgotten, so that soon the sentence remains unfinished or the air only half whistled through. The first words are the best fixed in the memory; these suggest others which stand next to them, and so on till the last, which have the least hold on the brain. These, however, as I have before mentioned, can be easily revived on repetition. This is also a very usual process in the human subject: for example, an Englishman speaking French will, in his own country, if no opportunity occur for conversation, apparently forget it; he no sooner, however, crosses the Channel and hears the language than it very soon comes back to him again. In trying to recall poems learned in childhood or in school days, although at that period hundreds of lines may have been known, it is found that in manhood we remember only the two or three first lines of the 'Iliad,' the 'Æneid,' or the 'Paradise Lost.'[1]

The following is communicated to me by Mr. Venn, of Cambridge, the well-known logician:—

I had a grey parrot, three or four years old, which had been taken from its nest in West Africa by those through whom I received it. It stood ordinarily by the window, where it could equally hear the front and back door bells. In the yard, by the back door, was a collie dog, who naturally barked violently at nearly all the comers that way. The parrot took to imitating the

[1] *Journal of Mental Science*, July 1879.

dog. After a time I was interested in observing the discriminative association between the back-door bell and the dog's bark in the parrot's mind. Even when the dog was not there, or for any other cause did not bark, the parrot would constantly bark when the back-door bell sounded, but never (that I could hear) when the front-door bell was heard.

This is but a trifle in the way of intelligence, but it struck me as an interesting analogous case to a law of association often noticed by writers on human psychology.

The celebrated parrot that belonged to the Buffon family and of which the Comte de Buffon wrote, exhibited in a strange manner the association of its ideas. For he was frequently in the habit of asking himself for his own claw, and then never failed to comply with his own request by holding it out, in the same way as he did when asked for his claw by anybody else. This, however, probably arose, not, as Buffon or his sister Madame Nadault supposed, from the bird not knowing its own voice, but rather from the association between the words and the gesture.

According to Margrave, parrots sometimes chatter their phrases in their dreams, and this shows a striking similarity of psychical processes in the operations of memory with those which occur in ourselves.

Similarly, Mr. Walter Pollock, writes me of his own parrot:—

In this parrot the sense of association is very strongly developed. If one word picked up at a former home comes into its head, and is uttered by it, it immediately follows this word up with all the other words and phrases picked up at the same place and period.

Lastly, parrots not only remember, but recollect; that is to say, they know when there is a missing link in a train of association, and purposely endeavour to pick it up. Thus, for instance, the late Lady Napier told me an interesting series of observations on this point which she had made upon an intelligent parrot of her own. They were of this kind. Taking such a phrase as 'Old Dan Tucker,' the bird would remember the beginning and the end, and try to recollect the middle. For it would say very slowly, 'Old—old—old—old' (and then very quickly) 'Lucy Tucker.' Feeling that this was not right, it would try again as before, 'Old—old—old—old—old Bessy Tucker,' substituting one word after another in the place of the sought-for word 'Dan.' And that the process was one of truly seeking for the desired word was proved by the fact that if, while the bird was saying, 'Old—old—old—old,' any one threw in the word 'Dan,' he immediately supplied the 'Tucker.'

Emotions.

As regards emotions, it is among birds that we first meet with a conspicuous advance in the tenderer feelings of affection and sympathy. Those relating to the sexes and the care of progeny are in this class proverbial for their intensity, offering, in fact, a favourite type for the poet and moralist. The pining of the 'love-bird' for its absent mate, and the keen distress of a hen on losing her chickens, furnish abundant evidence of vivid feelings of the kind in question. Even the stupid-looking ostrich has heart enough to die for love, as was the case with a male in the Rotund of the Jardin des Plantes in Paris, who, having lost his wife, pined rapidly away. It is remarkable that in some species—notably pigeons—conjugal fidelity should be so strongly marked; for this shows, not only what may be called a refinement of sexual feeling, but also the presence of an abiding image in the mind's eye of the lover. For instance,—

Referring to the habits of the mandarin duck (a Chinese species) Mr. Bennett says that Mr. Beale's aviary afforded a singular corroboration of the fidelity of the birds in question. Of a pair in that gentleman's possession, the drake being one night purloined by some thieves, the unfortunate duck displayed the strongest marks of despair at her bereavement, retiring into a corner, and altogether neglecting food and drink, as well as the care of her person. In this condition she was courted by a drake who had lost his mate, but who met with no encouragement from the widow. On the stolen drake being subsequently recovered and restored to the aviary, the most extravagant demonstrations of joy were displayed by the fond couple; but this was not all, for, as if informed by his spouse of the gallant proposals made to her shortly before his arrival, the drake attacked the luckless bird who would have supplanted him, beat out his eyes, and inflicted so many injuries as to cause his death.[1]

Similarly, to give an instance or two with regard to other birds, Jesse states the following as his own observation:—

A pair of swans had been inseparable companions for three years, during which time they had reared three broods of cygnets; last autumn the male was killed, and since that time the female has separated herself from all society with her own species; and, though at the time I am writing (the end of March) the breeding season for swans has far advanced, she remains in the same state of seclusion, resisting the addresses of a male swan, who has been making advances towards forming an acquaintance with her, either driving him away, or flying from him whenever he comes near her. How long she will continue in this state of widowhood I know not, but at present it is quite evident that she has not forgotten her former partner.

This reminds me of a circumstance which lately happened at Chalk Farm, near Hampton. A man, set to watch a field of peas which had been much preyed upon by pigeons, shot an old cock pigeon which had long been an inhabitant of the farm. His mate, around whom he had for many a year cooed, whom he had nourished from his own crop, and had assisted in rearing numerous young ones, immediately settled on the ground by his side, and showed her grief in the most expressive manner. The labourer took up the dead bird, and tied it to a short stake, thinking that it would frighten away the other depredators. In this situation, however, the widow did not forsake her deceased husband, but continued, day after day, walking slowly round the stick. The kind-hearted wife of the bailiff of the farm at last heard of the circumstance, and immediately went to afford what relief she could to the poor bird. She told me that, on arriving at the spot, she found the hen bird much exhausted, and that she had made a circular beaten track round the dead pigeon, making now and then a little spring towards him. On the removal of the dead bird the hen returned to the dovecote.[2]

As evidence of the intensity of the maternal instinct

---

[1] Couch, *Illustrations of Instinct*, p. 165.
[2] *Gleanings*, vol. i., pp. 112–13.

even in the case of barren birds, I may quote the following from the naturalist Couch. I do so because, although the instance is a trivial one, and also one of frequent occurrence, it is interesting as showing that a deeply rooted instinct or emotion may assert itself powerfully even in the absence of what may be termed its natural stimulus or object:—

I was once witness to a curious instance of the yearning for progeny in a diminutive bantam hen.

There was at this time a nest of the common hen in a secluded part of the garden, and the parent had been sitting on its eggs, till compelled by hunger she left them for a short time. This absence was fatal; for the bantam had in the meantime found its situation in a covered recess in the hedge, and I saw her creep into it with all the triumph of the discoverer of a treasure. The real mother now returned, and great was her agony at finding an intruder in her nest. The expression of her eye and the attitude of her head were emphatic of surprise at the impudence of the proceeding. But after many attempts to recover possession she was compelled to resign her rights, for the bantam was too resolute to be contended with; and though its body was not big enough to cover the whole of the eggs, and thus some of them were not hatched, yet in due season the pride of this audacious step-mother was gratified by strutting at the head of a company of robust chickens, which she passed off upon the feathered public as a brood of her own.[1]

As evidence of sympathy I shall quote *in extenso* an interesting case which has been communicated to me by a young lady, who desires her name withheld. There are several more or less corroborative cases in the anecdote-books,[2] so that I have no doubt as to the substantial accuracy of the account:—

My grandfather had a Swan River gander, which had been reared near the house, and had consequently attached himself to the members of the family; so much so that, on seeing any of them at a distance, he would run to meet them with all possible demonstrations of delight.

But 'Swanny' was quite an outcast from his own tribe; and as often as he made humble overtures to the other geese, so

---

[1] Couch, *Illustrations of Instinct*, p. 232.
[2] See especially Bingley, *Animal Biography*, vol. ii., pp. 327-29.

often was he driven away with great contempt, and on such occasions he would frequently run to some of his human friends, and laying his head on their laps, seem to seek for sympathy. At last, however, he found a friend among his own species. An old grey goose, becoming blind, was also discarded by her more fortunate companions, and Swanny lost no opportunity of recognising this comrade in distress. He at once took her under his protection and led her about. When he considered it well for her to have a swim, he would gently take her neck in his bill, and thus lead her, sometimes a considerable distance, to the water's edge. Having fairly launched her, he kept close by her side, and guided her from dangerous places by arching his neck over hers, and so turning her in the right direction. After cruising about a sufficient time, he would guide her to a convenient landing-place, and taking her neck in his bill as before, lead her to *terra firma* again. When she had goslings, he would proudly convoy the whole party to the water-side; and if any ill-fated gosling got into difficulties in a hole or deep cart-rut, Swanny with ready skill would put his bill under its body, and carefully raise it to the level ground.

My grandfather had also another gander who attached himself to him, and would follow him for hours through fields and lanes, pausing when he stood still, and waddling gravely by his side as he proceeded. This gander was not, like the other, discarded by his kind, but would leave them any time to walk with his master, and was exceedingly jealous of any one else who tried to share this privilege, excepting only his mistress. On one occasion, a gentleman venturing to place his hand on my grandfather's arm, the gander flew at him, and beat him severely with his wings, and it was with great difficulty that he was induced to let go.

The solicitude which most gregarious birds display when one of their number is wounded or captured, constitutes strong evidence of sympathy. As Jesse observes,—

There is one trait in the character of the rook which is, I believe, peculiar to that bird, and which does him no little credit; it is the distress which is exhibited when one of his fellows has been killed or wounded by a gun while they have been feeding in a field or flying over it. Instead of being scared away by the report of the gun, leaving their wounded or dead companion to his fate, they show the greatest anxiety and sympathy for him, uttering cries of distress, and plainly proving that they wish to render him assistance by hovering over him, or sometimes making a dart from the air close up to him, apparently to try and find out the reason why he did not follow them. . . . I have seen one of my labourers pick up a rook which he had shot at for the purpose of putting him up as a scarecrow in a field of wheat, and while the poor wounded bird was still fluttering in his hand, I have observed one of his companions make a wheel round in the air, and suddenly dart past him so as almost to touch him, perhaps with the last hope that he might still afford assistance to his unfortunate mate or companion. Even when the dead bird has been hung, *in terrorem*, to a stake in the field, he has been visited by some of his former friends, but as soon as they found that the case was hopeless, they have generally abandoned that field altogether.

When one considers the instinctive care with which rooks avoid any one carrying a gun, and which is so evident that I have often heard country people remark that a rook can smell gunpowder, one can more justly estimate the force of their love or friendship in thus continuing to hover round a person who has just destroyed one of their companions with an instrument the dangerous nature of which they seem fully capable of appreciating.[1]

The justice of these remarks may be better appreciated in the light of the following very remarkable observation, as an introduction to which I have quoted them.

Edward, the naturalist, having shot a tern, which fell winged into the sea, its companions hovered around the floating bird, manifesting much apparent solicitude, as terns and gulls always do under such circumstances. How far this apparent solicitude is real I have often speculated, as in the analogous case of the crows—wondering whether the emotions concerned were really those of sympathy or mere curiosity. The following observation, however, seems to set this question at rest. Having begun to make preparations for securing the wounded bird, Edward says: 'I expected in a few moments to have it in my possession, being not very far from the water's edge, and drifting shorewards with the wind.' He continues:—

While matters were in this position I beheld, to my utter astonishment and surprise, two of the unwounded terns take

[1] *Gleanings*, pp. 58–9.

hold of their disabled comrade, one at each wing, lift him out of the water, and bear him out seawards. They were followed by two other birds. After being carried about six or seven yards, he was let gently down again, when he was taken up in a similar manner by the two who had been hitherto inactive. In this way they continued to carry him alternately, until they had conveyed him to a rock at a considerable distance, upon which they landed him in safety. Having recovered my self-possession, I made toward the rock, wishing to obtain the prize which had been so unceremoniously snatched from my grasp. I was observed, however, by the terns; and instead of four, I had in a short time a whole swarm about me. On my near approach to the rock I once more beheld two of them take hold of the wounded bird as they had done already, and bear him out to sea in triumph, far beyond my reach. This, had I been so inclined, I could no doubt have prevented. Under the circumstances, however, my feelings would not permit me; and I willingly allowed them to perform without molestation an act of mercy, and to exhibit an instance of affection which man himself need not be ashamed to imitate.[1]

According to Clavigero,[2] the inhabitants of Mexico utilise the sympathy of the wild pelican for the procuring of fish. First a pelican is caught and its wing broken. The bird is then tied to a tree, and being both in pain and captivity, it utters cries of distress. Other pelicans are attracted by the cries, and finding their friend in such a sorry case, their bowels of compassion become moved in a very literal sense; for they disgorge from their stomachs and pouches the fish which they have caught, and deposit them within reach of the captive. As soon as this is done the men, who have been lying in wait concealed, run to the spot, drive off the friendly pelicans, and secure their fish, leaving only a small quantity for the use of the captive.

The parrot which belonged to the Buffon family showed much sympathy with a female servant to whom it was attached when the girl had a sore finger, which it displayed by its never leaving her sick room, and groaning as if itself in pain. As soon as the girl got better the bird again became cheerful.

[1] Smiles, *Life of Edward*, p. 240.   [2] *History of Mexico*, p. 220.

I shall conclude this brief demonstration of the keen sympathy which may exist in birds, by quoting the following very conclusive case in the words of its distinguished observer, Dr. Franklin:[1]—

I have known two parrots, said he, which had lived together four years, when the female became weak, and her legs swelled. These were symptoms of gout, a disease to which all birds of this family are very subject in England. It became impossible for her to descend from the perch, or to take her food as formerly, but the male was most assiduous in carrying it to her in his beak. He continued feeding her in this manner during four months, but the infirmities of his companion increased from day to day, so that at last she was unable to support herself on the perch. She remained cowering down in the bottom of the cage, making, from time to time, ineffectual efforts to regain the perch. The male was always near her, and with all his strength aided the feeble attempts of his dear better half. Seizing the poor invalid by the beak, or the upper part of the wing, he tried to raise her, and renewed his efforts several times.

His constancy, his gestures, and his continued solicitude, all showed in this affectionate bird the most ardent desire to relieve the sufferings and assist the weakness of his companion.

But the scene became still more interesting when the female was dying. Her unhappy spouse moved around her incessantly, his attention and tender cares redoubled. He even tried to open her beak to give some nourishment. He ran to her, then returned with a troubled and agitated look. At intervals he uttered the most plaintive cries; then, with his eyes fixed on her, kept a mournful silence. At length his companion breathed her last; from that moment he pined away, and died in the course of a few weeks.[1]

The jealousy of birds is proverbial; and that they also manifest the kindred passion of emulation, no one can doubt who has heard them singing against one another. Mr. Bold relates that a mule canary would always sing at his own image in a mirror, becoming more and more excited, till he ended by flying in rage against his supposed rival.

The late Lady Napier wrote me, among other 'anecdotes of a grey parrot left on a long visit to the family of General Sir William Napier, at the time residing in Ger-

---

[1] *Zoologist*, vol. ii.

many,' the following graphic description of the exultation displayed by the bird when it baffled the imitative powers of its master. The bird was the same as that already mentioned under the head of 'Memory':—

Sometimes when only two or three were in the room, at quiet occupations instead of talking, she would utter at short intervals a series of strong squalls or cries in an interjectional style, each more strange and grotesque than the previous one. My father on these occasions sometimes amused himself by imitating these cries as she uttered them, which seemed to excite her ingenuity in the production of them to the uttermost. As a last resource she always had recourse to a very peculiar one, which completely baffled him; upon which, with a loud ha! ha! ha! she made a somersault round her perch, swinging with her head downwards, sprung from one part of the cage to another, and tossed a bit of wood she used as a toy over her head in the most exulting triumph, repeating at intervals the inimitable cry, followed by peals of ha! ha! ha! to the great amusement of all present.

Allied to emulation is resentment, of which the following, communicated to me by a correspondent, may be taken as an example. If space permitted I could give confirmatory cases:—

One day the cat and the parrot had a quarrel. I think the cat had upset Polly's food, or something of that kind; however, they seemed all right again. An hour or so after, Polly was standing on the edge of the table; she called out in a tone of extreme affection, 'Puss, puss, come then—come then, pussy.' Pussy went and looked up innocently enough. Polly with her beak seized a basin of milk standing by, and tipped the basin and all its contents over the cat; then chuckled diabolically, of course broke the basin, and half drowned the cat.

Several strange but mutually corroborative stories seem to show cherished vindictiveness on the part of storks. Thus, in Captain Brown's book there occurs an account of a tame stork which lived in the college yard at Tübingen,—

And in a neighbouring house was a nest, in which other storks, that annually resorted to the place, used to hatch their eggs. At this nest, one day in autumn, a young collegian fired a shot, by which the stork that was sitting on it was probably wounded, for it did not fly out of the nest for some weeks afterwards. It was able, however, to take its departure at the usual time with the rest of the storks. But in the ensuing spring a strange stork was observed on the roof of the college, which, by clapping his wings and other gestures, seemed to invite the tame stork to come to him; but, as the tame one's wings were clipped, he was unable to accept the invitation. After some days the strange stork appeared again, and came down into the yard, when the tame one went out to meet him, clapping his wings as if to bid him welcome, but was suddenly attacked by the visitor with great fury. Some of the neighbours protected the tame bird, and drove off the assailant, but he returned several times afterwards, and incommoded the other through the whole summer. The next spring, instead of one stork only, four storks came together into the yard, and fell upon the tame one; when all the poultry present—cocks, hens, geese, and ducks—flocked at once to his assistance, and rescued him from his enemies. In consequence of this serious attack, the people of the house took precaution for the tame stork's security, and he was no more molested that year. But in the beginning of the third spring came upwards of twenty storks, which rushed at once into the yard and killed the tame stork before either man or any other animal could afford him protection.

A similar occurrence took place on the premises of a farmer near Hamburg, who kept a tame stork, and, having caught another, thought to make it a companion for the one in his possession. But the two were no sooner brought together than the tame one fell upon the other, and beat him so severely that he made his escape from the place. About four months afterwards, however, the defeated stork returned with three others, who all made a combined attack upon the tame one and killed him.[1]

The curiosity of birds is highly developed, so much so, indeed, that in this and other countries it is played upon by sportsmen and trappers. Unfamiliar objects being

---

[1] Watson, *Reasoning Power of Animals*, pp. 375-76, where see also some curious cases of male storks slaying their females upon the latter hatching out eggs of other birds. He gives an exactly similar case as having occurred with the domestic cock; and in Bingley (*loc. cit.*, vol. ii., p. 241) there is quoted from Dr. Percival another case of the same kind, in which a cock killed his hen as soon as she had hatched out a brood of young partridges from eggs which had been set to her.

placed within sight, say of ducks, the birds approach to examine them, and fall into the snares which have been prepared. Similarly, in oceanic islands unfrequented by man, the birds fearlessly approach to examine the first human beings that they have seen.

That birds exhibit pride might be considered doubtful if we had to rely only on the evidence supplied by the display of the peacock, and the strutting of the turkey-gobbler; for these actions, although so expressive of this emotion, may not really be due to it. But I think that the evident pleasure which is taken in achievement by talking birds can only be ascribed to the emotion in question. These birds regularly practise their art, and when a new phrase is perfected they show an unmistakable delight in displaying the result.

Play is exhibited by many species in various ways, and it seems to be this class of feelings in their most organised form which have led to the extraordinary instincts of the bower-birds of New South Wales. The 'playhouses' of the animals have been described by Mr. Gould in his 'History of the Birds of New South Wales.' Of course the play-instincts are here united with those of courtship, which are of such general occurrence among birds; but I think no one can read Mr. Gould's description of the bowers and the uses to which they are put without feeling that the love of sportive play must have been joined with the sexual instincts in producing the result. But, be this as it may, there can be no question that these bowers are highly interesting structures, as furnishing the most unexceptionable evidence of true æsthetic, if not artistic feeling on the part of the bird which constructs them; and, according to Mr. Herbert Spencer, the artistic feelings are physiologically allied with those of play. It is a matter of importance to obtain definite proof of an æsthetic sense in animals, because this constitutes the basis of Mr. Darwin's theory of sexual selection; but as he has treated the evidence on this subject in so exhaustive a manner, I shall not enter upon so wide a field further than to point out that the case of the bower-bird, even if it stood alone, would be amply sufficient to carry the general conclusion that some animals exhibit emotions of the beautiful. The following is Mr. Gould's description, *in extenso*, of the habits of the bird in question :—

The extraordinary bower-like structure, alluded to in my remarks on the genus, first came under my notice in the Sydney Museum, to which an example had been presented by Charles Cox, Esq. . . . On visiting the cedar bushes of the Liverpool range, I discovered several of these bowers or playing-houses on the ground, under the shelter of the branches of the overhanging trees, in the most retired part of the forest; they differed considerably in size, some being a third larger than others. The base consists of an extensive and rather convex platform of sticks firmly interwoven, on the centre of which the bower itself is built. This, like the platform on which it is placed, and with which it is interwoven, is formed of sticks and twigs, but of a more slender and flexible description, the tips of the twigs being so arranged as to curve inwards and nearly meet at the top; in the interior the materials are so placed that the forks of the twigs are always presented outwards, by which arrangement not the slightest obstruction is offered to the passage of the birds. The interest of this curious bower is much enhanced by the manner in which it is decorated with the most gaily coloured articles that can be collected, such as the blue tail-feathers of the Rose-hill and Pennantian parakeets, bleached bones and shells of snails, &c.; some of the feathers are inserted among the twigs, while others with the bones and shells are strewed near the entrances. The propensity of these birds to fly off with any attractive object is so well known to the natives that they always search the runs for any small missing article that may have been accidentally dropped in the bush. I myself found at the entrance of one of them a small neatly worked stone tomahawk of an inch and a half in length, together with some slips of blue cotton rag, which the birds had doubtless picked up at a deserted encampment of the natives.

It has now been clearly ascertained that these curious bowers are merely sporting-places in which the sexes meet, when the males display their finery, and exhibit many remarkable actions; and so inherent is this habit, that the living examples, which have been from time to time sent to this country, continue it even in captivity.[1] Those belonging to the Zoological Society have constructed

[1] See Darwin. *Descent of Man*, pp. 92, 381, 406, 413.

their bowers, decorated and kept them in repair, for several years. In a letter from the late Mr. F. Strange, it is said:—

My aviary is now tenanted by a pair of satin-birds, which for the last two months have been constantly engaged in constructing bowers. Both sexes assist in their erection, but the male is the principal workman. At times the male will chase the female all over the aviary, then go to the bower, pick up a gay feather or a large leaf, utter a curious kind of note, set all his feathers erect, run round the bower, and become so excited that his eyes appear ready to start from his head, and he continues opening first one wing and then another, uttering a low whistling note, and, like the domestic cock, seems to be picking up something from the ground, until at last the female goes gently towards him, when after two turns round her, he suddenly makes a dash, and the scene ends.'[1]

I have said that if this case stood alone it would constitute ample evidence that some animals possess emotions of the beautiful. But the case does not stand alone. Certain humming-birds, according to Mr. Gould, decorate the outsides of their nests 'with the utmost taste; they instinctively fasten thereon beautiful pieces of flat lichen, the larger pieces in the middle, and the smaller on the part attached to the branch. Now and then a pretty feather is intertwined or fastened to the outer sides, the stem being always so placed that the feather stands out beyond the surface.' Several other instances might be rendered of the display of artistic feeling in the architecture of birds; and, as Mr. Darwin so elaborately shows, there can scarcely be question that these animals take emotional pleasure in surveying beautiful plumage in the opposite sex, looking to the careful manner in which the males of many species display their fine colours to the females. Doubtless the evidence of æsthetic feeling is much stronger in the case of birds than it is in that of any other class; but if this feeling is accepted as a sufficient cause, through sexual selection, of natural decoration in the members of this class, we are justified in attributing to sexual selection, and so to æsthetic feeling, natural

[1] Gould, *Birds of Australia*, vol. i., pp. 442-45.

decoration in other classes, at least as low down in the scale as the Articulata. But, as I have said, Mr. Darwin has dealt with this whole subject in so exhaustive a manner that it is needless for me to enter upon it further than to say in general terms, that whatever we may think of his theory of sexual selection, his researches have unquestionably proved the existence of an æsthetic sense in animals.

The same fact appears to be shown in another way by the fondness of song-birds for the music of their mates. There can be no doubt that male birds charm their females with their strains, and that this, in fact, is the reason why song in birds has become developed. Of course it may be said that the vocal utterances of birds are not always, or even generally, musical; but this does not affect the fact that birds find some æsthetic pleasure in the sounds which they emit; it only shows that the standard of æsthetic taste differs in different species of birds as it does in different races of men. Moreover, the pleasure which birds manifest in musical sounds is not always restricted to the sounds which they themselves produce. Parrots seem certainly to take delight in hearing a piano play or a girl sing; and the following instance, published by the musician John Lockman, reveals in a remarkable manner the power of distinguishing a particular air, and of preferring it above others. He was staying at the house of a Mr. Lee in Cheshire, whose daughter used to play; and whenever she played the air of 'Speri si' from Handel's opera of 'Admetus,' a pigeon would descend from an adjacent dovecot to the window of the room where she sat, 'and listen to the air apparently with the most pleasing emotions,' always returning to the dovecot immediately the air was finished. But it was only this one air that would induce the bird to behave in this way.[1]

Special Habits.

Under this heading we shall have a number of facts to consider, which are more or less of a disconnected character.

[1] Bingley, *Animal Biography*, vol. ii., p. 220.

Taking first those special habits connected with the procuring of food, we may notice the instinct manifested by blackbirds and thrushes of conveying snails to considerable distances in order to hammer and break their shells against what may happen to be the nearest stone,[1] and the still more clever though somewhat analogous instinct exhibited by certain gulls and crows of flying with shell-fish to a considerable height and letting them fall upon stones for the purpose of smashing their shells.[2] Both these instincts manifest a high degree of intelligence, either on the part of the birds themselves, or on that of their ancestors; for neither of these instincts can be regarded as due to originally accidental adjustments favoured and improved by natural selection; they must at least originally have been intelligent actions purposely designed to secure the ends attained.

An interesting instinct is that of piracy, which in the animal kingdom reaches its highest or most systematic development among the birds. It is easy to see how it may be of more advantage to a species of strong bird that its members should become parasitic on the labours of other species than that they should forage for themselves, and so there is no difficulty in understanding the development of the plundering instinct by natural selection. We find all stages of this development among the sea-birds. Thus the gulls, although usually self-foragers, will, as I have often observed, congregate in enormous numbers where the guillemots have found a shoal of fish. Resting

[1] For full information, see Buckland, *Curiosities of Natural History*, p. 183.

[2] Of the crow (carrion and hooded), Edward says: 'He goes aloft with a crab, and lets it fall upon a stone or a rock chosen for the purpose. If it does not break, he seizes it again, goes up higher, lets it fall, and repeats his operation again and again until his object is accomplished. When a convenient stone is once met with, the birds resort to it for a long time. I myself know a pretty high rock, that has been used by successive generations of crows for about twenty years!' Also, as Handcock says, 'a friend of Dr. Darwin saw on the north coast of Ireland above a hundred crows preying upon mussels, which is not their natural food; each crow took a mussel up into the air, twenty or forty yards high, and let it fall on the stones, and thus breaking the shell, got possession of the animal. Ravens, we are told, often resort to the same contrivance.'

on or flying over the surface of the water, the gulls wait till a guillemot comes to the surface with a fish, and then wrest the latter from the beak of the former. In the robber-tern this instinct has proceeded further, so that the animal gains its subsistence entirely by plunder of other terns. I have often observed this process, and it is interesting that the common tern well knows the appearance of the robber; for no sooner does a robber-tern come up than the greatest consternation is excited among a flock of common terns, these flying about and screaming in a frantic manner. The white-headed eagle has also developed the plundering instinct in great perfection, as is shown by the following graphic account of Audubon:—

During spring and summer, the white-headed eagle, to procure sustenance, follows a different course, and one much less suited to a bird apparently so well able to supply itself without interfering with other plunderers. No sooner does the first hawk make its appearance along the Atlantic shore, or around the numerous and large rivers, than the eagle follows it, and, like a selfish oppressor, robs it of the hard-earned fruits of its labour. Perched on some tall summit, in view of the ocean or of some watercourse, he watches every motion of the osprey while on the wing. When the latter rises from the water, with a fish in its grasp, forth rushes the eagle in pursuit. He mounts above the fish-hawk, and threatens it by actions well understood; when the latter, fearing perhaps that its life is in danger, drops its prey. In an instant the eagle, accurately estimating the rapid descent of the fish, closes its wings, follows it with the swiftness of thought, and the next moment grasps it. The prize is carried off in silence to the woods, and assists in feeding the ever-hungry brood of the eagle.

The frigate pelican is likewise a professional thief, and attacks the boobies not only to make them drop the fish which they have newly caught, but also to disgorge those which are actually in their stomachs. The latter process is effected by strong punishment, which they continue until the unfortunate booby yields up its dinner. The punishment consists in stabbing the victim with its powerful beak. Catesby and Dampier have both observed and described these habits, and it seems from their account that the plunderer may either commit highway robbery in the air, or lie in wait for the boobies as they return to rest.

In antithesis to this habit of plundering other birds I may quote the following from 'Nature' (July 20, 1871), to show that the instinct of provident labour, so common among insects and rodents, is not altogether unrepresented in birds:—

The ant-eating woodpecker (*Melanerpes formicivorus*), a common Californian species, has the curious and peculiar habit of laying up provision against the inclement season. Small round holes are dug in the bark of the pine and oak, into each of which is inserted an acorn, and so tightly is it fitted or driven in, that it is with difficulty extricated. The bark of the pine trees, when thus filled, presents at a short distance the appearance of being studded with nails.

The following may also be quoted :—

It is the nature of this bird (guillemot), as well as of most of those birds which habitually dive to take their prey, to perform all their evolutions under water with the aid of their wings; but instead of dashing at once into the midst of the terrified group of small prey, by which only a few would be captured, it passes round and round them, and so drives them into a heap; and thus has an opportunity of snatching here one and there another as it finds it convenient to swallow them; and if any one pushes out to escape, it falls the first prey of the devourer. The manner in which this bird removes the egg of a gull or hen to some secure place to be devoured, when compared with that in which a like conveyance is made by the parent for the safety of its future progeny, affords a striking manifestation of the difference between appetite and affection. When influenced by affection, the brittle treasure is removed without flaw or fracture, and is replaced with tender care; but the plunderer at once plunges his bill into its substance, and carries it off on its point.[1]

Speaking of the feeding habits of the lapwing, Jesse says :—

When the lapwing wants to procure food, it seeks for a worm's cast, and stamps the ground by the side of it with its feet. After doing this for a short time, the bird waits for the

[1] Couch, *Illustrations of Instinct*, pp. 192-93.

issue of the worm from its hole, which, alarmed at the shaking of the ground, endeavours to make its escape, when it is immediately seized, and becomes the prey of the ingenious bird. The lapwing also frequents the haunts of moles, which, when in pursuit of worms on which they feed, frighten them, and the worm, in attempting to escape, comes to the surface of the ground, when it is seized by the lapwing.[1]

Again,—

A lady of Dr. E. Darwin's acquaintance saw a little bird repeatedly hop on a poppy stem, and shake the head with his bill, till many seeds were scattered, when it settled on the ground and picked up the seeds.[2]

It is a matter of common remark that in countries where vultures abound, these birds rapidly 'gather together where the carcass is,' although before the death of their prey no bird was to be seen in the sky. The question has always been asked whether the vultures are guided to the carcass by their sense of smell or by that of sight; but this question is really no longer an open one. When Mr. Darwin was at Valparaiso he tried the following experiment. Having tied a number of condors in a long row, and having folded up a piece of meat in paper, he walked backwards and forwards in front of the row, carrying the meat at a distance of three yards from them, 'but no notice whatever was taken.' He then threw the meat upon the ground, within one yard of an old male bird; 'he looked at it for a moment with attention, but then regarded it no more.' With a stick he next pushed the meat right under the beak of the bird. Then for the first time the bird smelled it, and tore open the paper 'with fury, and at the same moment every bird in the long row began struggling and flapping its wings.'[3] Thus there can be no doubt that vultures do not depend on their sense of smell for finding carrion at a distance. Nor is it mysterious why they should find it by their sense of sight. If over an area of many square miles there are a number of vultures flying as they do at a very high elevation, and if

---

[1] *Gleanings*, &c., vol. i., p. 71.
[2] *Ibid.*
[3] *Voyage of a Naturalist*, &c., p. 184

one of the number perceives a carcass and begins to descend, the next adjacent vultures would see the descent of the first one, and follow him as a guide, while the next in the series would follow these in the same way, and so on.

Coming now to special instincts relating to incubation and the care of offspring, a correspondent writes:—

Last spring I had a pair of canaries, in an ordinary breeding cage (with two small boxes for nests in a compartment at one end). In due course the first egg was laid, which I inspected through the little door made for that purpose. The next day I looked again; still only one egg, and so for four or five days. It being evident, from the appearance of the hen, that there were more eggs coming, and as she seemed in good health, I supposed she might have broken some; and I took out the box, and examined it carefully for the shells (but without pulling the nest to pieces), and found nothing, until towards the beginning of another week I went to take the one egg away, as the hen seemed preparing to sit upon it. There were two eggs! The next morning, to my surprise, she was sitting upon six eggs! She must therefore have buried four of them in the four corners of the box, and so deep that I had been unable to find them. At first I thought that she had done so merely from dislike at their being looked at, but on reflection it has occurred to me that she did it that all might be hatched at the same time (as they subsequently were); for she was perfectly tame, and would almost suffer herself to be handled when on her nest. Wild birds never seem to conceal their eggs before sitting; but then (having more amusements than cage birds) they do not revisit their eggs after laying, until they have laid their number, whereas a caged bird, having nothing to divert her attention from her nest, often sits on it the greater part of the day.

I am not aware that this curious display of forethought on the part of a caged bird has been hitherto recorded, and seeing, as my correspondent points out, that it has reference to the changed conditions of life brought about by domestication, it may be said to constitute the first step in the development of a new instinct, which, if the conditions were of sufficiently long continuance, might lead to an important and permanent change of the ancestral instinct.

I have several interesting facts, also communicated to

me by correspondents, similarly relating to individual variations of the ancestral instinct of incubation in order to meet the requirements of a novel environment. Thus Mr. J. F. Fisher tells me that while he was a commander in the East India trade he always took a quantity of fowls to sea for food. The laying-boxes being in a confined space, the hens used to quarrel over their occupancy; and one of the hens adopted the habit of removing the 'nest-eggs' which Mr. Fisher placed in one of the boxes to another box of the same kind not very far away. He watched the process through a chink of a door, and 'saw her curl her neck round the egg, thus forming a cup by which she lifted the egg,' and conveyed it to the other box. He adds :—

> I can give no information as to the more recondite question *why* the egg was removed, or the fastidious preference of the one box over the other, or the inventive faculty that suggested the neck as a makeshift hand; but from the despatch with which she effected the removal of the egg in the case I saw, I have no doubt that this hen was the one which had performed the feat so often before.

The explanation of the preference shown for the one box over the other may, I think, be gathered from another part of my correspondent's letter, for he there mentions incidentally that the box in which he placed the nest-egg, and from which the hen removed it, was standing near a door which was usually open, and thus situated in a more exposed position than the other box. But be this as it may, considering that among domestic fowls the habit of conveying eggs is not usual, such isolated cases are interesting as showing how instincts may originate. Jesse gives an exactly similar case ('Gleanings,' vol. i., p. 149) of the Cape goose, which removed eggs from a nest attacked by rats, and another case of a wild duck doing the same.

In the same connection, and with the same remarks, I may quote the following case in which a fowl adopted the habit of conveying, not her eggs, but her young chickens. I quote it from Houzeau ('Journ.,' i., p. 332), who gives the observation on the authority of his brother as eye-witness. The fowl had found good feeding-ground on the further side of a stream four metres wide. She adopted the habit of flying across with her chickens upon her back, taking one chicken on each journey. She thus transferred her whole brood every morning, and brought them back in a similar way to their nest every evening. The habit of carrying young in this way is not natural to Grallinaceæ, and therefore this particular instance of its display can only be set down as an intelligent adjustment by a particular bird.

Similarly, a correspondent (Mr. J. Street) informs me of a case in which a pair of blackbirds, after having been disturbed by his gardener looking into their nest at their young, removed the latter to a distance of twenty yards, and deposited them in a more concealed place. Partridges are well known to do this, and similarly, according to Audubon, the goatsucker, when its nest is disturbed, removes its eggs to another place, the male and female both transporting eggs in their beaks.[1]

Still more curiously, a case is recorded in 'Comptes Rendu' (1836) of a pair of nightingales whose nest was threatened by a flood, and who transported it to a safe place, the male and the female bearing the nest between them.

Now, it is easy to see that if any particular bird is intelligent enough, as in the cases quoted, to perform this adjustive action of conveying young—whether to feeding-grounds, as in the case of the hen, or from sources of danger, as in the case of partridges, blackbirds, and goat-suckers—inheritance and natural selection might develop the originally intelligent adjustment into an instinct common to the species. And it so happens that this has actually occurred in at least two species of birds—viz., the woodcock and wild duck, both of which have been repeatedly observed to fly with their young upon their backs to and from their feeding-ground.

Couch gives some facts of interest relating to the mode of escape practised by the water-rail, swan, and some other aquatic birds. This consists in sinking under water, with

[1] *Orn. Biog.*, i., p. 276.

only the bill remaining above the surface for respiration. When the swan has young, she may sink the head quite under water in order to allow the young to mount on it, and so be carried through even rapid currents.

The same author remarks that—

Many birds will carefully remove the meotings of the young from the neighbourhood of their nests, in order not to attract the attention of enemies; for while we find that birds which make no secret of their nesting-places are careless in such matters, the woodpecker and the marsh tit in particular are at pains to remove even the chips which are made in excavating the cavities where the nests are placed, and which might lead an observer to the sacred spot.

Similarly, Jesse observes:—

The excrement of the young of many birds who build their nests without any pretensions to concealment, such as the swallow, crow, &c., may at all times be observed about or under the nest; while that of some of those birds whose nests are more industriously concealed is conveyed away in the mouths of the parent birds, who generally drop it at a distance of twenty or thirty yards from the nest. Were it not for this precaution, the excrement itself, from its accumulation, and commonly from its very colour, would point out the place where the young were concealed. When the young birds are ready to fly, or nearly so, the old birds do not consider it any longer necessary to remove the excrement.

Sir H. Davy gives an account of a pair of eagles which he saw on Ben Nevis teaching their young ones to fly; and every one must have observed the same thing among commoner species of birds. The experiments of Spalding, however, have shown that flying is an instinctive faculty; so that when he reared swallows from the nest and liberated them only after they were fully fledged, they flew well immediately on being liberated. Therefore, the 'teaching to fly' by parent birds must be regarded as mere encouragement to develop instinctive powers, which in virtue of this encouragement are probably developed sooner than would otherwise be the case.

A few observations may here be offered on some

habits which do not fall under any particular heading.

The habit which many small birds display of mobbing carnivorous ones is probably due to a desire to drive off the enemy, and perhaps also to warn friends by the hubbub. It may therefore perhaps be regarded as a display of concerted action, of which, however, we shall have better evidence further on. I have seen a flock of common terns mob a pirate tern, which shows that this combined action may be directed as much against robbery as against murder. Couch says he has seen blackbirds mobbing a cat which was concealed in a bush, and here the motive would seem to be that of warning friends rather than that of driving away the enemy.

I have observed among the sea-gulls at the Zoological Gardens a curious habit, or mode of challenge. This consists in ostentatiously picking up a small twig or piece of wood, and throwing it down before the bird challenged, in the way that a glove used to be thrown down by the old knights. I observed this action performed repeatedly by several individuals of the glaucous and black-back species in the early spring-time of the year, and so it probably has some remote connection with the instinct of nest-building.

Nidification.

In connection with the habits and instincts peculiar to certain species of birds, I may give a short account of the more remarkable kinds of nidification that are met with in this class of animals. As the account must necessarily be brief, I shall only mention the more interesting of the usual types.

Petrels and puffins make their nests in burrows which they excavate in the earth. The great sulphur mountain in Guadaloupe is described by Wasser as 'all bored like a rabbit warren with the holes that these imps (*i.e.* petrels) excavate.' In the case of the puffin it is the male that does the work of burrowing. He throws himself upon his back in the tunnel which he has made, and digs it longer and longer with his broad bill, while casting out the mould with his webbed feet. The burrow when finished has several twists and turns in it, and is about ten feet deep. If a rabbit burrow is available, the puffin saves himself the trouble of digging by taking possession of the one already made. The kingfisher and land-martin also make their nests in burrows.

Certain auks lay their single egg on the bare rock while the stone curlew and goatsucker deposit theirs on the bare soil, returning, however, year after year to the same spot. Ostriches scrape holes in the sand to serve as extemporised nests for their eggs promiscuously dropped, which are then buried by a light-coating of sand, and incubated during the day by the sunbeams, and at night by the male bird. Sometimes a number of female ostriches deposit their eggs in a common nest, and then take the duty of incubation by turns. Similarly, gulls, sandpipers, plovers, &c., place their eggs in shallow pits hollowed out of the soil. The kingfisher makes a bed of undigested fish-bones ejected as pellets from her stomach, and 'some of the swifts secrete from their salivary glands a fluid which rapidly hardens as it dries on exposure to the air into a substance resembling isinglass, and thus furnish the "edible birds' nests" that are the delight of the Chinese epicures.'[1]

The house-martin builds its nest of clay, which it sticks upon the face of a wall, and renders more tenacious by working into it little bits of straw, splinters of wood, &c. According to Mr. Gilbert White:—

That this work may not, while it is soft and green, pull itself down by its own weight, the provident architect has prudence and forbearance enough not to advance her work too fast; but by building only in the morning, and by dedicating the rest of the day to food and amusement, gives it sufficient time to dry and harden. About half an inch seems a sufficient layer for a day. Thus careful workmen, when they build mud walls (informed at first perhaps by these little birds), raise but a moderate layer at a time, and then desist, lest the work should become top-heavy, and ruined by its own weight. By this method, in about ten or twelve days is formed a hemispheric nest, with a small aperture towards the top, strong, compact,

[1] Newton, *Encycl. Brit.*, art. 'Birds.'

and warm, and perfectly fitted for all the purposes for which it was intended.

Other birds build in wood. The tomtit and the woodpecker excavate a hole in a tree, and carefully carry away the chips, so as not to give any indication of the whereabouts of their nests. Wilson says that the American woodpecker makes an excavation five feet in depth, of a tortuous form, to keep out wind and rain.

The orchard starling suspends its nest from the branches of a tree, and uses for its material tough kinds of grass, the blades of which it weaves together. Wilson found one of these blades to be thirteen inches long, and to be woven in and out thirty-four times.

We may next notice the weaver (*Ploceus textor*) and tailor (*Prinia, Orthotomus,* and *Sylvia*). The former intertwines slender leaves of grass so as to produce a web sufficiently substantial for the protection of its young. The tailor-birds sew together leaves wherewith to make their nests, using for the purpose cotton and thread where they can find it, and natural vegetable fibres where they cannot obtain artificial. Colonel Sykes says that he has found the threads thus used for sewing knotted at the ends.[1]

Forbes saw the tailor-bird of the East Indies constructing its nest, and observed it to choose a plant with large leaves, gather cotton which it regularly spun into a thread by means of its bill and claws, and then sew the leaves together, using its beak as a needle, or rather awl.

This instinct is rendered particularly interesting to evolutionists from the fact that it is exhibited by three distinct genera. For, as the instinct is so peculiar and unique, it is not likely to have originated independently in the three genera, but must be regarded as almost certainly derived from a common ancestral type—thus showing that an instinct may be perpetuated unaltered after the differentiation of structure has proceeded beyond a specific distinction. The genus *Sylvia* inhabits Italy, the other two inhabit India. *Sylvia* uses for thread spiders' web col-

[1] *Catalogue of Birds,* &c., p. 16.

lected from the egg-pouches, which is stitched through holes made in the edges of leaves, presumably with the beak.

The baya bird of India 'hangs its pendulous dwelling from a projecting bough, twisting it with grass into a form somewhat resembling a bottle with a prolonged neck, the entrance being inverted, so as to baffle the approaches of its enemies, the tree snakes and other reptiles.'

Sir E. Tennent, from whom this account is taken, adds :—

The natives assert that the male bird carries fire-flies to the nest, and fastens them to its sides by particles of soft mud. Mr. Layard assures me that although he has never succeeded in finding the fire-fly, the nest of the male bird (for the female occupies another during incubation) invariably contains a patch of mud on each side of the perch.

Dr. Buchanan confirms the report of the natives here alluded to, and says :—

At night each of the habitations is lighted up by a fire-fly stuck on the top with a bit of clay. The nest consists of two rooms; sometimes there are three or four fire-flies, and their blaze in the little cells dazzles the eyes of the bats, which often kill the young of these birds.

While this work is passing through the press I meet with the following, which appears to refer to some independent, and therefore corroborative observation concerning the above-stated fact, and in any case is worth adding, on account of the observation concerning the rats, which, if trustworthy, would furnish a sufficient reason for the instinct of the birds. The extract is taken from a letter to 'Nature' (xxiv., p. 165), published by Mr. H. A. Severn :

I have been informed on safe authority that the Indian bottle-bird protects his nest at night by sticking several of these glow-beetles around the entrance by means of clay; and only a few days back an intimate friend of my own was watching three rats on a roof rafter of his bungalow when a glow-fly lodged very close to them; the rats immediately scampered off.

The Talegallus of Australia is, in the opinion of Gould,—

Among the most important of the ornithological novelties which the exploration of Western and Southern Australia has unfolded to us, and this from the circumstance of its not hatching its own eggs, which, instead of being incubated in the usual way, are deposited in mounds of mixed sand and herbage, and there left for the heating of the mass to develop the young, which, when accomplished, force their way through the sides of the mound, and commence an active life from the moment they see the light of day.[1]

Sir George Grey measured one of these mounds, and found it to be 'forty-five feet in circumference, and if rounded in proportion on the top (it being at the time unfinished) would have been full five feet high.' The heat round the eggs was taken to be 89°.

A curious aberration of the nest-building instinct is sometimes shown by certain birds—particularly the common wren—which consists in building a supernumerary nest. That is to say, after one nest is completed, another is begun and finished before the eggs are laid, and the first nest is not used, though sometimes it is used in preference to the second.

As showing at once the eccentricity which birds sometimes display in the choice of a site, and also the determination of certain birds to return to the same site in successive years, I may allude to the case published by Bingley, of a pair of swallows which built their nest upon the wings and body of a dead owl, which was hanging from the rafters of a barn, and so loosely as to sway about with every gust of wind. The owl with the nest upon it was placed as a curiosity in the museum of Sir Ashton Lever, and he directed that a shell should be hung upon the rafters in the place which had been previously occupied by the dead owl. Next year the swallows returned and constructed their new nest in the cavity of the shell.[2]

The following is quoted from Thompson's 'Passions of Animals,' p. 205:—

The sociable grosbeak of Africa is one of the few instances of birds living in community and uniting in constructing one

---

[1] Gould, *Birds of Australia*, vol. ii., p. 155, where see for further description.
[2] *Animal Biography*, vol. ii., p. 204.

huge nest for the whole society. L. Valiant's account has been fully confirmed by other travellers. He says: 'I observed on the way a tree with an enormous nest of these birds, which I have called republicans; and as soon as I arrived at my camp I despatched a few men with a waggon to bring it to me, that I might open and examine the hive. When it arrived, I cut it in pieces with a hatchet, and saw that the chief portion of the structure consisted of a mass of Boshman's grass, without any mixture, but so compact and firmly basketed together as to be impenetrable to the rain. This is the commencement of the structure, and each bird builds its particular nest under this canopy. But the nests are formed only beneath the eaves, the upper surface remaining void, without, however, being useless; for as it has a projecting rim, and is a little inclined, it serves to let the water run off, and preserves each little dwelling from the rain. Figure to yourself a huge irregular sloping roof, all the eaves of which are covered with nests, crowded one against another, and you will have a tolerably accurate idea of these singular edifices. Each individual nest is three or four inches in diameter, which is sufficient for the bird; but, as they are all in contact with one another around the eaves, they appear to the eye to form but one building, and are distinguishable from each other only by a little external aperture which serves as an entrance to the nest; and even this is sometimes common to three different nests, one of which is situated at the bottom and the other two at the sides. This large nest, which was one of the most considerable I had anywhere seen in the course of my journey, contained 320 inhabited cells, which, supposing a male and female to each, would form a society of 640 individuals; but as these birds are polygamous, such a calculation would not be exact.'

The following is quoted from Couch ('Illustrations of Instinct,' p. 227 *et seq.*):—

Mr. Waterton says there is a peculiarity in the nidification of the domestic swan too singular to be passed over without notice. At the time it lays its first egg the nest which it has prepared is of very moderate size; but as incubation proceeds we see it increase vastly in height and breadth. Every soft material, such as pieces of grass and fragments of sedges, is laid hold of by the sitting swan as they float within her reach, and are added to the nest. This work of accumulation is performed by her during the entire period of incubation, be the weather wet or dry, settled or unsettled; and it is perfectly astonishing to see with what assiduity she plies her work of aggrandisement to a nest already sufficient in strength and size to answer every end. My swans generally form their nest on an island quite above the reach of a flood; and still the sitting bird never appears satisfied with the quantity of materials which are provided for her nest. I once gave her two huge bundles of oaten straw, and she performed her work of apparent supererogation by applying the whole of it to her nest, already very large, and not exposed to destruction had the weather become ever so rainy.

This same author continues:—

It is probable that this disposition to accumulation, in its general bearing, has reference to heat rather than the flood; but that the wild swan has a foresight regarding danger, and a quick perception as to the means of securing safety, appears from an instance mentioned by Captain Parry, in his Northern voyage. When everything was deeply involved in ice, the voyagers were obliged to pay much attention to discern whether they were travelling over water or land; but some birds, which formed their nest at no great distance from the ships, were under no mistake in so important a matter; and when the thaw took place it was seen that the nest was situated on an island in the lake.

The following cases are likewise taken from Couch (*loc. cit.*, p. 225):—

This swan was eighteen or nineteen years old, had brought up many broods, and was highly valued by the neighbours. She exhibited, some eight or nine years past, one of the most remarkable powers of instinct ever recorded. She was sitting on four or five eggs, and was observed to be very busy in collecting weeds, grasses, &c., to raise her nest; a farming man was ordered to take down half a load of haulm, with which she most industriously raised her nest and the eggs two feet and a half; that very night there came down a tremendous fall of rain, which flooded all the malt-shops and did great damage. Man made no preparation, the bird did; instinct prevailed over reason. Her eggs were above, and only just above, the water.

During the early part of the summer of 1835, a pair of water-hens built their nest by the margin of the ornamental pond at Bell's Hill, a piece of water of considerable extent, and ordinarily fed by a spring from the height above, but into which the contents of another large pond can occasionally be admitted. This was done while the female was sitting; and as the nest had been built when the water level stood low, the sudden influx of this large body of water from the second pond caused a rise of several inches, so as to threaten the speedy immersion and consequent destruction of the eggs. This the birds seem to have been aware of, and immediately took precautions against so imminent a danger; for when the gardener, upon whose veracity I can safely rely, seeing the sudden rise of the water, went to look after the nest, expecting to find it covered and the eggs destroyed, or at least forsaken by the hen, he observed, whilst at a distance, both birds busily engaged about the brink where the nest was placed; and when near enough he clearly perceived that they were adding, with all possible despatch, fresh materials to raise the fabric beyond the level of the increased contents of the pond; and that the eggs had by some means been removed from the nest by the birds, and were then deposited upon the grass about a foot or more from the margin of the water. He watched them for some time, and saw the nest rapidly increase in height; but I regret to add that he did not remain long enough, fearing he might create alarm, to witness the interesting act of replacing the eggs which must have been effected shortly after; for, upon his return in less than an hour, he found the hen quietly sitting upon them in the newly raised nest. In a few days afterwards the young were hatched, and, as usual, soon quitted the nest and took to the water with their parents. The nest was shown to me *in situ* shortly after, and I could then plainly discern the formation of the new with the older part of the fabric.

We must not conclude these remarks on nidification without alluding to Mr. Wallace's chapters on the 'Philosophy of Birds' Nests,' in his work on 'Natural Selection.' This writer is inclined to suppose that birds do not build their nests distinctive of their various species by the teachings of hereditary instinct, but by the young birds intelligently observing the construction of the nests in which they are hatched, and purposely imitating this construction when in the following season they have occasion to build nests of their own. With reference to this theory it is only needful to say that it is antecedently improbable, and not well substantiated by facts. It is antecedently improbable because, when any habit has been continued for a number of generations—especially when the habit is of a peculiar and detailed character—the probability is that it has become instinctive; we should have almost as much reason to anticipate that the nest of the little crustacean *Podocerus*, or the cell of the hive-bee, is constructed by a process of conscious imitation, as that this is the case with the nests of birds. And this theory is not well substantiated by facts because, if the theory were true, we should expect considerable differences to be usually presented by nests of the same species. Unless the construction of the nest of any given species were regulated by a common instinct, numberless idiosyncratic peculiarities would necessarily require to arise, and there would only be a very general uniformity of type presented by the nests of the same species.

A more valuable contribution to the 'Philosophy of Birds' Nests' is furnished by this able naturalist when he directs attention to a certain general correlation between the form of the nest and the colour of the female. For, on reviewing the birds of the world, he certainly makes good the proposition that, as a general rule, liable however to frequent exceptions, dull-coloured females sit on open nests, while those that are conspicuously coloured sit in domed nests. But Mr. Darwin, in a careful review of all the evidence, clearly shows that this interesting fact is to be attributed, not, as Mr. Wallace supposed, to the colour of the female having been determined through natural selection by the form of the nest, but to the reverse process of the form of the nest having been determined by the colour of the female.[1]

Another general fact of interest connected with nidification must not be omitted. This is that the instincts of nidification, although not so variable as the theory of Mr. Wallace would require, are nevertheless highly plastic. The falcon, which usually builds on a cliff, has been known to lay its eggs on the ground in a marsh; the golden eagle sometimes builds in trees or on the ground while the heron varies its site between trees, cliffs, and open fen.[2] Again, Audubon, in his 'Ornithological Biography,' gives many cases of conspicuous local variations in the nests of the same species in the northern and

---

[1] See *Descent of Man*, p. 452 *et seq.*
[2] See Newton, *Ency. Brit.*, art. 'Birds.'

southern United States; and, as Mr. Wallace truly observes,—

> Many facts have already been given which show that birds do adapt their nests to the situations in which they place them; and the adoption of eaves, chimneys, and boxes by swallows, wrens, and many other birds, shows that they are always ready to take advantage of changed conditions. It is probable, therefore, that a permanent change of climate would cause many birds to modify the form or materials of their abode, so as better to protect their young.[1]

In America the change of habits in this respect undergone by the house-swallow has been accomplished within the last three hundred years.

Closely connected, if not identical, with this fact is another, namely, that in some species which have been watched closely for a sufficient length of time, a steady improvement in the construction of nests has been observed. Thus C. G. Leroy, who filled the post of Ranger of Versailles about a century ago, and therefore had abundant opportunities of studying the habits of animals, wrote an essay on 'The Intelligence and Perfectibility of Animals from a Philosophical Point of View.' In this essay he has anticipated the American observer Wilson in noticing that the nests of young birds are distinctly inferior to those of older ones, both as regards their situation and construction. As we have here independent testimony of two good observers to a fact which in itself is not improbable, I think we may conclude that the nest-making instinct admits of being supplemented, at any rate in some birds, by the experience and intelligence of the individual. M. Pouchet has also recorded that he has found a decided improvement to have taken place in the nests of the swallows at Rouen during his own lifetime; and this accords with the anticipation of Leroy that if our observation extended over a sufficient length of time, and in a manner sufficiently close, we should find that the accumulation of intelligent improvements by individuals of successive generations would begin to tell upon the in-

---

[1] *Natural Selection*, pp. 232-3.

herited instinct, so that all the nests in a given locality would attain to a higher grade of excellence.

Leroy also says that when swallows are hatched out too late to migrate with the older birds, the instinct of migration is not sufficiently imperative to induce them to undertake the journey by themselves. 'They perish, the victims of their ignorance, and of the tardy birth which made them unable to follow their parents.'

Cuckoo.

Perhaps the strangest of the special instincts manifested by birds is that of the cuckoo laying its eggs in the nests of other birds. As the subject is an important one from several points of view, I shall consider it at some length.

It must first be observed that the parasitic habit in question is not practised by all species of the genus—the American cuckoo, for instance, being well known to build its nest and rear its young in the ordinary manner. The Australian species, however, manifests the same instinct as the European. The first observer of the habit practised by the European cuckoo was the illustrious Jenner, who published his account in the 'Philosophical Transactions.'[1] From this account the following is an extract:—

The cuckoo makes choice of the nests of a great variety of small birds. I have known its eggs entrusted to the care of the hedge-sparrow, water-wagtail, titlark, yellowhammer, green linnet, and winchat. Among these it generally selects the three former, but shows a much greater partiality to the hedge-sparrow than to any of the rest; therefore, for the purpose of avoiding confusion, this bird only, in the following account, will be considered as the foster-parent of the cuckoo, except in instances which are particularly specified.

When the hedge-sparrow has sat her usual time, and disengaged the young cuckoo and some of her own offspring from the shell,[2] her own young ones, and any of her eggs that remain unhatched, are soon turned out, the young cuckoo remaining

---

[1] *Phil. Trans.*, vol. lxxviii., p. 221 *et seq.*
[2] The young cuckoo is generally hatched first.

possessor of the nest, and sole object of her future care. The young birds are not previously killed, nor are the eggs demolished, but all are left to perish together, either entangled about the bush which contains the nest, or lying on the ground under it.

On June 18, 1787, I examined the nest of a hedge-sparrow, which then contained a cuckoo's and three hedge-sparrow's eggs. On inspecting it the day following, I found the bird had hatched, but that the nest now contained a young cuckoo and only one young hedge-sparrow. The nest was placed so near the extremity of a hedge, that I could distinctly see what was going forward in it; and, to my astonishment, saw the young cuckoo, though so newly hatched, in the act of turning out the young hedge-sparrow.

The mode of accomplishing this was very curious. The little animal, with the assistance of its rump and wings, contrived to get the bird upon its back, and making a lodgment for the burden by elevating its elbows, clambered backward with it up the side of the nest till it reached the top, when, resting for a moment, it threw off its load with a jerk, and quite disengaged it from the nest. It remained in this situation a short time, feeling about with the extremities of its wings, as if to be convinced whether this business was properly executed, and then dropped into the nest again. With these (the extremities of its wings) I have often seen it examine, as it were, an egg and nestling before it began its operations; and the sensibility which these parts appeared to possess seemed sufficiently to compensate the want of sight, which as yet it was destitute of. I afterwards put in an egg, and this by a similar process was conveyed to the edge of the nest and thrown out. These experiments I have since repeated several times in different nests, and have always found the young cuckoo disposed to act in the same manner. In climbing up the nest it sometimes drops its burden, and thus is foiled in its endeavours; but after a little respite the work is resumed, and goes on almost incessantly till it is effected. It is wonderful to see the extraordinary exertions of the young cuckoo, when it is two or three days old, if a bird be put into the nest with it that is too weighty for it to lift out. In this state it seems ever restless and uneasy. But this disposition for turning out its companions begins to decline from the time it is two or three till it is about twelve days old, when, as far as I have hitherto seen, it ceases. Indeed, the disposition for throwing out the egg appears to cease a few days sooner; for I have frequently seen the young cuckoo, after it had been hatched nine or ten days, remove a nestling that had been placed in the nest with it, when it suffered an egg, put there at the same time, to remain unmolested. The singularity of its shape is well adapted to these purposes; for, different from other newly hatched birds, its back from the scapulæ downwards is very broad, with a considerable depression in the middle. This depression seems formed by nature for the design of giving a more secure lodgment to the egg of the hedge-sparrow, or its young one, when the young cuckoo is employed in removing either of them from the nest. When it is about twelve days old this cavity is quite filled up, and then the back assumes the shape of nestling birds in general. . . . The circumstance of the young cuckoo being destined by nature to throw out the young hedge-sparrows seems to account for the parent cuckoo dropping her egg in the nests of birds so small as those I have particularised. If she were to do this in the nest of a bird which produced a large egg, and consequently a large nestling, the young cuckoo would probably find an insurmountable difficulty in solely possessing the nest, as its exertions would be unequal to the labour of turning out the young birds. (I have known a case in which a hedge-sparrow sat upon a cuckoo's egg and one of her own. Her own egg was hatched five days before the cuckoo's, when the young hedge-sparrow had gained such a superiority in size that the young cuckoo had not powers sufficient to lift it out of the nest till it was two days old, by which time it had grown very considerably. This egg was probably laid by the cuckoo several days after the hedge-sparrow had begun to sit; and even in this case it appears that its presence had created the disturbance before alluded to, as all the hedge-sparrow's eggs had gone except one.) . . . June 27, 1787.—Two cuckoos and a hedge-sparrow were hatched in the same nest this morning; one hedge sparrow's egg remained unhatched. In a few hours after, a contest began between the cuckoos for the possession of the nest, which continued undetermined till the next afternoon; when one of them, which was somewhat superior in size, turned out the other, together with the young hedge-sparrow and the unhatched egg. This contest was very remarkable. The combatants alternately appeared to have the advantage, as each carried the other several times nearly to the top of the nest, and then sunk down again oppressed with the weight of its burden; till at length, after various efforts, the strongest prevailed, and was afterwards brought up by the hedge-sparrows.

To what cause, then, may we attribute the singularities of

the cuckoo? May they not be owing to the following circumstances,—the short residence this bird is allowed to make in the country where it is destined to propagate its species, and the call that nature has upon it, during that short residence, to produce a numerous progeny? The cuckoo's first appearance here is about the middle of April, commonly on the 17th. Its egg is not ready for incubation till some weeks after its arrival, seldom before the middle of May. A fortnight is taken up by the sitting bird in hatching the egg. The young bird generally continues three weeks in the nest before it flies, and the foster-parents feed it more than five weeks after this period; so that, if a cuckoo should be ready with an egg much sooner than the time pointed out, not a single nestling, even one of the earliest, would be fit to provide for itself before its parent would be instinctively directed to seek a new residence, and be thus compelled to abandon its young one; for old cuckoos take their final leave of this country the first week in July.

Had nature allowed the cuckoo to have stayed here as long as some other migrating birds, which produce a single set of young ones (as the swift or nightingale, for example), and had allowed her to have reared as large a number as any bird is capable of bringing up at one time, there might not have been sufficient to have answered her purpose; but by sending the cuckoo from one nest to another, she is reduced to the same state as the bird whose nest we daily rob of an egg, in which case the stimulus for incubation is suspended.

A writer in 'Nature' (vol. v., p. 383; and vol. ix., p. 123), to whom Mr. Darwin refers in the latest edition of 'The Origin of Species' as an observer that Mr. Gould has found trustworthy, precisely confirms, from observations of his own, the above description of Jenner. So far, therefore, as the observations are common I shall not quote his statements; but the following additional matter is worth rendering:—

But what struck me most was this: the cuckoo was perfectly naked, without a vestige of a feather or even a hint of future feathers; its eyes were not yet opened, and its neck seemed too weak to support the weight of its head. The pipits (in whose nest the young cuckoo was parasitic) had well-developed quills on the wings and back, and had bright eyes partially open; yet they seemed quite helpless under the manipulations of the cuckoo, which looked a much less developed creature. The cuckoo's legs, however, seemed very muscular, and it appeared to feel about with its wings, which were absolutely featherless, as with hands—the 'spurious wing' (unusually large in proportion) looking like a spread-out thumb. The most singular thing of all was the direct purpose with which the blind little monster made for the open side of the nest, the only part where it could throw its burden down the bank. [The latter remark has reference to the position of the nest below a heather bush, on the declivity of a low abrupt bank, where the only chance of dislodging the young birds was to eject them over the side of the nest remote from its support upon the bank.] As the young cuckoo was blind, it must have known the part of the nest to choose by feeling from the inside that that part was unsupported.

Such being the facts, we have next to ask how they are to be explained on the principles of evolution. At first sight it seems that although the habit saves the bird which practises it much time and trouble, and so is clearly of benefit to the individual, it is not so clear how the instinct is of benefit to the species; for as cuckoos are not social birds, and therefore cannot in any way depend on mutual co-operation, it is difficult to see that this saving of time and trouble to the individual can be of any use to the species. But Jenner seems to have hit the right cause in the concluding part of the above quotation. If it is an advantage that the cuckoo should migrate early, it clearly becomes an advantage, in order to admit of this, that the habit should be formed of leaving her eggs for other birds to incubate. At any rate, we have here a sufficiently probable explanation of the *raison d'être* of this curious instinct; and whether it is the true reason or the only reason, we are justified in setting down the instinct to the creating influence of natural selection.

Mr. Darwin, in his 'Origin of Species,' has some interesting remarks to make on this subject. First, he is informed by Dr. Merrell that the American cuckoo, although as a rule following the ordinary custom of birds in incubating her own eggs, nevertheless occasionally deposits them in the nests of other birds.

Now let us suppose that the ancient progenitor of our European cuckoo had the habits of the American cuckoo, and that she occasionally laid her egg in another bird's nest. If the old bird profited by this occasional habit through being able to migrate earlier, or through any other cause; or if the young were made more vigorous by advantage being taken of the mistaken instinct of another species than when reared by their own mother, encumbered as she could hardly fail to be by having eggs and young at the same time;[1] then the old birds or the fostered young would gain an advantage.[2]

The instinct would seem to be a very old one, for there are two great changes of structure in the European cuckoo which are manifestly correlated with the instinct. Thus, the shape of the young bird's back has already been noted; and not less remarkable than this is the small size of the egg from which the young bird is hatched. For the egg of the cuckoo is not any larger than that of the skylark, although an adult cuckoo is four times the size of an adult skylark. And 'that the small size of the egg is a real case of adaptation (in order to deceive the small birds in whose nests it is laid), we may infer from the fact of the non-parasitic American cuckoo laying full-sized eggs.' Yet, although the instinct in question is doubtless of high antiquity, there have been occasional instances observed in cuckoos of reversion to the ancestral instinct of nidification; for, according to Adolf Müller, 'the cuckoo occasion-

---

[1] Allusion is here made to the fact that the cuckoo lays her eggs at intervals of two or three days, and therefore that if all were incubated by the mother, they would hatch out at different times—a state of things which actually obtains in the case of the American cuckoo, whose nest contains eggs and young at the same time.

[2] It is worth while to observe, as bearing on this theory of the origin of this parasitic habit, that even non-parasitic birds occasionally deposit their eggs in nests of other birds. Thus, Professor A. Newton writes in his admirable essay on 'Birds' in the Encyclopædia Britannica, 'Certain it is that some birds, whether by mistake or stupidity, do not unfrequently lay their eggs in the nests of others. It is within the knowledge of many that pheasants' eggs and partridges' eggs are often laid in the same nest; and it is within the knowledge of the writer that gulls' eggs have been found in the nests of eider-ducks, and *vice versâ*; that a redstart and a pied flycatcher will lay their eggs in the same convenient hole—the forest being rather deficient in such accommodation; that an owl and a duck will resort to the same nest-hole, set up by the scheming woodman for his own advantage; and that the starling, which constantly dispossesses the green woodpecker, sometimes discovers that the rightful heir of the domicile has to be brought up by the intruding tenant.'

ally lays her eggs on the bare ground, sits on them, and feeds her young.'

In 'Nature' for November 18, 1869, Professor A. Newton, F.R.S., has published an article on a somewhat obscure point connected with the instincts of the cuckoo. He says that Dr. Baldamus has satisfied him, by an exhibition of sixteen specimens of cuckoos' eggs found in the nests of different species of birds, 'that the egg of the cuckoo is approximately coloured and marked like those of the bird in whose nest it is found,' for the purpose, no doubt, of deceiving the foster-parents. Professor Newton adds, however :—

Having said this much, and believing as I do the Doctor to be partly justified in the carefully worded enunciation of what he calls a 'law of nature,' I must now declare that it is only 'approximately,' and by no means *universally* true that the cuckoo's egg is coloured like those of the victims of her imposition, &c.

Still, when so great an authority as Professor Newton expresses himself satisfied that there is a marked *tendency* to such imitation, which in some cases leads to extraordinary variations in the colouring of the cuckoo's egg, the alleged fact becomes one which demands notice. The question, of course, immediately arises, How is it conceivable that the fact, if it is a fact, can be explained? We cannot imagine the cuckoo to be able consciously to colour her egg during its formation in order to imitate the eggs among which she is about to lay it; nor can we suppose that having laid an egg and observed its colouring, she then carries it to the nest of the bird whose eggs it most resembles. Professor Newton suggests another theory, which he seems to think sufficient, but which I confess seems to me little more satisfactory than the impossible theories just stated. He says :—

Only one explanation of the process can, to my mind, be offered. Every person who has studied the habits of birds with sufficient attention will be conversant with the tendency which certain of those habits have to become hereditary. It is, I am sure, no violent hypothesis to suppose that there is a very reasonable probability of each cuckoo most commonly placing her eggs in the nest of the same bird, and of this habit being transmitted to her posterity.

Now it will be seen that it requires but only an application to this case of the principle of 'natural selection,' or 'survival of the fittest,' to show that if my argument be sound, nothing can be more likely than that, in the course of time, that principle should operate so as to produce the facts asserted, the eggs which best imitated those of particular foster-parents having the best chance of duping the latter, and so of being hatched out.

Now, granting to this hypothesis the assumption that individual cuckoos have special predilections as to the species in whose nests they are to lay their eggs, and that some of these species require to be deceived by imitative colouring of the egg to prevent their tilting it out, there is still an enormous difficulty to be met. Supposing that one cuckoo out of a hundred happens to lay eggs sufficiently like those of the North African magpies (a species alluded to by Professor Newton) to deceive the latter into supposing the egg to be one of their own. This I cannot think is too small a proportion to assume, seeing that, *ex hypothesi*, the resemblance must be tolerably close, and that the egg of the magpie does not resemble the great majority of eggs of the cuckoo. Now, in order to sustain the theory, we must suppose that the particular cuckoo which happens to have the peculiarity of laying eggs so closely resembling those of the magpie, must also happen to have the peculiarity of desiring to lay its eggs in the nest of a magpie. The conjunction of these two peculiarities would, I should think, at a moderate estimate reduce the chances of an approximately coloured egg being laid in the appropriate nest to at least one thousand to one. But supposing the happy accident to have taken place, we have next to suppose that the peculiarity of laying these exceptionably coloured eggs is not only constant for the same individual cuckoo, but is inherited by innumerable generations of her progeny; and, what is much more difficult to grant, that the fancy for laying eggs in the nest of a magpie is similarly inherited. I think, therefore, notwithstanding Professor Newton's strong opinion upon the subject, that the ingenious hypothesis must be dismissed as too seriously encumbered by the difficulties which I have mentioned. We may with philosophical safety invoke the influence of natural selection to explain all cases of protective colouring when the *modus operandi* need only be supposed simple and direct; but in a case such as this the number and complexity of the conditions that would require to meet in order to give natural selection the possibility of entrance, seem to me much too considerable to admit of our entertaining the possibility of its action—at all events in the way that Professor Newton suggests. Therefore, if the facts are facts, I cannot see how they are to be explained.

Cuckoos are not the only birds which manifest the parasitic habit of laying their eggs in other birds' nests.

Some species of *Melothrus*, a widely distinct genus of American birds, allied to our starlings, have parasitic habits like those of the cuckoo; and the species present an interesting gradation in the perfection of their instincts. The sexes of *Melothrus cadius* are stated by an excellent observer, Mr. Hudson, sometimes to live promiscuously together in flocks and sometimes to pair. They either build a nest of their own, or seize on one belonging to some other bird, occasionally throwing out the nestlings of the stranger. They either lay their eggs in the nest thus appropriated, or oddly enough build one for themselves on the top of it. They usually sit on their own eggs and rear their own young; but Mr. Hudson says it is probable that they are occasionally parasitic, for he has seen the young of this species feeding old birds of a distinct kind and clamouring to be fed by them. The parasitic habits of another species of *Melothrus*, the *M. Canariensis*, are much more highly developed than those of the last, but are still far from perfect. This bird, as far as it is known, invariably lays its eggs in the nests of strangers, but it is remarkable that several together sometimes commence to build an irregular untidy nest of their own, placed in singularly ill-adapted situations, as on the leaves of a large thistle. They must, however, as far as Mr. Hudson has ascertained, complete a nest for themselves. They often lay so many eggs, from fifteen to twenty, in the same foster-nest, that few or none can possibly be hatched. They have, moreover, the extraordinary habit of pecking holes in the eggs, whether of their own species or of their foster-parents, which they find in the appropriated nests. They drop also many eggs on the bare ground, which are thus wasted. A third species, the *M. precius* of North America, has acquired instincts as perfect as those of the cuckoo, for it never lays more than an egg in a foster-nest, so that the young bird is securely reared. Mr. Hudson is a strong disbeliever in evolution, but he appears to have been so much struck by the imperfect instincts of the *Melothrus Canariensis* that he quotes my words, and asks, 'Must we consider these habits not as especially endowed or created instincts, but as small consequences of one general law, namely transition?'[1]

Such are all the facts and considerations which I have to present with reference to the curious instinct in question. It will be seen that—with one doubtful or not sufficiently investigated exception, viz., that of cuckoos adapting the colour of their eggs to that of the eggs of the foster-parents—there is nothing connected with these instincts that presents any difficulty to the theory of evolution. We may, perhaps, at first sight wonder why some counteracting instinct should not have been developed by the same agency in the birds which are liable to be thus duped; but here we must remember that the deposition of a parasitic egg is, comparatively speaking, an exceedingly rare event, and therefore not one that is likely to lead to the development of a special instinct to meet it.

General Intelligence.

Under this heading I shall here, as in the case of this heading elsewhere, string together all the instances which I have met with, and which I deem trustworthy, of the display of unusually high intelligence in the class, family, order, or species of animals under consideration—the object of this heading in all cases being that of supplying, by the facts mentioned beneath it, a general idea of the upper limit of intelligence which is distinctive of each group of animals.

That birds recognise their own images in mirrors as birds there can be no question. Houzeau, who records observations of his own in this connection with parrots,[2] adds that dogs are more difficult to deceive by mirrors in

---

[1] *Origin of Species*, p. 215. [2] Tom. i., p. 130.

this way than birds, on account of their depending so much upon smell for their information. No doubt individual differences are to be met with in animals of both classes, and much depends on previous experience. Young dogs, or dogs which have never seen a mirror before, are not, as a rule, difficult to deceive, even though they have good noses. I myself had a setter with an excellent nose, who on many repeated occasions tried to fight his own image, till he found by experience that it was of no use. As to birds, I have seen canaries suppose their own images to be other canary birds, and also the reflection of a room to be another room—the birds flying against a large mirror and falling half stunned. I mention the latter circumstance because it afforded evidence of the superior intelligence of a linnet, which on the same occasion dashed itself against the mirror once, but never a second time, while the canaries did so repeatedly.

Mrs. Frankland, in 'Nature'(xxi., p. 82), gives the following account of a bullfinch paying more attention to a portrait of a bullfinch than to his own image in a mirror, which is certainly remarkable; and as the fact seems to have been observed repeatedly, it can scarcely be discredited:

> The following is a curious instance of discrimination which I have observed in my bullfinch. He is in the habit of coming out of his cage in my room in the morning. In this room there is a mirror with a marble slab before it, and also a very cleverly executed water-colour drawing of a hen bullfinch, life size. The first thing that my bullfinch does on leaving his cage is to fly to the picture (perching on a vase just below it) and pipe his tune in the most insinuating manner, accompanied with much bowing to the portrait of the hen bullfinch. After having duly paid his addresses to it, he generally spends some time on the marble slab in front of the looking-glass, but without showing the slightest emotion at the sight of his own reflection, or courting it with a song. Whether this perfect coolness is due to the fact of the reflection being that of a cock bird, or whether (since he shows no desire to fight the reflected image) he is perfectly well aware that he only sees himself, it is difficult to say.

That birds possess considerable powers of imagination, or forming mental pictures of absent objects, may be inferred from the fact of their pining for absent mates, parrots calling for absent friends, &c. The same fact is further proved by birds dreaming, a faculty which has been noticed by Cuvier, Jerdon, Thompson, Bennet, Houzeau, Bechstein, Lindsay, and Darwin.[1]

The facility with which birds lend themselves to the education of the show-man is certain evidence of considerable docility, or the power of forming novel associations of ideas. Thus, according to Bingley,—

Some years ago the Sieur Roman exhibited in this country the wonderful performances of his birds. These were goldfinches, linnets, and canary birds. One appeared dead, and was held up by the tail or claw without exhibiting any signs of life. A second stood on its head, with its claws in the air, &c., &c.[2]

And many years ago there was exhibited a very puzzling automaton, which, although of very small size and quite isolated from any possibly mechanical connection with its designer, performed certain movements in any order that the fancy of the observers might dictate. The explanation turned out to be that within the mechanism of the figure there was a canary bird which had been taught to run in different directions at different words or tones of command, so by its weight starting the mechanism to perform the particular movement required.

The rapidity with which birds learn not to fly against newly erected telegraph wires, displays a large amount of observation and intelligence. The fact has been repeatedly observed. For instance, Mr. Holden says:—

About twelve years ago I was residing on the coast of county Antrim, at the time the telegraph wires were set up along that charming road which skirts the sea between Larne and Cushendall. During the winter months large flocks of starlings always migrated over from Scotland, arriving in the early morning. The first winter after the wires were stretched along the coast I frequently found numbers of starlings lying dead or wounded on the road-side, they having evidently in their flight in the dusky morn struck against the telegraph wires, not

---

[1] See *Birds of India*, i., p. 21; *Passions of Animals*, p. 60; *Fac. Men. des Ani.*, tom. ii., p. 183; *Mind in Lower Animals*, vol. ii., p. 96; and *Descent of Man*, p. 74.

[2] *Animal Biography*, vol. ii., p. 173.

blown against them, as these accidents often occurred when there was but little wind. I found that the peasantry had come to the conclusion that these unusual deaths were due to the flash of the telegraph messages killing any starlings that happened to be perched on the wires when working. Strange to say that throughout the following and succeeding winters hardly a death occurred among the starlings on their arrival. It would thus appear that the birds were deeply impressed, and understood the cause of the fatal accidents among their fellow-travellers the previous year, and hence carefully avoided the telegraph wires; not only so, but the young birds must also have acquired this knowledge and perpetuated it, a knowledge which they could not have acquired by experience or even instinct, unless the instinct was really inherited memory derived from the parents whose brains were first impressed by it.[1]

Similar facts are given in Buckland's 'Curiosities of Natural History,'[2] and I have myself known of a case in Scotland where a telegraph was erected across a piece of moorland. During the first season some of the grouse were injured by flying against the wires, but never in any succeeding season. Why the young birds should avoid them without having had individual experience may, I think, be explained by the consideration that in birds which fly in flocks or coveys, it is the older ones that lead the way. This explanation would not, of course, apply to birds which fly singly; but I am not aware that any observations have gone to show that the young of such birds avoid the wires.

I quote the following exhibition of intelligence in an eagle from Menault:—

The following account of the patience with which a golden eagle submitted to surgical treatment, and the care which it showed in the gradual use of the healing limb, must suggest the idea that something very near to prudence and reason existed in the bird. This eagle was caught in a fox-trap set in the forest of Fontainebleau, and its claw had been terribly torn. An operation was performed on the limb by the conservators of the Zoological Gardens at Paris, which the noble bird bore with a rational patience. Though his head was left loose, he made no attempts to interfere with the agonising extraction of the

[1] *Nature*, xx., p. 266. [2] Vol. i., p. 216. See also *Descent of Man*, p. 80

splinters, or to disturb the arrangements of the annoying bandages. He seemed really to understand the nature of the services rendered, and that they were for his good.[1]

Speaking of the Urubu vultures, Mr. Bates says:—

They assemble in great numbers in the villages about the end of the wet season, and are then ravenous with hunger. My cook could not leave the open kitchen at the back of the house for a moment whilst the dinner was cooking, on account of their thievish propensities. Some of them were always loitering about, watching their opportunity, and the instant the kitchen was left unguarded, the bold marauders marched in and lifted the lids of the saucepans with their beaks to rob them of their contents. The boys of the village lie in wait, and shoot them with bow and arrow; and vultures have consequently acquired such a dread of these weapons, that they may be often kept off by hanging a bow from the rafters of the kitchen.[2]

Mrs. Lee, in her 'Anecdotes,' says that one day her gardener was struck by the strange conduct of a robin, which the man had often fed. The bird fluttered about him in so strange a manner—now coming close, then hurrying away, always in the same direction—that the gardener followed its retreating movements. The robin stopped near a flower-pot, and fluttered over it in great agitation. It was soon found that a nest had been formed in the pot, and contained several young. Close by was a snake, intent, doubtless, upon making a meal of the brood.

The following appeared in the 'Gardener's Chronicle' for Aug. 3, 1878, under the initials 'T. G.' I wrote to the editor requesting him to supply me with the name of his correspondent, and also to state whether he knew him to be a trustworthy man. In reply the editor said that he knew his correspondent to be trustworthy, and that his name is Thomas Guring:—

About thirty years ago the small market town in which I reside was skirted by an open common, upon which a number of geese were kept by cottagers. The number of the birds was very great. . . . Our corn market at that time was held in the street in front of the principal inn, and on the market day a good deal of corn was scattered from sample bags by millers. Somehow the geese found out about the spilling of corn, and they appear to have held a consultation upon the subject. . . .

---

[1] Menault, *Wonders of Instinct*, p. 132.
[2] *Nat. on Amazons* p. 177; *Anecdotes*, p. 135.

From this time they never missed their opportunity, and the entry of the geese was always looked for and invariably took place. On the morning after the market, early, and always on the proper morning, fortnightly, in they came cackling and gobbling in merry mood, and they never came on the wrong day. The corn, of course, was the attraction, but in what manner did they mark the time? One might have supposed that their perceptions were awakened on the market day by the smell of corn, or perhaps by the noise of the market traffic; but my story is not yet finished, and its sequel is against this view. It happened one year that a day of national humiliation was kept, and the day appointed was that on which our market should have been held. The market was postponed, and the geese for once were baffled. There was no corn to tickle their olfactory organs from afar, no traffic to appeal to their sense of hearing. I think our little town was as still as it usually is on Sundays. . . . The geese should have stopped away; but they knew their day, and came as usual. . . . I do not pretend to remember under what precise circumstances the habit of coming into the street was acquired. It may have been formed by degrees, and continued from year to year; but how the old birds, who must have led the way, marked the time so as to come in regularly and fortnightly, on a particular day of the week, I am at a loss to conceive.

Livingstone's 'Expedition to the Zambesi, 1865,' p. 209, gives a conclusive account of the bird called the honey-guide, which leads persons to bees' nests. 'They are quite as anxious to lure the stranger to the bees' hive as other birds are to draw him away from their own nests.' The object of the bird is to obtain the pupæ of the bees which are laid bare by the ravaging of the nest. The habits of this bird have long been known and described in books on popular natural history; but it is well that the facts have been observed by so trustworthy a man as Livingstone. He adds, 'How is it that members of this family have learned that all men, white and black, are fond of honey?' We can only answer, by intelligent observation in the first instance, passing into individual and hereditary habit, and so eventually into a fixed instinct.

Brehm relates an instance of cautious sagacity in a pewit. He had placed some horsehair snares over its nest, but the bird seeing them, pushed them aside with her bill. Next day he set them thickly round the nest; but now the bird, instead of running as usual to the nest along the ground, alighted directly upon it. This shows a considerable appreciation of mechanical appliances, as does also the following.

Mrs. G. M. E. Campbell writes to me:—

At Ardglass, co. Down, Ireland, is a long tract of turf coming to the edge of the rocks overhanging the sea, where cattle and geese feed; at a barn on this tract there was a low enclosure, with a door fastening by a hook and staple to the side-post: when the hook was out of the staple, the door fell open by its own weight. I one day saw a goose with a large troop of goslings coming off the turf to this door, which was secured by the hook being in the staple. The goose waited for a minute or two, as if for the door to be opened, and then turned round as if to go away, but what she did was to make a rush at the door, and making a dart with her beak at the point of the hook nearly threw it out of the staple; she repeated this manœuvre, and succeeded at the third attempt, the door fell open, and the goose led her troop in with a sound of triumphant chuckling. How had the goose learned that the force of the rush was needful to give the hook a sufficient toss?

Mrs. K. Addison sends me the following instance of the use of signs on the part of an intelligent jackdaw. The bird was eighteen months old, and lived in some bushes in Mrs. Addison's garden. She writes:—

I generally made a practice of filling a large basin which stands under the trees every morning for Jack's bath. A few days ago I forgot this duty, and was reminded of the fact in a very singular manner. Another of my daily occupations is to open my dressing-room shutters about eleven o'clock of a morning. Now these said shutters open almost on to the trees where Jack lives. The day I forgot his bath, when I opened the shutters I found my little friend waiting just outside them, as though he knew that he should see me there; and when he did he placed himself immediately in front of me, and then shook himself and spread out his wings just as he always does in his bath. The action was so suggestive and so unmistakable, that I spoke just as I would have done to a child—'Oh yes, Jack, of course you shall have some water.'

Mr. W. W. Nichols writes to 'Nature:'—

The Central Prison at Agra is the roosting-place of great numbers of the common blue pigeon; they fly out to the neighbouring country for food every morning, and return in the evening, when they drink at a tank just outside the prison walls. In this tank are a large number of fresh-water turtles, which lie in wait for the pigeons just under the surface of the water and at the edge of it. Any bird alighting to drink near one of these turtles has a good chance of having its head bitten off and eaten; and the headless bodies of pigeons have been picked up near the water, showing the fate which has sometimes befallen the birds. The pigeons, however, are aware of the danger, and have hit on the following plan to escape it. A pigeon comes in from its long flight, and, as it nears the tank, instead of flying down at once to the water's edge, will cross the tank at about twenty feet above its surface, and then fly back to the side from which it came, apparently selecting for alighting a safe spot which it had remarked as it flew over the bank; but even when such a spot has been selected the bird will not alight at the edge of the water, but on the bank about a yard from the water, and will then run down quickly to the water, take two or three hurried gulps of it, and then fly off to repeat the same process at another part of the tank till its thirst is satisfied. I had often watched the birds doing this, and could not account for their strange mode of drinking till told by my friend the superintendent of the prison, of the turtles which lay in ambush for the pigeons.

As a still more remarkable instance of the display of intelligence by a bird of this species, I shall quote the following observation of Commander R. H. Napier, also published in 'Nature' (viii., p. 324):—

A number of them (pouters) were feeding on a few oats that had been accidentally let fall while fixing the nose-bag on a horse standing at bait. Having finished all the grain at hand, a large 'pouter' rose, and flapping its wings furiously, flew directly at the horse's eyes, causing the animal to toss his head, and in doing so, of course shake out more corn. I saw this several times repeated—in fact, whenever the supply on hand had been exhausted. . . . Was not this something more than instinct?

The following display of intelligence on the part of swallows is communicated to me by Mr. Charles Wilson.

It can scarcely be attributed to accident, and does not admit of mal-observation. My informant says:—

Two swallows were building a nest in the verandah of a house in Victoria, but as their nest was resting partly on a bell-wire, it was by this means twice pulled down. They then began afresh, making a tunnel through the lower part of the nest, through which the wire was able to act without doing damage.

Another gentleman writes me of another use to which he has observed swallows put the artifice of building tunnels. Being molested by sparrows which desired to take forcible possession of their nest, a pair of swallows modified the entrance of the latter, so that instead of opening by a simple hole under the eaves of a house, it was carried on in the form of a tunnel.

Linnæus says that the martin, when it builds under the eaves of houses, sometimes is molested by sparrows taking possession of the nest. The pair of martins to which the nest belongs are not strong enough to dislodge the invaders; but they convoke their companions, some of whom guard the captives, whilst others bring clay, close up the entrance of the nest, and leave the sparrows to die miserably. This account has been to a large extent independently confirmed by Jesse, who seems not to have been acquainted with the statement of Linnæus. He writes:—

Swallows seem to entertain the recollection of injury, and to resent it when an opportunity offers. A pair of swallows built their nest under the ledge of a house at Hampton Court. It was no sooner completed than a couple of sparrows drove them from it, notwithstanding the swallows kept up a good resistance, and even brought others to assist them. The intruders were left in peaceable possession of the nest, till the two old birds were obliged to quit it to provide food for their young. They had no sooner departed than several swallows came and broke down the nest; and I saw the young sparrows lying dead on the ground. As soon as the nest was demolished, the swallows began to rebuild it.[1]

The same author gives the following and somewhat similar case:—

[1] *Gleanings*, vol. ii., p. 96.

A pair of swallows built their nest against one of the first-floor windows of an uninhabited house in Merrion Square, Dublin. A sparrow, however, took possession of it, and the swallows were repeatedly seen clinging to the nest, and endeavouring to gain an entrance to the abode they had erected with so much labour. All their efforts, however, were defeated by the sparrow, who never once quitted the nest. The perseverance of the swallows was at length exhausted: they took flight, but shortly afterwards returned, accompanied by a number of their congeners, each of them having a piece of dirt in its bill. By this means they succeeded in stopping up the hole, and the intruder was immured in total darkness. Soon afterwards the nest was taken down and exhibited to several persons, with the dead sparrow in it. In this case there appears to have been not only a reasoning faculty, but the birds must have been possessed of the power of communicating their resentment and their wishes to their friends, without whose aid they could not thus have avenged the injury they had sustained.[1]

That birds sometimes act in concert may also be gathered from the following observations recorded by Mr. Buck:—

I have constantly seen a flock of pelicans, when on the feed, form a line across a lake, and drive the fish before them up its whole length, just as fishermen would with a net.[2]

The following is extracted from Sir E. Tennent's 'Natural History of Ceylon,' and displays remarkable intelligence on the part of the crows in that island:—

One of these ingenious marauders, after vainly attitudinising in front of a chained watch-dog, that was lazily gnawing a bone, and after fruitlessly endeavouring to divert his attention by dancing before him, with head awry and eye askance, at length flew away for a moment, and returned bringing a companion which perched itself on a branch a few yards in the rear. The crow's grimaces were now actively renewed, but with no better success, till its confederate, poising itself on its wings, descended with the utmost velocity, striking the dog upon the spine with all the force of its strong beak. The *ruse* was successful; the dog started with surprise and pain, but not quickly enough to seize his assailant, whilst the bone he had been gnawing was snatched away by the first crow the

[1] *Ibid*, p. 99.   [2] *Nature*, vol. xiii., p. 303.

instant his head was turned. Two well-authenticated instances of the recurrence of this device came within my knowledge at Colombo, and attest the sagacity and powers of communication and combination possessed by these astute and courageous birds.

This account, which would be difficult of credence if narrated by a less competent author, is strikingly confirmed by an independent observation on the crows of Japan, which has recently been published by Miss Bird, in whose words I shall render it. She writes:—

In the inn garden I saw a dog eating a piece of carrion in the presence of several of these covetous birds. They evidently said a great deal to each other on the subject, and now and then, one or two of them tried to pull the meat away from him, which he resented. At last a big strong crow succeeded in tearing off a piece, with which he returned to the pine where the others were congregated, and after much earnest speech they all surrounded the dog, and the leading bird dexterously dropped the small piece of meat within reach of his mouth, when he immediately snapped at it, letting go the big piece unwisely for a second, on which two of the crows flew away with it to the pine, and with much fluttering and hilarity they all ate, or rather gorged it, the deceived dog looking vacant and bewildered for a moment, after which he sat under the tree and barked at them inanely. A gentleman told me that he saw a dog holding a piece of meat in like manner in the presence of three crows, which also vainly tried to tear it from him, and after a consultation they separated, two going as near as they dared to the meat, while the third gave the tail a bite sharp enough to make the dog turn round with a squeak, on which the other villains seized the meat, and the three fed triumphantly upon it on the top of a wall.[1]

These two independent statements by competent observers of such similar exhibitions of intelligence by crows, justifies us in accepting the fact, remarkable though it be. As further corroboration, however, I shall quote still another independent and closely similar observation, which I find in a letter to me from Sir J. Clarke Jervoise, who says, while writing of rooks which he has observed in England:—

[1] *Unbeaten Tracks in Japan*, vol. ii., pp. 149-50.

A pheasant used to come very boldly and run off with large pieces of food, which he could only divide by shaking, and he was closely watched by the rooks for the pieces that flew out of his reach. He learned to run off into the shrubs, followed by the rooks, who pulled his tail to make him drop his food.

I shall next quote a highly interesting observation which seems to have been well made, and which displays remarkable intelligence on the part of the birds described. These are Turnstones, which, as their name implies, turn over stones, &c., in order to obtain as food the sundry small creatures concealed beneath. In this case the observer was Edward. Being concealed in a hollow, and unnoticed by the birds, he saw a pair trying to turn over the body of a stranded cod-fish, three and a half feet long, and buried in the sand to a depth of several inches. He thus describes what he saw:—

Having got fairly settled down in my pebbly observatory, I turned my undivided attention to the birds before me. They were boldly pushing at the fish with their bills, and then with their breasts. Their endeavours, however, were in vain: the object remained immovable. On this they both went round to the opposite side, and began to scrape away the sand from beneath the fish. After removing a considerable quantity, they again came back to the spot which they had left, and went once more to work with their bills and breasts, but with as little apparent success as formerly. Nothing daunted, however, they ran round a second time to the other side, and recommenced their trenching operations with a seeming determination not to be baffled in their object, which evidently was to undermine the dead animal before them, in order that it might be the more easily overturned.

While they were thus employed, and after they had laboured in this manner at both sides alternately for nearly half an hour, they were joined by another of their own species, which came flying with rapidity from the neighbouring rocks. Its timely arrival was hailed with evident signs of joy. I was led to this conclusion from the gestures which they exhibited, and from a low but pleasant murmuring noise to which they gave utterance so soon as the new-comer made his appearance. Of their feelings he seemed to be perfectly aware, and he made his reply to them in a similar strain. Their mutual congratulations being over, they all three set to work; and after labouring vigorously for a few minutes in removing the sand, they came round to the other side, and putting their breasts simultaneously to the fish, they succeeded in raising it some inches from the sand, but were unable to turn it over. It went down again into its sandy bed, to the manifest disappointment of the three. Resting, however, for a space, and without leaving their respective positions, which were a little apart the one from the other, they resolved, it appears, to give the work another trial. Lowering themselves, with their breasts pressed close to the sand, they managed to push their bills underneath the fish, which they made to rise about the same height as before. Afterwards, withdrawing their bills, but without losing the advantage which they had gained, they applied their breasts to the object. This they did with such force, and to such purpose, that at length it went over, and rolled several yards down a slight declivity. It was followed to some distance by the birds themselves before they could recover their bearing.[1]

I shall now bring this chapter to a close by presenting all the evidence that I have been able to collect with regard to the punishment of malefactors among rooks.

Goldsmith, who used constantly to observe a rookery from his window, says that the selection of a site for the building of a nest is a matter of much anxious deliberation on the part of a young crow couple; the male and female 'examining all the trees of a grove very attentively, and when they have fixed upon a branch that seems fit for their purpose, they continue to sit upon it, and observe it very sedulously for two or three days longer:'—

It often happens that the young couple have made choice of a place too near the mansion of an older pair, who do not choose to be incommoded by such troublesome neighbours; a quarrel, therefore, instantly ensues, in which the old ones are always victorious. The young couple, thus expelled, are obliged again to go through their fatigues—deliberating, examining, and choosing; and, having taken care to keep their due distance, the nest begins again, and their industry deserves commendation. But their activity is often too great in the beginning; they soon grow weary of bringing the materials of their nests from distant places, and they very early perceive that sticks may be provided nearer home, with less honesty indeed, but some degree of address. Away they go, therefore, to pilfer as fast as they can, and, whenever they see a nest unguarded, they

[1] Smiles, *Life of Edward*, pp. 244-6.

take care to rob it of the very choicest sticks of which it is composed. But these thefts never go unpunished, and probably, upon complaint being made, there is a general punishment inflicted. I have seen eight or ten rooks come upon such occasions, and, setting upon the new nest of the young couple, all at once tear it to pieces in a moment.

At length, however, the young pair find the necessity of going more regularly to work. While one flies to fetch the materials, the other sits upon the tree to guard it; and thus in the space of three or four days, with a skirmish now and then between, the pair have filled up a commodious nest, composed of sticks without, and of fibrous roots and long grass within. From the instant the female begins to lay, all hostilities are at an end; not one of the whole grove, that a little before treated her so rudely, will now venture to molest her, so that she brings forth her brood with perfect tranquillity. Such is the severity with which even native rooks are treated by each other; but if a foreign rook should attempt to make himself a denizen of their society, he would meet with no favour, the whole grove would at once be up in arms against him, and expel him without mercy.

Couch says ('Illustrations of Instinct,' p. 334 *et seq.*): —

The wrong-doers being discovered, the punishment is appropriate to the offence; by the destruction of their dishonest work they are taught that they who build must find their own bricks or sticks, and not their neighbours', and that if they wish to live in the enjoyment of the advantages of the social condition, they must endeavour to conform their actions to the principles of the rookery of which they have been made members.

It is not known what enormities led to the institution of another tribunal of the same kind, called the Crow Court, but according to Dr. Edmonson, in his 'View of the Shetland Islands,' its proceedings are as authoritative and regular, and it is remarkable as occurring in a species (*Corvus Cornice*) so near akin to the rook. The Crow Court is a sort of general assembling of birds who, in their usual habits, are accustomed to live in pairs, scattered at great distances from each other; when they visit the south or west of England, as they do in severe winters, they are commonly solitary. In their summer haunts in the Shetland Islands, numbers meet together from different points on a particular hill or field; and on these occasions the assembly is not complete, and does not begin its business for a day or two, till, all the deputies having arrived, a general clamour or croaking ensues, and the whole of the court, judges, barristers, ushers, audience, and all, fall upon the two or three prisoners at the bar, and beat them till they kill them. When this is accomplished the court breaks up and quietly disperses.

In the northern parts of Scotland (says Dr. Edmonson), and in the Faroe Islands, extraordinary meetings of crows are occasionally known to occur. They collect in great numbers, as if they had all been summoned for the occasion; a few of the flock sit with drooping heads, and others seem as grave as judges, while others again are exceedingly active and noisy; in the course of about one hour they disperse, and it is not uncommon, after they have flown away, to find one or two left dead on the spot. These meetings will sometimes continue for a day or two before the object, whatever it may be, is completed. Crows continue to arrive from all quarters during the session. As soon as they have all arrived, a very general noise ensues; and, shortly after, the whole fall upon one or two individuals, and put them to death. When the execution has been performed, they quietly disperse.

Similarly, the Bishop of Carlisle writes in the 'Nineteenth Century' for July 1881:—

I have seen also a jackdaw in the midst of a congregation of rooks, appa ently being tried for some misdemeanour. First Jack made a speech, which was answered by a general cawing of the rooks; this subsiding, Jack again took up his parable, and the rooks in their turn replied in chorus. After a time the business, whatever it was, appeared to be settled satisfactorily: if Jack was on his trial, as he seemed to be, he was honourably acquitted by acclamation; for he went to his home in the towers of Ely Cathedral, and the rooks also went their way.

Lastly, Major-General Sir George Le Grand Jacob, K.C.S.I., C.B., writes to me that while sitting in a verandah in India, he saw three or four crows come and perch on a neighbouring house. They then cawed continuously with such peculiar sound and vigour as to attract his attention. His account proceeds:—

Soon a gathering of crows from all quarters took place, until the roof of the guard-house was blackened by them. Thereupon a prodigious clatter ensued; it was plain that a 'palaver' was going forward. Some of its members, more eager than others, skipping about, I became much interested, and narrowly watched the proceedings, all within a dozen yards of me. After much cawing and clamour, the whole group suddenly rose into the air, and kept circling round half a dozen of their fellows, one of whom had been clearly told off for punishment, for the five repeatedly attacked it in quick succession, allowing no opportunity for their victim to escape, which he was trying to do, until they had cast him fluttering on the ground about thirty yards from my chair. Unfortunately I rushed forward to pick up the bird, prostrate but fluttering on the grass which was like a lawn before the building. I succeeded only in touching it, for it wriggled away from my grasp, and flew greatly crippled and close to the ground into the neighbouring bushes, where I lost sight of it. All the others, after circling round me and chattering, angrily as I thought, flew away, on my resuming my seat, in the direction taken by their victim.

[Since going to press I have seen, through the kindness of Mr. Seebohm, some specimens of cuckoo's eggs coloured in imitation of those belonging to the birds in the nests of which they are laid. There can be no question about the imitation, and I add this note to mitigate the criticism which I have passed upon Professor Newton's theory of the cause. For Mr. Seebohm has pointed out to me that the theory becomes more probable if we consider that a cuckoo reared in the nest of any particular bird is likely afterwards to choose a similar nest for the deposition of its own eggs. Whether or not the memory of a bird would thus act could only, of course, be certainly proved by experiment; but in view of the possibility that it may, Professor Newton's theory becomes more probable than it is if the selection of the appropriate nest is supposed to depend only on inheritance.

I must also add that Dr. Sclater has been kind enough to draw my attention to a remarkable description of a species of Bower-bird, published by Dr. Beccari in the *Gardener's Chronicle* for March 16, 1879. This species is called the Gardener Bower-bird (*Amblyornis niornata*), and inhabits New Guinea. The animal is about the size of a turtle-dove, and its bower—or rather hut—is built round the stem of a tree in the shape of a cone, with a space between the stem of the tree and the walls of the hut. The latter are composed of stems of an orchid with their leaves on—this particular plant being chosen by the birds apparently because its leaves remain long fresh. But the most extraordinary structure is the garden, which is thus described by Dr. Beccari:—' Before the cottage there is a meadow of moss. This is brought to the spot and left free from grass, stones, or anything which would offend the eye. On this green turf flowers and fruits of pretty colour are placed, so as to form an elegant little garden. The greater part of the decoration is collected round the entrance to the nest, and it would appear that the husband offers these his daily gifts to his wife. The objects are very various, but always of a vivid colour. There were some fruits of a Garcinia like a small-sized apple. Others were the fruits of Gardencias of a deep yellow colour in the interior. I saw also small rosy fruits, probably of a Scitamineous plant, and beautiful rosy flowers of a new Vaccinium. There were also fungi and mottled insects placed on the turf. As soon as the objects are faded they are moved to the back of the hut.' There is a fine-coloured plate of this bird in its garden, published in the *Birds of New Guinea*, by Mr. Gould Part ix., 1879.]

CHAPTER XI.

MAMMALS.

I SHALL devote this chapter to the psychology of all the Mammalia which present any features of psychological interest, with the exception of the rodents, the elephant, the dog and cat tribe among Carnivora, and the Primates —all of which I shall reserve for separate treatment.

Marsupials.

In the 'Transactions of the Linnean Society,' Major Mitchell gives an interesting account of the structure reared by a small Australian marsupial (*Conilurus constructor*) for the purposes of defence against the dingo dog. It consists of a large pile of dry sticks and brushwood, 'big enough to make two or three good cart-loads.' Each stick and fragment is closely intertwined or woven with the rest, so that the whole forms a solid, compact mass. In the middle of this large structure is the nest of the animal.

The marsupials are as low in the scale of mammalian intelligence as they are in that of mammalian structure: so that, except the above, I have met with no fact connected with the psychology of this group that is worth quoting, except, perhaps, the following, which appears to show deliberation and decision on the part of the kangaroo. Jesse writes:—

A gentleman who had resided for several years in New South Wales related the following circumstance, which he assured me he had frequently witnessed while hunting the kangaroo: it furnishes a strong proof of the affection of that animal for her young, even when her own life has been placed in the most imminent danger. He informed me that, when a female kangaroo has been hard pressed by dogs, he has seen her, while she has been making her bounds, put her fore-paws into her pouch, take a young one from it, and then throw it as far on one side as she possibly could out of the way of the dogs. But for this manœuvre her own life and that of her young one would have been sacrificed. By getting rid of the latter she has frequently effected her escape, and probably returned afterwards to seek for her offspring.

Cetaceans.

The following is quoted from Thompson:—

In 1811, says Mr. Scoresby, one of my harpooners struck a sucker, with the hope of leading to the capture of the mother. Presently she arose close to the 'fast boat,' and seizing the young one, dragged about 600 feet of line out of the boat with remarkable force and velocity. Again she rose to the surface, darted furiously to and fro, frequently stopped short or suddenly changed her direction, and gave every possible intimation of extreme agony. For a length of time she continued thus to act, though pursued closely by the boats; and, inspired with courage and resolution by her concern for her young, seemed regardless of the dangers which surrounded her. At length one of the boats approached so near that a harpoon was hove at her; it hit, but did not attach itself. A second harpoon was struck, but this also failed to penetrate; but a third was more successful, and held. Still she did not attempt to escape, but allowed other boats to approach; so that in a few minutes three more harpoons were fastened, and in the course of an hour afterwards she was killed.[1]

Mr. Saville Kent communicates an article to 'Nature' (vol. viii., p. 229) on 'Intellect of Porpoises.' He says:--

The keeper in charge of these interesting animals is now in the habit of summoning them to their meals by the call of a whistle; his approaching footsteps, even, cause great excitement in their movements. . . . The curiosity attributed to these creatures, as illustrated by the experiences of Mr. Matthew Williams, receives ample confirmation from their habits and confinement. A new arrival is at once subjected to the most importunate attention, and, advancing from familiarity to con-

[1] *Passions of Animals*, p. 154.

tempt if disapproved of, soon becomes the object of attack and persecution. A few dog-fish (*Acanthias* and *Mastelus*), three or four feet long, now fell victims to their tyranny, the porpoises seizing them by their tails, and swimming off with and shaking them in a manner scarcely conducive to their comfort or dignified appearance, reminding the spectator of a large dog worrying a rat. . . . On one occasion I witnessed the two *Cetacea* acting evidently in concert against one of these unwieldy fish (skates), the latter swimming close to the top of the water, and seeking momentary respite from its relentless enemies by lifting its unfortunate caudal appendage high above its surface —the peculiar tail of the skate being the object of sport to the porpoises, which seized it in their mouths as a convenient handle whereby to pull the animal about, and worry it incessantly.

In a subsequent number of 'Nature' (vol. ix., p. 42) Mr. C. Fox writes:—

Several years ago a herd of porpoises was scattered by a net which I had got made to enclose some of them. . . . The whole 'sculle' was much alarmed, and two were secured. I conclude that their companions retained a vivid remembrance of the sea-fight, as these *Cetacea*, although frequent visitants in this harbour (Falmouth) previously, and often watched for, were not seen in it again for two years or more.

Horse and Ass.

The horse is not so intelligent an animal as any of the larger Carnivora, while among herbivorous quadrupeds his sagacity is greatly exceeded by that of the elephant, and in a lesser degree by that of his congener the ass. On the other hand, his intelligence is a grade or two above that of perhaps any ruminant or other herbivorous quadruped.

The emotional life of this animal is remarkable, in that it appears to admit of undergoing a sudden transformation in the hands of the 'horse-tamer.' The celebrated results obtained by Rarey in this connection have since been repeated with more or less success by many persons in various parts of the world, and the 'method' appears to be in all cases essentially the same. The untamed and apparently untamable animal has its fore-leg or legs strapped up, is cast on its side and allowed to struggle for a while. It is then subjected to various manipulations, which, without necessarily causing pain, make the animal feel its helplessness and the mastery of the operator. The extraordinary fact is that, after having once felt this, the spirit or emotional life of the animal undergoes a complete and sudden change, so that from having been 'wild' it becomes 'tame.' In some cases there are subsequent relapses, but these are easily checked. Even the truly 'wild' horse from the prairie admits of being completely subdued in a marvellously short time by the Gauchos, who employ an essentially similar method, although the struggle is here much more fierce and prolonged.[1] The same may be said of the taming of wild elephants, although in this case the facts are not nearly so remarkable from a psychological point of view, seeing that the process of taming is so much more slow.

Another curious emotional feature in the horse is the liability of all the other mental faculties of the animal to become abandoned to that of terror. For I think I am right in saying that the horse is the only animal which, under the influence of fear, loses the possession of every other sense in one mad and mastering desire to run. With its entire mental life thus overwhelmed by the flood of a single emotion, the horse not only loses, as other animals lose, 'presence of mind,' or a due balance among the distinctively intellectual faculties, but even the avenues of special sense become stopped, so that the wholly demented animal may run headlong and at terrific speed against a stone wall. I have known a hare come to grief in a somewhat similar fashion when hotly pursued by a dog; this, however, was clearly owing to the hare looking behind instead of before, in a manner not, under the circumstances, unwise; but, as I have said, there is no animal except the horse whose whole psychology is thus liable to be completely dominated by a single emotion.

As for its other emotions, the horse is certainly an affectionate animal, pleased at being petted, jealous of

[1] See Mr. Darwin's account in *Naturalist's Voyage round the World*, pp. 151-2.

companions receiving favour, greatly enjoying play with others of its kind, and also the sport of the hunting-field. Lastly, horses exhibit pride in a marked degree, as do also mules. Such animals, when well kept, are unmistakably pleased with gay trappings, so that 'in Spain, as a punishment for disobedience, it is usual to strip the animal of its gaudy coronal and bells, and to transfer them to another' (Thompson).

The memory of the horse is remarkably good, as almost every one must have had occasion to observe who has driven one over roads which the animal may have only once traversed a long time before. As showing the duration of memory I may quote the following letter to Mr. Darwin from the Rev. Rowland H. Wedgwood, which I find among the MSS. of the former:—

I want to tell you of an instance of long memory in a horse. I have just driven my pony down from London here, and though she has not been here for eight years, she remembered her way quite well, and made a bolt for the stables where I used to keep her.

A few instances of the display of intelligence by members of the horse tribe may bring this section to a close.

Mr. W. J. Fleming writes me concerning a vicious horse he had which, while being groomed, frequently used to throw a ball of wood attached to his halter at the groom. He did so by flexing his fetlock and jamming the ball between the pastern and the leg, then throwing the ball backwards 'with great force.'

I myself had a horse which was very clever at slipping his halter after he knew that the coachman was in bed. He would then draw out the two sticks in the pipe of the oat-bin, so as to let all the oats run down from the bin above upon the stable floor. Of course he must have observed that this was the manner in which the coachman obtained the oats, and desiring to obtain them, did what he had observed to be required. Similarly, on other occasions he used to turn the water-tap to obtain a drink, and pull the window cord to open the window on hot nights.

The anecdote books contain several stories very much alike concerning horses spontaneously visiting blacksmiths' shops when they require shoeing, or feel their shoes uncomfortable. The appended account, vouched for as it is by a good authority, may be taken as corroborative of these stories. I quote the account from 'Nature' (May 19, 1881):—

The following instance of animal intelligence is sent to us by Dr. John Rae, F.R.S., who states that the Mr. William Sinclair mentioned is respectable and trustworthy. The anecdote is taken from the 'Orkney Herald' of May 11 :—"A well-authenticated and extraordinary case of the sagacity of the Shetland pony has just come under our notice. A year or two ago Mr. William Sinclair, pupil-teacher, Holm, imported one of these little animals from Shetland on which to ride to and from school, his residence being at a considerable distance from the school buildings. Up to that time the animal had been unshod, but some time afterwards Mr. Sinclair had it shod by Mr. Pratt, the parish blacksmith. The other day Mr. Pratt, whose smithy is a long distance from Mr. Sinclair's house, saw the pony, without halter or anything upon it, walking up to where he was working. Thinking the animal had strayed from home, he drove it off, throwing stones after the beast to make it run homewards. This had the desired effect for a short time; but Mr. Pratt had only got fairly at work once more in the smithy when the pony's head again made its appearance at the door. On proceeding a second time outside to drive the pony away, Mr. Pratt, with a blacksmith's instinct, took a look at the pony's feet, when he observed that one of its shoes had been lost. Having made a shoe he put it on, and then waited to see what the animal would do. For a moment it looked at the blacksmith as if asking whether he was done, then pawed once or twice to see if the newly-shod foot was comfortable, and finally gave a pleased neigh, erected its head, and started homewards at a brisk trot. The owner was also exceedingly surprised to find the animal at home completely shod the same evening, and it was only on calling at the smithy some days afterwards that he learned the full extent of his pony's sagacity."

In 'Nature,' also (vol. xx., p. 21), Mr. Claypole, of Antioch Cottage, Ohio, writes as follows:—

A friend of mine is employed on a farm near Toronto, Ontario, where a horse, belonging to the wife of the farmer, is never required to work, but is allowed to live the life of a gentleman, for the following reason. Some years ago the lady above mentioned fell off a plank bridge into a stream when the water was deep. The horse, which was feeding in a field close by, ran to the spot, and held her up with his teeth till assistance arrived, thus probably saving her life. Was this reason or instinct?

Mr. Strickland, also writing to 'Nature' (vol. xix.. p. 410), says:—

A mare here had her first foal when she was ten or twelve years old. She was blind of one eye. The result was, she frequently trod upon the foal or knocked it over when it happened to be on the blind side of her, in consequence of which the foal died when it was three or four months old. The next year she had another foal, and we fully expected the result would be the same. But no; from the day it was born she never moved in the stall without looking round to see where the foal was, and she never trod upon it or injured it in any way. You see that reason did not teach her that she was killing her first foal; her care for the second was the result of memory, imagination, and thought after the foal was dead, and before the next one was born. The only difference that I can see between the reasoning power of men and animals is that the latter is applied only to the very limited space of providing for their bodily wants, whereas that of men embraces a vast amount of other objects besides this.

Houzeau (vol. ii., p. 207) says that the mules used in the tramways at New Orleans prove that they are able to count five; for they have to make five journeys from one end of the tramway to the other before they are released, and they make four of these journeys without showing that they expect to be released, but bray at the end of the fifth. This observation, however, requires to be confirmed, for unless carefully made we must suppose that the fact may be due to the mules seeing the ostler waiting to take them out.

Mr. Samuel Goodbehere, solicitor, writes me from Birmingham the following instance as having fallen under his own observation:—

We had a Welsh cob pony or Galloway about 14 hands high, who was occasionally kept in a shed (in a farmyard), partly closed at the front by a gate which was secured by a bolt inside and a drop latch outside. The pony (who was able to put his head and neck over the gate, but could not reach the outside latch) was constantly found loose in the yard, which was considered quite a mystery until it was solved one day by my observing the pony first pushing back the inside bolt, and then neighing until a donkey, who had the run of the yard and an adjoining paddock, came and pushed up the outside latch with his nose, thus letting the pony at liberty, when the two marched off together.

The following is the only instance that I have met with in any of the horse tribe of that degree of sagacity which leads to the intentional concealment of wrong-doing. In the case of elephants, dogs, and monkeys we find abundant evidence on this head, which therefore renders the following instance more antecedently credible, and, as it is also narrated on good authority, I do not hesitate to quote it.

Professor Niphon, of Washington University, St. Louis, U.S., says:—

A friend of mine living at Iowa City had a mule, whose ingenuity in getting into mischief was more than ordinarily remarkable. This animal had a great liking for the company of an oat-bin, and lost no opportunity, when the yard gate and barn door were open, to secure a mouthful of oats. Finally the mule was found in the barn in the morning, and for a long time it was found impossible to discover how he had come there. This went on for some time, until the animal was 'caught in the act.' It was found that he had learned how to open the gate, reaching over the fence to lift the latch, and that he then effectually mystified his masters by turning round and backing against it until it was latched. He then proceeded to the barn door, and pulling out the pin which held the door, it swung open of its own accord. From the intelligence which this animal displayed on many occasions, I am of the opinion that had not discovery of his trick prevented, it would soon have occurred to him to retrace his steps before daylight, in order to avoid the clubbing which the stable boys gave him in the morning. It may be added that this animal had enjoyed no unusual educational advantages, and his owners found it to their interest to discourage his intellectual efforts as much as possible.'

[1] *Nature*, vol. xx., p. 21.

Ruminants.

Concerning sympathy, Major-General Sir George Le Grand Jacob, C.B., &c., writes me of instances which he observed of doe ibexes raising with their heads the bucks which he shot, and supporting them during flight.

A vivid and intelligent class of emotions, in which sympathy and rational fear are blended, seem to be exhibited by cattle in slaughterhouses. Many years ago a pamphlet was written upon the subject, and more recently Mr. Robert Hamilton, F.C.S., without apparently knowing of this previous publication, wrote another pamphlet, conveying precisely similar statements. These are too long to quote *in extenso*; but from a letter which the latter gentleman writes to me I may make the following extract:—

> The animal witnessing the process of killing, flaying, &c., repeated on one after another of its fellows, gets to comprehend to the full extent the dreadful ordeal, and as it mentally grasps the meaning of it all, the increasing horror depicted in its condition can be clearly seen. Of course some portray it much more vividly than others; the varying intelligence manifested in this respect is only another link which knits them in oneness with the human family.

Pride is well marked in sheep and cattle, as shown by the depressing effects produced on a 'bell-wether' or leading cow by transferring the bell to another member of the herd; and it is said that in Switzerland the beasts which on show days are provided with garlands, are evidently aware of the distinction thus placed upon them. With some amount of poetic exaggeration this fact is noted by Schiller, who says in 'Wilhelm Tell,'—

> See with what pride your steer his garland wears;
> He knows himself the leader of the herd;
> But strip him of it, and he'd die of grief.

With regard to the general intelligence of ruminants I may first quote the following:—

> The sagacity with which the bisons defend themselves

against the attack of wolves is admirable. When they scent the approach of a drove of these ravenous creatures, the herd throws itself into the form of a circle, having the weakest and the calves in the middle, and the strongest ranged on the outside; thus presenting an impenetrable front of horns.[1]

The buffalo of the Old World manifests sagacity very similar. As Sir J. E. Tennent informs us,—

The temper of the wild buffalo is morose and uncertain; and such is its strength and courage, that in the Hindu epic of the 'Ramayana' its onslaught is compared with that of the tiger. It is never quite safe to approach them if disturbed in their pasture, or alarmed from their repose in the shallow lakes. On such occasions they hurry into line, draw up in defensive array, with a few of the oldest bulls in advance; and, wheeling in circles, their horns clashing with a loud sound as they clank them together in their rapid evolutions, they prepare for attack: but generally, after a menacing display, the herd betake themselves to flight; then forming again at a safer distance, they halt as before, elevating their nostrils, and throwing back their heads to take a defiant survey of the intruders.[2]

When tamed this animal is used for sporting purposes in a manner which displays the spirit of curiosity of deer, hogs, and other animals. Thus, Sir J. E. Tennent continues:—

A bell is attached to its neck, and a box or basket with one side open is securely strapped on its back. This at night is lighted with flambeaux of wax, and the buffalo bearing it is slowly driven into the jungle. The huntsmen with their fowling-pieces keep close under the darkened side, and as it moves slowly onwards, the wild animals, startled by the sound and bewildered by the light, steal cautiously towards it in stupefied fascination. Even the snake, I am assured, will be attracted by this extraordinary object; and the leopard, too, falls a victim to curiosity.[3]

Livingstone says of the African buffalo, that he has known the animal, when pursued by hunters, to 'turn back to a point a few yards from its own trail, and then lie down in a hollow for the hunter to come up,'—a fact

---

[1] Thompson, *Passions of Animals*, p. 308.
[2] *Natural History of Ceylon*, p. 54.
[3] *Ibid.*, p. 56.

which displays a level of intelligence in this animal surpassing that which is met with in most Carnivora.[1]

Livingstone also says:—

It is curious to observe the intelligence of game; in districts where they are much annoyed by fire-arms they keep out on the most open spots of country they can find, in order to have a widely extended range of vision, and a man armed is carefully shunned.... But here, where they are killed by the arrows of the Balonda, they select for safety the densest forest, where the arrow cannot be easily shot.[2]

Jesse, who had many opportunities of observing the fact, says:—

I have been much delighted with watching the manner in which some of the old bucks in Bushey Park continue to get the berries from the fine thorn trees there. They will raise themselves on their hind legs, give a spring, entangle their horns in the lower branches of the tree, give them one or two shakes, and they will then quietly pick them up.[3]

The same author elsewhere says:—

Few things, indeed, can show more forcibly the powerful instinct which is implanted in animals for their self-preservation than the means which they take to avoid danger. I saw an instance of this lately in a stag. It had been turned out before a pack of hounds, and, when somewhat pressed by them, I observed it twice to go amongst a flock of sheep, and in both cases to double back, evidently, I should imagine, with the intention of baffling the pursuit of the dogs. It would thus seem that the animal was aware of its being followed by the scent, and not by sight. If this be the case, it affords another proof that animals are possessed of something more than common instinct.[4]

This author also says that he has 'frequently observed the buffalo at the Zoological Farm on Kingston Hill' display the following proof of intelligence. Being of a ferocious disposition, a strong iron ring was fixed through the septum of his nose, to which a chain about two feet long was attached. At the free end of the chain there was another ring about four inches in diameter. 'In grazing the buffalo must have put his feet on this ring,

---

[1] *Missionary Travels*, p. 328.　[2] *Ibid.*, p. 280.
[3] *Gleanings*, &c., vol. i., p. 20.　[4] *Ibid.*, vol. ii., p. 20.

and in raising his head the jerk would have produced considerable pain. In order to avoid this the animal has the sense to put his horn through the lower ring, and thus avoid the inconvenience he is put to. I have seen him do this in a very deliberate manner, putting his head on one side while he got his horn through the ring, and then shaking his head till the ring rested at the bottom of the horn.'[1]

The following is quoted from Mrs. Lee's 'Anecdotes' (p. 366), and is rendered credible not only because her own observations are generally good, but also because we shall subsequently find unquestionable evidence of the display of similar intelligence by cats:—

A goat and her kids frequented a square in which I once lived, and were often fed by myself and servants—a circumstance which would have made no impression, had I not heard a thumping at the hall door, which arose from the buttings of the goat when the food was not forthcoming, and whose example was followed by the two little things. After a time this remained unheeded, and, to our great astonishment, one day the area bell used by the tradespeople, the wire of which passed by the side of one of the railings, was sounded. The cook answered it, but no one was there save the goat and kids, with their heads bent down towards the kitchen window. It was thought that some boy had rung for them; but they were watched, and the old goat was seen to hook one of her horns into the wire and pull it. This is too much like reason to be ascribed to mere instinct.

P. Wakefield, in his 'Instinct Displayed,'[2] gives two separate cases of an intelligent manœuvre performed by goats. On both occasions two goats met on a ridge of rock with a precipice on each side, and too narrow to admit of their passing one another. One of these cases occurred on the ramparts of Plymouth Citadel, and was witnessed by 'many persons;' the other took place at Ardenglass, in Ireland. 'In both these instances the animals looked at each other for some time, as if they were considering their situation, and deliberating what was best to be done in the emergency.' In each case one of the goats then 'knelt down with great caution, and

[1] *Ibid.*, pp. 226-7.   [2] Pp. 66 and 97.

crouched as close as it could lie, when the other walked over its back.' This manœuvre on the part of goats has also been recorded by other writers, and is not so incredible as it may at first sight appear, if we remember that in their wild state these animals must not unfrequently find themselves in this predicament.

Mr. W. Forster, writing from Australia, gives me the following account of the intelligence of a bull :—

A rather tame bull, bred of a milch cow, used to puzzle me by being found inside a paddock used for cultivation, and enclosed by a two-railed fence, of which the lower rail was unusually high. At last I saw the animal lie down close to the fence, and roll over on his back, with four legs in the air, by which proceeding he was inside the paddock. I never knew another beast perform this feat; and although it must have been often done in the presence of a number of cows, not one of them ever imitated it, though they would all have unquestionably followed the bull through an opening in the fence, or by the slip-rails.

Mr. G. S. Erb, writing from Salt Lake City, gives me an interesting account of the sagacity displayed by the wild deer of the United States in avoiding gun-traps, which, except for the cutting of the string, to which the teeth of the animal are not so well adapted, is strikingly similar to the sagacity which we shall see to be displayed in this respect by sundry species of Carnivora. He says :—

My method was this: I would fell or cut down a maple tree, the top of which they are very partial to; and as the ground was invariably covered with snow to the depth of 12 inches, food was scarce, and the deer would come and browse, probably from hearing the tree fall. I would place a loaded gun 20 feet from the top of the tree at which it was pointing; I would attach a line the size of an ordinary fish-line to a lever that pressed against the trigger; the other end of the line I would fasten to the tree-top. By this means the deer could not pass between the tree and the gun without getting shot, or at least shot at; but I never succeeded in killing one when my line was as large as a fish-line, *i.e.* about one-sixteenth of an inch in thickness. Commencing at the body of the tree on one side, the deer would eat all the tops to within 12 inches of the line, and then go around the gun and eat all on the other side, never touching the line. I tried this at least sixty times, always with the same result. Then I took a black linen thread, and had no difficulty in killing them, as it was so small and black that they could not distinguish it.

Pigs.

There can be no doubt that pigs exhibit a degree of intelligence which falls short only of that of the most intelligent Carnivora. The tricks taught the so-called 'learned pigs' would alone suffice to show this; while the marvellous skill with which swine sometimes open latches and fastenings of gates, &c., is only equalled by that of the cat. The following account of pigs in their wild state shows that they manifest the same kind of sagacious co-operation in facing an enemy as that which we have just seen to be manifested by the bison and the buffalo, although here it seems to be displayed in a manner still more organised:—

Wild swine associate in herds and defend themselves in common. Green relates that in the wilds of Vermont a person fell in with a large herd in a state of extraordinary restlessness; they had formed a circle with their heads outwards, and the young ones placed in the middle. A wolf was using every artifice to snap one, and on his return he found the herd scattered, but the wolf was dead and completely ripped up. Schmarda recounts an almost similar encounter between a herd of tame swine and a wolf, which he witnessed on the military positions of Croatia. He says that the swine, seeing two wolves, formed themselves into a wedge, and approached the wolves slowly, grunting and erecting their bristles. One wolf fled, but the other leaped on to the trunk of a tree. As soon as the swine reached it they surrounded it with one accord, when, suddenly and instantaneously, as the wolf attempted to leap over them, they got him down and destroyed him in a moment.[1]

In Bingley's 'Memoirs of British Quadrupeds' (page 452) there is an account drawn up at his request by Sir Henry Mildmay, concerning the docility of the pig. The Toomer brothers were King's keepers in the New Forest, and they conceived the idea of training a sow to point game. This they succeeded in doing within a fort-

[1] Thompson, *Passions of Animals*, p 308.

night, and in a few more weeks it also learnt to retrieve. Her scent was exceedingly good, and she stood well at partridges, black game, pheasants, snipes, and rabbits, but never pointed hares. She was more useful than a dog, and afterwards became the property of Sir Henry Mildmay. According to Youatt,[1] Colonel Thornton also had a sow similarly trained. The same author says that a sow belonging to Mr. Craven had a litter of pigs, one of which, when old enough, was taken and roasted, then a second and a third. These were necessarily taken when the mother returned in the evening from the woods for supper. But the next time she came she was alone, and, 'as her owners were anxious to know what was become of her brood, she was watched on the following evening, and observed driving back her pigs at the extremity of the wood, with much earnest grunting, while she went off to the house, leaving them to wait for her return. It was evident that she had noticed the diminution of her family, and had adopted this method to save those that remained.'[2]

Mr. Stephen Harding sends me the following as an observation of his own:—

On the 15th ult. (Nov. 1879) I saw an intelligent sow pig about twelve months old, running in an orchard, going to a young apple tree and shaking it, pricking up her ears at the same time, as if to listen to hear the apples fall. She then picked the apples up and ate them. After they were all down she shook the tree again and listened, but as there were no more to fall she went away.

The proverbial indifference to dirt attributed to the pig seems scarcely to be justified; the worst that can be said is that the animal prefers cool mud to dry heat, and the filth which swine often exhibit in their sty is the fault of the farmers rather than of the animals. Or, to quote from Thompson's 'Passions,'—

A washed sow in the hot season of our temperate climate, and in almost every season of such a climate as that of Palestine, 'returns to her wallowing in the mire' simply because she feels scorched, and blistered, and sickened under the ardent

[1] *On the Pig*, p. 17.   [2] *Ibid.*

sunshine; and hence, when she receives from man the aid which is due to her as a domesticated animal, she demands not dirt all the year through, nor any day at all, but shade in summer, shelter in winter, and a clean, dry bed in every season.

Cheiroptera.

Mr. Bates says of bats: 'The fact of their sucking the blood of persons sleeping is now well established; but it is only a few persons who are subject to this bloodletting. ... I am inclined to think many different kinds of bats have this propensity' ('Nat. on Amaz.,' p. 91). The particular species of bat, however, which has been most universally accredited with this habit, viz., the vampire, is perfectly harmless.

Mr. G. Clark ('A Brief Notice of the Fauna of Mauritius') gives an account of the intelligence displayed by a tame bat (*Pteropus vulgaris*). As soon as its master came into the room, it welcomed him with cries; and if not at once taken up to be petted, it climbed up his dress, rubbed its head against him, and licked his hands. If Mr. Clark took anything in his hand, the bat would carefully examine it by sight and smell, and when he sat down the bat would hang upon the back of his chair, following all his movements with its eyes.

Carnivora.

I shall here run together a few facts relating to the intelligence of carnivorous animals other than those to be considered in subsequent chapters.

Seals.—In their wild state these animals have not much opportunity for the display of intelligence; but when tamed it is seen that the latter is considerable. They are then affectionate animals, liking to be petted, and showing attachment to their homes. The most remarkable species of the order from a psychological point of view are the so-called Pinnipeds, whose habits during the breeding season are so peculiar that I think it is worth while to quote the best account that has hitherto been published on the subject. This is the elaborate work of Mr. Joel Asaph Allen:[1] —

From the time of the first arrivals in May up to the 1st of June, as late as the middle of this month if the weather be clear, is an interval in which everything seems quiet; very few seals are added to the pioneers. By the 1st of June, however, or thereabouts, the foggy, humid weather of summer sets in, and with it the bull-seals come up by hundreds and thousands, and locate themselves in advantageous positions for the reception of the females, which are from three weeks to a month later, as a rule. The labour of locating and maintaining a position in the rookery is really a serious business for those bulls which come in last, and for those that occupy the water-line, frequently resulting in death from severe wounds in combat sustained. It appears to be a well-understood principle among the able-bodied bulls that each one shall remain undisturbed on his ground, which is usually about ten feet square, provided he is strong enough to hold it against all comers; for the crowding in of fresh bulls often causes the removal of those who, though equally able-bodied at first, have exhausted themselves by fighting earlier, and are driven by the fresher animals back further and higher up on the rookery. Some of these bulls show wonderful strength and courage. I have marked one veteran, who was among the first to take up his position, and that one on the water-line, when at least fifty or sixty desperate battles were fought victoriously by him with nearly as many different seals who coveted his position; and when the fighting season was over (after the cows have mostly all hauled up) I saw him covered with scars and gashes, raw and bloody, an eye gouged out, but holding it bravely over his harem of fifteen or twenty cows, all huddled together on the same spot he had first chosen. The fighting is mostly or entirely done with the mouth, the opponents seizing each other with the teeth and clenching the jaws; nothing but sheer strength can shake them loose, and that effort almost always leaves an ugly wound, the sharp canines tearing out deep gutters in the skin and blubber, or shredding the flippers into ribbon-strips. They usually approach each other with averted heads and a great many false passes before either one or the other takes the initiative by gripping; the heads are darted out and back as quick as flash, their hoarse roaring and shrill piping whistle never ceases, while their fat

[1] *History of the North American Pinnipeds.* The quotations are taken from pp. 348 to 361.

bodies writhe and swell with exertion and rage, fur flying in air and blood streaming down—all combined make a picture fierce and savage enough, and, from its great novelty, exceedingly strange at first sight. In these battles the parties are always distinct, the offensive and the defensive; if the latter proves the weaker he withdraws from the position occupied, and is never followed by his conqueror, who complacently throws up one of his hind flippers, fans himself, as it were, to cool himself from the heat of the conflict, uttering a peculiar chuckle of satisfaction and contempt, with a sharp eye open for the next covetous bull or 'sea-catch' (native name for the bulls on the rookeries, especially those who are able to maintain their position).

* * * * * * *

All the bulls, from the very first, that have been able to hold their positions have not left them for an instant, night or day; nor do they do so until the end of the rutting season, which subsides entirely between the 1st and 10th of August, beginning shortly after the coming of the cows in June. Of necessity, therefore, this causes them to fast, to abstain entirely from food of any kind, or water for at least three months; and a few of them stay four months before going into the water for the first time after hauling up in May. This alone is remarkable enough, but it is simply wonderful when we come to associate the condition with unceasing activity, restlessness, and duty devolved upon the bulls as heads and fathers of large families. They do not stagnate like bears in caves; it is evidently accomplished or due to the absorption of their own fat, with which they are so liberally supplied when they take their positions on the breeding-ground, and which gradually diminishes while they remain on it.

* * * * * * *

They are noticed and received by the bulls on the water-line station with much attention; they are alternately coaxed and urged up on the rocks, and are immediately under the most jealous supervision; but owing to the covetous and ambitious nature of the bulls which occupy the stations reaching some way back from the water-line, the little cows have a rough-and-tumble time of it when they begin to arrive in small numbers at first; for no sooner is the pretty animal fairly established on the station of bull No. 1 who has installed her there, than he perhaps sees another one of her style down in the water from which she has just come, and in obedience to his polygamous feeling, he devotes himself anew to coaxing the later arrival in the same winning manner so successful in her case, when bull

No. 2, seeing bull No. 1 off his guard, reaches out his long strong neck, and picks the unhappy but passive creature up by the scruff of hers, just as a cat does a kitten, and deposits her on his seraglio-ground; then bulls Nos. 3, 4, 5, and so on in the vicinity, seeing this high-handed operation, all assail one another, and especially bull No. 2, and have a tremendous fight perhaps for half a minute or so; and during this commotion the cow is generally moved or moves farther back from the water two or three stations more, where, when all gets quiet, she usually remains in peace. Her late lord and master, not having the exposure to such diverting temptation as had her first, gives her such care that she not only is unable to leave did she wish, but no other bull can seize upon her. This is only one instance of the many different trials and tribulations which both parties on the rookery subject themselves to before the harems are filled. Far back, fifteen or twenty stations deep from the water-line sometimes, but generally not more than, on an average, ten or fifteen, the cows crowd in at the close of the season for arriving, July 10 to 14, and then they are able to go about pretty much as they please, for the bulls have become greatly enfeebled by this constant fighting and excitement during the past two months, and are quite content with even only one or two partners.

* * * * * *

I have found it difficult to ascertain the average number of cows to one bull on the rookery, but I think it will be nearly correct to assign to each male from twelve to fifteen females occupying the stations nearest the water, those back in the rear from five to nine. I have counted forty-five cows all under the charge of one bull, which had them penned up on a flat table-rock near Kestaire Point; the bull was enabled to do this quite easily, as there was but one way to go to or come from this seraglio, and on this path the old Turk took his stand and guarded it well. At the rear of all these rookeries there is always a large number of able-bodied bulls, who wait patiently, but in vain, for families, most of them having had to fight as desperately for the privilege of being there as any of their more fortunately located neighbours, who are nearer the water than themselves; but the cows do not like to be in any outside position, when they are not in close company lying most quiet and content in the largest harems; and these large families pack the surface of the ground so thickly that there is hardly moving or turning room until the females cease to come up from the sea; but the inaction on the part of the bulls in the rear during the rutting season only serves to qualify them to move into the places vacated by those males who are obliged to leave from exhaustion, or to take the position of fearless and jealous protectors for the young pups in the fall. The courage with which the fur-seal holds his position as the head and guardian of a family is of the very highest order compared with that of other animals. I have repeatedly tried to drive them when they have fairly established themselves, and have almost always failed, using every stone at my command, making all the noise I could, and finally, to put their courage to the full test, I walked up to within twenty feet of a bull at the rear and extreme end of Tolstoi Rookery, who had four cows in charge, and commenced with my double-barrelled breech-loading shot-gun to pepper him all over with mustard-seed or dust-shot. His bearing in spite of the noise, smell of powder, and pain, did not change in the least from the usual attitude of determined defence which nearly all the bulls assume when attacked with showers of stones and noise; he would dart out right and left and catch the cows which timidly attempted to run after each report, fling and drag them back to their places; then, stretching up to his full height, look me directly and defiantly in the face, roaring and spitting most vehemently. The cows, however, soon got away from him, but he still stood his ground, making little charges on me of ten or fifteen feet in a succession of gallops or lunges, spitting furiously and then retreating to the old position, back of which he would not go, fully resolved to hold his own or die in the attempt.

This courage is all the more noteworthy from the fact that, in regard to man, it is invariably of a defensive character. The seal, if it makes you turn when you attack it, never follows you much farther than the boundary of its station, and no aggravation will compel it to become offensive, as far as I have been able to observe.

* * * * * *

The apathy with which the young are treated by the old on the breeding-grounds is somewhat strange. I have never seen a cow caress or fondle her offspring, and should it stray but a short distance from the harem, it can be picked up and killed before the mother's eyes, without causing her to show the slightest concern. The same indifference is exhibited by the bull to all that takes place outside of the boundary of his seraglio. While the pups are, however, within the limits of his harem-ground he is a jealous and fearless protector; but if the little animals pass beyond this boundary, then they may be carried off without the slightest attention in their behalf from their guardian.

* * * * * *

Early in August (8th) the pups that are nearest the water on the rookeries essay swimming, but make slow and clumsy progress, floundering about, when over head in depth, in the most awkward manner, thrashing the water with their fore-flippers, not using the hinder ones. In a few seconds, or a minute at the most, the youngest is so wary that he crawls out upon the rocks or beach, and immediately takes a recuperative nap, repeating the lesson as quick as he awakes and is rested. They soon get familiar with the water and delight in it, swimming in endless evolutions, twisting, turning, diving; and when exhausted, they draw up on the beach again, shake themselves as young dogs do, either going to sleep on the spot, or having a lazy frolic among themselves.

In this matter of learning to swim, I have not seen any 'driving' of the young pups into the water by the old in order to teach them this process, as has been affirmed by writers on the subject of seal life.

Otter.—The fact that otters admit of being taught to catch fish and bring them to their masters, shows no small degree of docility on the part of these animals. 'I have seen,' says Dr. Goldsmith, 'an otter go to a gentleman's pond at word of command, drive the fish into a corner, and, seizing upon the largest of the whole, bring it off in his mouth to his master.' And several other cases of the same kind are given by Bingley.[1]

Weasel.—' Mdlle. de Faister described her tame weasel to Buffon as playing with her fingers like a kitten, jumping on her head and neck; and if she presented her hands at the distance of three feet, it jumped into them without ever missing. It distinguished her voice amidst twenty people, and sprang over everybody to get at her. She found it impossible to open a drawer or a box, or even to look at a paper, without his examining it also. If she took up a paper or book, and looked attentively at it, the weasel immediately ran upon her hand, and surveyed with an inquisitive air whatever she happened to hold.'[2]

---

[1] *Animal Biography*, vol. iii., pp. 301–2.
[2] Thompson, *Passions in Animals*, p. 337.

Polecat.—Professor Alison, in his article on 'Instinct,' in Todd's 'Cyclopædia of Anatomy,' quotes the following account from the 'Magazine of Natural History' (vol. iv., p. 206) touching a remarkable instinct manifested by polecats. 'I dug out five young polecats, comfortably embedded in dry, withered grass; and in a side-hole, of proper dimensions for such a larder, I picked out forty large frogs and two toads, all alive, but merely capable of sprawling a little. On examination, I found that the whole number, toads and all, had been purposely and dexterously bitten through the brain.' The analogy of this instinct to that which has already been mentioned as having been much more recently observed by M. Fabre in the sphex insect is noteworthy.

Ferret.—I once kept a ferret as a domestic pet. He was a very large specimen, and my sister taught him a number of tricks, such as begging for food (which he did quite as well and patiently as any terrier), leaping over sticks, &c. He became a very affectionate animal, delighting much in being petted, and following like a dog when taken out for walk. He would, however, only follow those persons whom he well knew. That his memory was exceedingly good was shown by the fact that after an absence of many months, during which he was never required to beg, or to perform any of his tricks, he went through all his paces perfectly the first time that we again tried him.

I strongly suspect that ferrets dream, as I have frequently seen them when fast asleep moving their noses and twitching their claws as if in pursuit of rabbits. Another fact I may mention as bearing on the intelligence of these animals. On one occasion, while ferreting rabbits, I lost the ferret about a mile away from home. Some days afterwards the animal returned to his home. Similar cases have been communicated to me by several sporting friends, but certainly the return of a ferret under such circumstances is the exception, and not the rule.

Wolverine.—Amazing tales are told concerning the intelligence of this animal, which for the most part are certainly exaggerations. Still there is no doubt that the creature does display a degree of sagacious cunning unsurpassed, if not unequalled, in the animal kingdom. This may be shown by the two following quotations from the statements of trustworthy writers. The first is a letter kindly sent me by Dr. J. Rae, F.R.S., in reply to my request for information concerning the intelligence of this animal:—

The narratives of most travellers in America tell wonderful stories of the glutton or wolverine, but I do not know that any of my experiences of this extremely acute animal indicate what I call reasoning powers. They are very suspicious, and can seldom or never be taken with poisoned bait, trap, or gun. The poisoned baits are usually found broken up, but not eaten by them; traps are destroyed or entered, but not where the trapper desired; and guns, except when concealed after the Eskimo fashion by a covering of snow, are avoided.

In 1853, on the Arctic coast, when about to change our domicile from a tent to the warmer snow hut, my man had carried over about 100 lbs. or more of fine venison steaks to the snow houses about a quarter of a mile from our tents; and as there were at the time no traces either of foxes, wolves, or wolverines about, the meat was placed overnight in one of the huts, and the door left open. During the night two wolverines came, but, evidently dreading some trap or danger in the open door, would not enter that way, but cut a hole for themselves through the wall of the snow hut, and carried off all our fine steaks, a considerable quantity of which was picked up close to our house when the thaw took place in the spring, it having been hid in the snow, but completely spoilt for use, by a well-known filthy habit.

Dr. Rae has also drawn my attention to the following account contained in the Miscellaneous Publications of the Geological Survey of the United States.[1] The writer of this account is Captain Elliot Cones:—

To the trapper the wolverines are equally annoying. When they have discovered a line of marten traps they will never abandon the road, and must be killed before the trapping can be successfully carried on. Beginning at one end, they proceed from trap to trap along the whole line, pulling them successively to pieces, and taking out the baits from behind. When they can eat no more, they continue to steal the baits and cache

[1] Vol. viii., Washington, 1877: 'A Monograph of the North American *Mustelidæ*.'

them. If hungry they may devour two or three of the martens they find captured, the remainder being carried off and hidden in the snow at a considerable distance. The work of demolition goes on as fast as the traps can be renewed.

The propensity to steal and hide things is one of the strongest traits of the wolverine. To such an extent is it developed that the animal will often secrete articles of no possible use to itself. Besides the wanton destruction of marten traps, it will carry off the sticks and hide them at a distance, apparently in sheer malice. Mr. Ross, in the article above quoted, has given an amusing instance of the extreme of this propensity. The desire for accumulating property seems so deeply implanted in this animal, that, like tame ravens, it does not appear to care much what it steals so that it can exercise its favourite propensity to commit mischief. An instance occurred within my own knowledge, in which a hunter and his family having left their lodge unguarded during their absence, on their return found it completely gutted—the walls were there, but nothing else. Blankets, guns, kettles, axes, cans, knives, and all the other paraphernalia of a trapper's tent had vanished, and the tracks left by the beast showed who had been the thief. The family set to work, and by carefully following up all his paths recovered, with some trifling exceptions, the whole of the lost property.

* * * * * *

At Peel's River, on one occasion, a very old carcajou discovered my marten road, on which I had nearly a hundred and fifty traps. I was in the habit of visiting the line about once a fortnight, but the beast fell into the way of coming oftener than I did, to my great annoyance and vexation. I determined to put a stop to his thieving and his life together, cost what it might. So I made six strong traps at as many different points, and also set three steel traps. For three weeks I tried my best to catch the beast without success; and my worst enemy would allow that I am no green hand in these matters. The animal carefully avoided the traps set for his own benefit, and seemed to be taking more delight than ever in demolishing my marten traps and eating the martens, scattering the poles in every direction, and câching what baits or martens he did not devour on the spot. As we had no poison in those days, I next set a gun on the bank of a little lake. The gun was concealed in some low bushes, but the bait was so placed that the carcajou must see it on his way up the bank. I blockaded my path to the gun with a small pine tree, which completely hid it. On my first visit afterwards I found that the beast had gone up to the bait and smelled it, but had left it untouched. He had next pulled up the pine tree that blocked the path, and gone around the gun and cut the line which connected the bait with the trigger, just behind the muzzle. Then he had gone back and pulled the bait away, and carried it out on the lake, where he lay down and devoured it at his leisure. There I found my string. I could scarcely believe that all this had been done designedly, for it seemed that faculties fully on a par with human reason would be required for such an exploit if done intentionally. I therefore rearranged things, tying the string where it had been bitten. But the result was exactly the same for three successive occasions, as I could plainly see by the footprints; and what is most singular of all, each time the brute was careful to cut the line a little back of where it had been tied before, as if actually reasoning with himself that even the knots might be some new device of mine, and therefore a source of hidden danger he would prudently avoid. I came to the conclusion that that carcajou ought to live, as he must be something at least human, if not worse. I gave it up, and abandoned the road for a period.

* * * * * *

With so much for the tricks and the manners of the beast behind our backs, roaming at will in his vast solitudes, what of his actions in the presence of man? It is said that if one only stands still, even in full view of an approaching carcajou, he will come within fifty or sixty yards, provided he be to windward, before he takes the alarm. Even then, if he be not warned by sense of smell, he seems in doubt, and will gaze earnestly several times before he finally concludes to take himself off. On these and similar occasions he has a singular habit—one not shared, so far as I am aware, by any other beast whatever. He sits on his haunches and shades his eyes with one of his fore-paws, just as a human being would do in scrutinising a dim or distant object. The carcajou, then, in addition to his other and varied accomplishments, is a perfect sceptic—to use this word in its original signification. A sceptic, with the Greeks, was simply one who would shade his eyes to see more clearly.

Bears.—There is no doubt that the intelligence of these animals stands very high in the psychological scale, although the actual instances which I have met of the display of their intelligence are few. The tricks which are taught performing bears do not count for much as proof of high sagacity, as they for the most part consist in teaching the animals to assume unnatural positions, or display grotesque antics—performances which speak indeed for the general docility of the creatures, but scarcely for their high intelligence. Still even here it is worth while to remark that all species of bears would probably not lend themselves to this kind of education, for the emotional temperament manifested by the different species is unquestionably diverse. Thus, making all allowances for exaggeration, it seems certain that the grizzly bear displays a courage and ferocity which are foreign to the disposition of the brown bear, and indeed to that of most other animals. The polar bear likewise displays much bravery under the influence of hunger or maternal feeling, although under other circumstances it usually deems discretion the better part of valour. The following incident displays considerable intelligence on the part of this animal.

Scoresby, in his 'Account of the Arctic Regions,' gives the instance to which I allude:—

The animal with two cubs was being pursued by a party of sailors over an ice-field. She urged her young to an increase of speed by running before them, turning round, and manifesting, by a peculiar action and voice, her anxiety for their progress; but finding that her pursuers were gaining upon them, she carried, or pushed, or pitched them alternately forward, until she effected their escape. In throwing them before her, the little creatures placed themselves across her path to receive the impulse; and when projected some yards in advance, they ran onwards until she overtook them, when they alternately adjusted themselves for a second throw.

As the polar bear is not exposed to any enemies except man, this method of escaping is not likely to be instinctive, but was probably an intelligent adaptation to the particular circumstances of the case.

Mr. S. J. Hutchinson writes me as follows with regard to this same species:—

One Sunday, at the 'Zoo,' some one threw a bun to the bears, but it fell in the water in that quadrant-shaped pond you will remember. The bun fell just at the angle, and the bear seemed disinclined to enter the water, but stood on the edge of the pond, and commenced *stirring* the water with its paw, so that it established a sort of rotatory current, which eventually brought the bun within reach. When one leg got tired it used the other, but in the same direction. I watched the whole performance with the greatest interest myself.

In corroboration of this most remarkable observation I quote the following from Mr. Darwin's 'Descent of Man' (p. 76), which is so precisely similar, that the fact of bears reaching the high level of intelligence which the fact implies can scarcely be doubted. 'A well-known entomologist, Mr. Westropp, informs me that he observed in Vienna a bear deliberately making with his paw a current in some water which was close to the bars of his cage, so as to draw a piece of floating bread within his reach.'

CHAPTER XII.

RODENTS.

THE rodents, psychologically considered, are, of all orders in the animal kingdom, most remarkable for the differences presented by constituent species. For while the group contains many animals, such as the guinea-pig, whose instincts and intelligence cannot be said to rise above the lowest level that obtains among mammalian forms, it also contains other animals with instincts as remarkable as those of the squirrel, intelligence as considerable as that of the rat, and a psychological development as unique as that of the beaver. In no other group of animals do we meet with nearly so striking an exemplification of the truth that zoological or structural affinity is only related in a most loose and general way to psychological or mental similarity. Up to a certain point, however, even here we meet with an exemplification of what I may call a complementary truth, namely, that similarity of organisation and environment is in a general way related to similarity of instincts (though not necessarily of intelligence). This is obviously the case with the habit from which the order takes its name; for whether the instinct of gnawing is here the cause or the result of peculiar organisation, the instinct is unquestionably correlated with the peculiarity. And similarly, though less obviously, is this the case with the instinct of storing food for winter consumption, which is more prevalent among the rodents than in any other order of mammals—rats, mice, squirrels, harvesters, beavers, &c., all manifesting it with remarkable vigour and persistency. Here we probably have a case of similar organisation and environment determining the same instinct; for the latter is not of sufficiently general occurrence among all species of rodents to allow us to suppose that the species in which it does occur have derived it from a common ancestry.

Rabbit.

Rabbits are somewhat stupid animals, exhibiting but small resources under novel circumstances, although inheriting several clever instincts, such as that of rapidly deciding upon the alternative of flight or crouching, which is usually done with the best judgment. I have, however, often observed that the animal does not seem to have sense enough to regard the colour of the surface on which it crouches, so that if this happens to be inappropriate, the rabbit may become conspicuous, and so its crouching a source of danger. I have been particularly struck with the fact that black rabbits inherit the crouching instinct as strongly as do normally coloured ones, with the effect of rendering themselves highly conspicuous. This shows that the instinct is not necessarily correlated with the colour which alone renders the instinct useful, but that both have developed simultaneously and independently, and by natural selection. The fact also shows that the crouching of rabbits is purely instinctive, and not due to any conscious process of comparing their own colour with that of the surfaces on which they crouch. No doubt the instinct began and was developed by natural selection placing a premium upon the better judgment of those individuals which know when best to seek safety in flight and when by crouching—protective colouring being added at the same time by the same agency.

Another fact, which every one who shoots must have observed, goes to show the stupidity of rabbits, or their inability to learn by experience. When alarmed they run for their burrows, and when they reach them, instead of entering they very frequently squat down to watch the enemy. Now, although they well know the distance at which it is safe to allow a man with a gun to approach, excess of curiosity, or a mistaken feeling of security in being so near their homes, induces the animals to allow a man to approach within easy shooting distance. Yet that in other respects rabbits can learn much by experience must be evident to all who are accustomed to shoot with ferrets. From burrows which have not been much ferreted, rabbits will bolt soon after the ferret is put in; but this is not the case where rabbits have had previous experience of the association between ferrets and sportsmen. Rather than bolt under such circumstances, and so face the known danger of the waiting gun, rabbits will often allow themselves to be torn with the ferrets' claws and mutilated by their teeth. This is the case, no matter how silently the sportsmen may conduct their operations; the mere fact of a ferret entering their burrows seems to be enough to assure the rabbits that sportsmen are waiting outside.[1]

In its emotions the rabbit is for the most part a very timid animal, although the males fight severely with one another—having more strongly developed than any other animal the strange but effectual instinct of castrating their rivals. Moreover, even against other animals, rabbits will, when compelled to do so, stand upon the defensive. To show this I may quote a letter which several years ago I published in 'Nature:'—

I have occasion just now to keep over thirty Himalayan rabbits in an outhouse. A short time ago it was observed that some of these rabbits had been attacked and slightly bitten by rats. Next day the person who feeds the rabbits observed, upon entering the outhouse, that nearly all the inmates were congregated in one corner; and upon going to ascertain the cause, found one rat dead, and another so much injured that it could scarcely run. Both rats were of an unusually large size, and their bodies were much mangled by the rabbits' teeth.

I never before knew that domestic rabbits would fight with any carnivorous antagonist. That wild rabbits never do so I infer from having several times seen ferrets turn out from the most crowded burrow in a warren young stoats and weasels not more than four inches long.

[1] It is particularly remarkable that if under these circumstances a rabbit bolts and, seeing the sportsman, doubles back into its burrow, being then certain that the sportsman is waiting, it will usually allow itself to be slowly and painfully killed by the ferret rather than bolt a second time. This is remarkable because it proves the strength of an abiding image or idea in the mind of the animal.

It is evident that the show-fight instinct cannot have been developed in Himalayan rabbits by means of natural selection, but it is no less evident that if it ever arose in wild rabbits it would be preserved and intensified by such means.

The following observation of my own on a previously unnoticed instinct displayed by wild rabbits is, I think, of sufficient interest to render. Most people are aware that if a rabbit is shot near the mouth of its burrow, the animal will employ the last remnant of its life in struggling into it. Having several times observed that wounded rabbits which had thus escaped appeared again several days afterwards above ground, lying dead a few feet from the mouth of the burrow, I wished to ascertain whether the wounded animals had themselves come out before dying, possibly for air, or had been taken out by their companions. I therefore shot numerous rabbits while they were sitting near their burrows, taking care that the distance between the gun and the animal should be such as to insure a speedy, though not an immediate death. Having marked the burrows at which I shot rabbits in this manner I returned to them at intervals for a fortnight or more, and found that about one-half of the bodies appeared again on the surface in the way described. That this reappearance above ground is not due to the victim's own exertions, I am now quite satisfied; for not only did two or three days generally elapse before the body thus showed itself—a period much too long for a severely wounded rabbit to survive—but in a number of cases decomposition had set in. Indeed, on one occasion scarcely anything of the animal was left save the skin and bones. This was in a large warren.

It is a curious thing that I have hitherto been unable to get any bodies returned to the surface, of rabbits which I *inserted* into their burrows *after death*. I account for this by supposing that the stench of the decomposing carcass is not so intolerable to the other occupants of the burrow when it is near the orifice as it is when further in. Similarly, I find that there is not so good a chance of bodies being returned from an extensive warren of intercommunicating holes, as there is from smaller warrens or blind holes; the reason probably being that in the one case the living inhabitants are free to vacate the offensive locality, while in the other case they are not so. Anyhow, there can be no reasonable doubt that the instinct of removing their dead has arisen in rabbits from the necessity of keeping their confined domiciles in a pure state.

Hare.

The hare is a more intelligent animal than the rabbit. Possibly its much greater powers of locomotion may be one cause of its mental superiority to its nearest congener. I have never myself observed a hare commit the mistake already mentioned in the case of the rabbit, viz., that of crouching for concealment upon an inappropriately coloured surface. But the best idea of the comparatively high intelligence of the hare will be gained by the following quotations. The first of these is taken from Loudoun's 'Magazine of Natural History' (vol. iv., p. 143):—

It is especially conscious of the scent left by its feet, and of the danger which threatens it in consequence; a reflection which implies as much knowledge of the habits of its enemies as of its own. When about to enter its seat for the purpose of rest, it leaps in various directions, and crosses and recrosses its path with repeated springs; and at last, by a leap of greater energy than it has yet used, it effects a lodgment in the selected spot, which is chosen rather to disarm suspicion than to protect it from injury. In the 'Manuel du Chasseur' some instances are quoted from an ancient volume on hunting by Jaques du Fouillouse. A hare intending to mislead its pursuers has been seen spontaneously to quit its seat and to proceed to a pond at the distance of nearly a mile, and having washed itself, push off again through a quantity of rushes. It has, too, been known, when pursued to fatigue by dogs, to thrust another hare from its seat and squat itself down in its place. This author has seen hares swim successively through two or three ponds, of which the smallest was eighty paces round. He has known it, after a long chase, to creep under the door of a sheep-house and rest among the cattle, and when the hounds were in pursuit, it would get into the middle of a flock of sheep and accompany them in all their motions round the field, refusing by any means to quit the shelter they afforded. The stratagem of its passing forward on one side of a hedge and returning by the other, with only the breadth of the hedge between itself and its enemies, is of frequent occurrence, and it has even been known to select its seat close to the walls of a dog-kennel. This latter circumstance, however, is illustrative of the principles of reflection and reasoning; for the fox, weasel, and polecat are to the hare more dangerous enemies than the hound; and the situations chosen were such as those ferocious creatures were not likely to approach. A gentleman was engaged in the amusement of coursing, when a hare, closely pressed, passed under a gate, while the dogs followed by leaping over it. The delay caused to her pursuers by this manœuvre seems to have taught a sudden and useful lesson to the persecuted creature; for as soon as the dogs had cleared the gate and overtaken her, she doubled and returned under the gate as before, the dogs again following and passing over it. And this flirtation continued backwards and forwards until the dogs were fairly tired of the amusement; when the hare, taking advantage of their fatigue, quietly stole away.

The following note, by Mr. Yarrell, is significant of a process of reasoning derived from observations of the course of nature, such as would do no discredit to a higher race of creatures :—

A harbour of great extent on our northern coast has an island near the middle of considerable size, the nearest point of which is a mile distant from the mainland at high water, and with which point there is frequent communication by a ferry. Early one morning in spring two hares were observed to come down from the hills of the mainland towards the sea-side; one of which from time to time left its companion, and proceeding to the very edge of the water, stopped there a minute or two, and then returned to its mate. The tide was rising, and after waiting some time, one of them, exactly at high water, took to the sea, and swam rapidly over, in a straight line, to the opposite projecting point of land. The observer on this occasion, who was near the spot, but remained unperceived by the hares, had no doubt they were of different sexes, and that it was the male (like another Leander) which swam across the water, as he had probably done many times before. It was remarkable that the hares had remained on the shore nearly half an hour; one of them occasionally examining, as it would seem, the state of the current, and ultimately taking to the sea at that precise period of the tide called slackwater, when the passage across could be effected without being carried by the force of the stream either above or below the desired point of landing. The other hare then cantered back to the hills. (Loudoun's 'Magazine of Natural History,' vol. v., p. 99.)

According to Couch ('Illustrations of Instinct,' p. 177)—

When followed by dogs, it will not run through a gate, though this is obviously the most ready passage; nor in crossing a hedge will it prefer a smooth and even part, but the roughest, where thorns and briars abound; and when it mounts an eminence it proceeds obliquely, and not straightforward. And whether we suppose these actions to proceed from a desire to avoid those places where traps may probably have been laid, or from knowing that his pursuers will exactly follow his footsteps, and he has resolved to lead them through as many obstacles as possible, in either case an estimation of causes and consequences is to be discovered.

It is a remarkable thing that both hares and rabbits should allow themselves to be overtaken in the open field by weasels. I have myself witnessed the process, and am at a loss to account for it. The hare or rabbit seems perfectly aware of the dangerous character of the weasel, and yet does not put forth its powers of escape. It merely toddles along with the weasel toddling behind, until tamely allowing itself to be overtaken. This anomalous case may perhaps be akin to the alleged phenomena of the fascination of birds and small rodents by snakes; but in any case there seems to have been here a remarkable failure of natural selection in doing duty to the instincts of these swift-footed animals.

We must not close this account of the intelligence of the hare genus without alluding to the classical case of Cowper's hares. The following abstract is taken from Tegg's edition of 'The Life and Works of William Cowper,' p. 633:—

Puss was ill three days, during which time I nursed him, kept him apart from his fellows, . . . and by constant care, &c., restored him to perfect health. No creature could be more grateful than my patient after his recovery, a sentiment which he most significantly expressed by licking my hand, first the back of it, then the palm, then every finger separately, then between all the fingers, as if anxious to leave no part of it unsaluted; *a ceremony which he never performed but once again upon a similar occasion.* Finding him extremely tractable, I made it my custom to carry him always after breakfast into the garden.... I had not long habituated him to this taste of liberty before he began to be impatient for the return of the time when he might enjoy it. He would invite me to the garden by drumming upon my knee, and by a look of such expression as it was not possible to misinterpret. If this rhetoric did not immediately succeed, *he would take the skirt of my coat between his teeth and pull it with all his force.* He seemed to be happier in human society than when shut up with his natural companions.

Rats and Mice

Rats are well known to be highly intelligent animals. Unlike the hare or rabbit, their shyness seems to proceed from a wise caution rather than from timidity; for, when circumstances require, their boldness and courage in combat is surprising. Moreover, they never seem to lose their presence of mind; for, however great their danger, they seem always ready to take advantage of any favouring circumstances that may arise. Thus, when matched with so formidable an opponent as a ferret in a closed room, they have been known to display wonderful cunning in taking advantage of the light—keeping close under the window so as to throw the glare into the eyes of the enemy, darting forwards time after time to deliver a bite, and then as often retiring to their vantage-ground.[1] But the emotions of rats do not appear to be of an entirely selfish character. There are so many accounts in the anecdote books of blind rats being led about by their seeing companions, that it is difficult to discredit an observation so frequently confirmed.[2] Moreover, rats have been frequently known to assist one another in defending themselves from dangerous enemies. Several observations of this kind are recorded by the trustworthy writer Mr. Rodwell, in his somewhat elaborate work upon this animal.

[1] See Watson's *Reasoning Power in Animals*, and *Quarterly Review*, c. i., p. 135.
[2] See especially Jesse, *Gleanings*, &c., iii., p. 206; and *Quarterly Review*, c. i., p. 135.

Again, as showing affection for human beings, I may quote the following:—' The mouse which had been tamed by Baron Trench in his prison having been taken from him, watched at the door and crept in when it was opened; being removed again, it refused all food, and died in three days.'[1]

With regard to general intelligence, every one knows the extraordinary wariness of rats in relation to traps, which is only equalled in the animal kingdom by that of the fox and the wolverine. It has frequently been regarded as a wonderful display of intelligence on the part of rats that while gnawing through the woodwork of a ship, they always stop before they completely perforate the side; but, as Mr. Jesse suggests, this is probably due to their distaste of the salt water. No such disparaging explanation, however, is possible in some other instances of the display of rat-intelligence. Thus, the manner in which they transport eggs to their burrows has been too frequently observed to admit of doubt. Rodwell gives a case in which a number of eggs were carried from the top of a house to the bottom by two rats devoting themselves to each egg, and alternately passing it down to each other at every step of the staircase.[2] Dr. Carpenter also received from an eye-witness a similar account of another instance.[3] According to the article in the *Quarterly Review*, already mentioned, rats will not only convey eggs from the top of the house to the bottom, but from bottom to top. ' The male rat places himself on his fore-paws, with his head downwards, and raising up his hind legs and catching the egg between them, pushes it up to the female, who stands on the step above, and secures it with her fore-paws till he jumps up to her; and this process is repeated from step to step till the top is reached.'

' The captain of a merchantman,' says Mr. Jesse, ' trading to the port of Boston, in Lincolnshire, had constantly missed eggs from his sea stock. He suspected that he was robbed by his crew, but not being able to dis-

---

- Thompson, *Passions of Animals*, p. 368.
- [2] *The Rat, its Natural History*, p. 102.
- [3] Mrs. Lee, *Anecdotes of Animals*, p. 264.

cover the thief, he was determined to watch his storeroom. Accordingly, having laid in a fresh stock of eggs, he seated himself at night in a situation that commanded a view of his eggs. To his great astonishment he saw a number of rats approach; they formed a line from his egg baskets to their hole, and handed the eggs from one to another in their fore-paws.'[1]

Another device to which rats resort for the procuring of food is mentioned in all the anecdote books, and it seemed so interesting that I tried some direct experiments upon the subject. I shall first state the alleged facts in the words of Watson:—

As to oil, rats have been known to get oil out of a narrow-necked bottle in the following way:—One of them would place himself, on some convenient support, by the side of the bottle, and then, dipping his tail into the oil, would give it to another to lick. In this act there is something more than what we call instinct; there is reason and understanding.[2]

Jesse also gives the following account:—

A box containing some bottles of Florence oil was placed in a store-room which was seldom opened; the box had no lid to it. On going to the room one day for one of the bottles, the owner found that the pieces of bladder and cotton at the mouth of each bottle had disappeared, and that much of the contents of the bottles had been consumed. The circumstance having excited suspicion, a few bottles were refilled with oil, and the mouths of them secured as before. Next morning the coverings of the bottles had been removed, and some of the oil was gone. However, upon watching the room, which was done through a little window, some rats were seen to get into the box, and insert their tails into the necks of the bottles, and then withdrawing them, they licked off the oil which adhered to them.[3]

Lastly, Rodwell gives another case similar in all essential respects, save that the rat licked its own tail instead of presenting it to a companion.

The experiment whereby I tested the truth of these

---

[1] Jesse, *Gleanings*, &c., ii., p. 281.
[2] *Reasoning Power in Animals*, p. 293.
[3] *Loc. cit.*

statements was a very simple one. I recorded it in
'Nature' as follows:—

It is, I believe, pretty generally supposed that rats and mice use their tails for feeding purposes when the food to be eaten is contained in vessels too narrow to admit the entire body of the animal. I am not aware, however, that the truth of this supposition has ever been actually tested by any trustworthy person, and so think the following simple experiments are worth publishing. Having obtained a couple of tall-shaped preserve bottles with rather short and narrow necks, I filled them to within three inches of the top with red currant jelly which had only half stiffened. I covered the bottles with bladder in the ordinary way, and then stood them in a place infested by rats. Next morning the bladder covering each of the bottles had a small hole gnawed through it, and the level of the jelly was reduced in both bottles to the same extent. Now, as this extent corresponded to about the length of a rat's tail if inserted at the hole in the bladder, and as this hole was not much more than just large enough to admit the root of this organ, I do not see that any further evidence is required to prove the manner in which the rats obtained the jelly, viz., by repeatedly introducing their tails into the viscid matter, and as repeatedly licking them clean. However, to put the question beyond doubt, I refilled the bottles to the extent of half an inch above the jelly level left by the rats, and having placed a circle of moist paper upon each of the jelly surfaces, covered the bottles with bladder as before. I now left the bottles in a place where there were no rats or mice, until a good crop of mould had grown upon one of the moistened pieces of paper. The bottle containing this crop of mould I then transferred to the place where the rats were numerous. Next morning the bladder had again been eaten through at one edge, and upon the mould there were numerous and distinct tracings of the rats' tails, resembling marks made with the top of a pen-holder. These tracings were evidently caused by the animals sweeping their tails about in a fruitless endeavour to find a hole in the circle of paper which covered the jelly.

With regard to mice, the Rev. W. North, rector of Ashdown, in Essex, placed a pot of honey in a closet, in which a quantity of plaster rubbish had been left by builders. The mice piled up the plaster in the form of a heap against the sides of the pot, in order to constitute an inclined plane whereby to reach the rim. A quantity of the rubbish had also been thrown into the pot, with the effect of raising the level of the honey that remained to near the rim of the pot; but, of course, the latter fact may have been due to accident, and not to design.[1] This is a case in which mal-observation does not seem to have been likely.

Powelsen, a writer on Iceland, has related an account of the intelligence displayed by the mice of that country, which has given rise to a difference of competent opinion, and which perhaps can hardly yet be said to have been definitely settled. What Powelsen said is that the mice collect in parties of from six to ten, select a flat piece of dried cow-dung, pile berries or other food upon it, then with united strength drag it to the edge of any stream they wish to cross, launch it, embark, and range themselves round the central heap of provisions with their heads joined over it, and their tails hanging in the water, perhaps serving as rudders. Pennant afterwards gave credit to this account, observing that in a country where berries were scarce, the mice were compelled to cross streams for distant forages.[2] Dr. Hooker, however, in his 'Tour in Iceland,' concludes that the account is a pure fabrication. Dr. Henderson, therefore, determined on trying to arrive at the truth of the matter, with the following result :—' I made a point of inquiring of different individuals as to the reality of the account, and am happy in being able to say that it is now established as an important fact in natural history by the testimony of two eye-witnesses of unquestionable veracity, the clergyman of Briamslaek, and Madame Benedictson of Stickesholm, both of whom assured me that they had seen the expedition performed repeatedly. Madame Benedictson, in particular, recollected having spent a whole afternoon, in her younger days, at the margin of a small lake on which these skilful navigators had embarked, and amusing herself and her companions by driving them away from the sides of the lake as they approached them. I was also informed

---

[1] Jesse, *Gleanings*, iii., p. 176.
[2] *Introduction to Arctic Zoology*, p 70.

that they make use of dried mushrooms as sacks, in which they convey their provisions to the river, and thence to their homes.'[1]

Before leaving the mice and rats I may say a few words upon certain mouse- and rat-like animals which scarcely require a separate section for their consideration. Of the harvesting mouse Gilbert White says:—

One of their nests I procured this autumn, most artificially plaited and composed of blades of wheat, perfectly round, and about the size of a cricket-ball, with the aperture so ingeniously closed that there was no discovering to what part it belonged. It was so compact and well filled that it would roll across the table without being discomposed, though it contained eight little mice that were naked and blind. As the nest was perfectly full, how could the dam come at her litter respectively, so as to administer a teat to each? Perhaps she opens different places for that purpose, adjusting them again when the business is over; but she could not possibly be contained herself in the ball with the young ones, which, moreover, would be daily increasing in size. This wonderful procreant cradle, an elegant instance of the efforts of instinct, was found in a wheat-field, suspended on the head of a thistle.

Pallas has described the provident habits of the so-called 'rat-hare' (*Lagomys*), which lays up a store of grass, or rather hay, for winter consumption. These animals, which occur in the Altai Mountains, live in holes or crevices of rock. About the middle of the month of August they collect grass, and spread it out to dry into hay. In September they form heaps or stacks of the hay, which may be as much as six feet high, and eight feet in diameter. It is stored in their chosen hole or crevice, protected from the rain.

The following is quoted from Thompson's 'Passions of Animals,' pp. 235-6:—

The life of the harvester rat is divided between eating and fighting. It seems to have no other passion than that of rage, which induces it to attack every animal that comes in its way, without in the least attending to the superior strength of its enemy. Ignorant of the art of saving itself by flight, rather

[1] Dr. Henderson, *Journal of a Residence in Iceland in* 1814 *and* 1815, vol. ii., p. 187.

than yield, it will allow itself to be beaten to pieces with a stick. If it seizes a man's hand, it must be killed before it will quit its hold. The magnitude of the horse terrifies it as little as the address of the dog, which last is fond of hunting it. When a harvester perceives a dog at a distance, it begins by emptying its cheek-pouches, if they happen to be filled with grain; it then blows them up so prodigiously, that the size of the head and neck greatly exceeds that of the rest of the body. It rears itself upon its hind legs, and thus darts upon the enemy. If it catches hold, it never quits it but with the loss of its life; but the dog generally seizes it behind, and strangles it. This ferocious disposition prevents it from being at peace with any animal whatever. It even makes war against its own species. When two harvesters meet, they never fail to attack each other, and the stronger always devours the weaker. A combat between a male and a female commonly lasts longer than between two males. They begin by pursuing and biting each other, then each of them retires aside, as if to take breath. After a short interval they renew the combat, and continue to fight till one of them falls. The vanquished uniformly serves as a repast to the conqueror.

If we contrast the fearless disposition of the harvester with the timidity of the hare or rabbit, we observe that in respect of emotions, no less than in that of intelligence, the order Rodentia comprises the utmost extremes.

The so-called 'prairie-dog' is a kind of small rodent, which makes burrows in the ground, and a slight elevation above it. The animals being social in their habits, their warrens are called 'dog-towns.' Prof. Jillson, Ph.D., kept a pair in confinement (see 'American Naturalist,' vol. v., pp. 24–29), and found them to be intelligent and highly affectionate animals. These burrows he found to contain a 'granary,' or chambers set apart for the reception of stored food. With regard to the association said to exist between this animal and the owl and rattle-snake, Prof. Jillson says, 'I have seen many dog-towns, with owls and dogs standing on contiguous, and in some cases on the same mound, but never saw a snake in the vicinity.' The popular notion that the owl acts the part of sentry to the dog requires, to say the least, confirmation.

Beaver.

Most remarkable among rodents for instinct and intelligence unquestionably stands the beaver. Indeed, there is no animal—not even excepting the ants and bees—where instinct has risen to a higher level of far-reaching adaptation to certain constant conditions of environment, or where faculties, undoubtedly instinctive, are more puzzlingly wrought up with faculties no less undoubtedly intelligent. So much is this the case that, as we shall presently see, it is really impossible by the closest study of the pyschology of this animal to distinguish the web of instinct from the woof of intelligence ; the two principles seem here to have been so intimately woven together, that in the result, as expressed by certain particular actions, it cannot be determined how much we are to attribute to mechanical impulse, and how much to reasoned purpose.

Fortunately, the doubt that for many years shrouded the facts has been dispelled by the conscientious and laborious observations of the late Mr. Lewis H. Morgan,[1] whose work throughout displays the judicious accuracy of a scientific mind. As this is much the most trustworthy, as well as the most exhaustive essay upon the subject, I shall mainly rely upon it for my statement of facts, and while presenting these I shall endeavour to point out the psychological explanation, or difficulty of explanation, to which they are severally open.

The beaver is a social animal, the male living with his single female and progeny in a separate burrow or 'lodge.' Several of these lodges, however, are usually built close together, so as to form a beaver colony. The young quit the lodge of their parents when they enter upon the summer of their third year, seek mates, and establish new lodges for themselves. As each litter numbers three or four, and breeding is annual, it follows that a beaver lodge never or rarely contains more than twelve individuals, while the number usually ranges from four to eight. Every season, and particularly when a district becomes

---
[1] *The American Beaver and his Works* (Lippincott & Co. 1868).

overstocked, some of the beavers migrate. The Indians say that in their local migrations the old beavers go up stream, and the young down; assigning as a reason that in the struggle for existence greater advantages are afforded near the source than lower down a stream, and therefore that the old beavers appropriate the former. But although lodges may thus be vacated by the old beavers, they are not left tenantless; their lease is, as it were, transferred to another beaver couple. This process of transference of ownership goes on from generation to generation, so that the same lodges are continuously occupied for centuries.

These lodges, which are always constructed in or near water, are of three kinds—the island, bank, and lake-lodge. The first are formed on small islands which may happen to occur in the ponds made by the beaver-dams. The floor of the lodge is a few inches above the level of the water, and into it there open two, or sometimes more entrances : —

> These are made with great skill, and in the most artistic manner. One is straight, or as nearly so as possible, with its floor, which is of course under water, an inclined plane, rising gradually from the bottom of the pond into the chamber; while the other is abrupt in its descent, and often sinuous in its course. The first we shall call the 'wood entrance,' from its evident design to facilitate the admission into the chamber of their wood cuttings, upon which they subsist during the season of winter. These cuttings, as will elsewhere be shown, are of such size and length that such an entrance is absolutely necessary for their free admission into the lodge. The other, which we shall call the 'beaver entrance,' is the ordinary run-way for their exit and return. It is usually abrupt, and often winding. In the lodge under consideration, the wood entrance descended from the outer run of the chamber entrance about ten feet to the bottom of the pond in a straight line, and upon an inclined plane; while the other, emerging from the line of the chamber at the side, descended quite abruptly to the bottom of the moat or trench, through which the beavers must pass, in open water, out into the pond. Both entrances were rudely arched, with a roof of interlaced sticks filled in with mud intermixed with vegetable fibre, and were extended to the bottom

of the pond or trench, with the exception of the opening at their ends. At the places where they were constructed through the floor they were finished with neatness and precision; the upper parts and sides forming an arch more or less regular, while the bottom and floor edges were formed with firm and compacted earth, in which small sticks were embedded. It is difficult to realise the artistic appearance of some of these entrances without actual inspection.

Upon the floor of the lodge there is constructed a house of sticks, brushwood, and mud, in the form of a circular or oval chamber, the size of which varies with the age of the lodge; for by a continuous process of repair (which consists in removing the decayed sticks, &c., from the interior and working them up with new material upon the exterior) the whole lodge progressively increases in size: eventually in this way the interior chamber may attain a diameter of seven or eight feet.

The 'bank lodges' are of two kinds:—

One is situated upon the bank of the stream or pond, a few feet back from its edge, and entered by an underground passage from the bed of the stream, excavated through the natural earth up into the chamber. The other is situated upon the edge of the bank, a portion of it projecting over and resting upon the bed of the channel, so as to have the floor of the chamber rest upon the bank as upon solid ground, while the external wall on the pond side projects beyond it, and is built up from the bottom of the pond.

Lastly, the 'lake lodges' are constructed on the shores of lakes, which, being usually shelving and hard, require some further variation in the structure of the lodges. These, therefore, are of interest 'as illustrations of the capacity of the beavers to vary the mode of construction of their lodges in accordance with the changes of situation.' One-half or two-thirds of the lodge is in this case 'built out upon the lake for the obvious purpose of covering the entrance, as well as for its extension into deep water.'

All these forms of lodge are, historically regarded, modified burrows.

The beaver is a burrowing animal. Indulging this propensity, he excavates chambers underground, and constructs artificial lodges upon its surface, both of which are indispensable to his security and happiness. The lodge is but a burrow above ground, covered with an artificial roof, and possesses some advantages over the latter as a place for rearing young.

There are reasons for believing that the burrow is the normal residence of the beavers, and that the lodge grew out of it, in the progress of their experience, by a process of natural suggestion. ... In addition to the lodge, the same beavers who inhabit it have burrows in the banks surrounding the pond. They never risk their personal safety upon their lodge alone, which, being conspicuous to their enemies, is liable to attack. ... As the entrances are always below the surface level of the pond, there are no external indications to mark the site of the burrow,

except occasionally a small pile of beaver-cuttings a foot or more high. These, the trappers affirm, are purposely left there by the beavers to keep the snow loose over the ends of their burrows during winter for the admission of air.

Mr. Morgan adds the very probable suggestion that this habit of piling up cuttings for purposes of ventilation may have constituted the origin of lodge-building.

It is but a step from such a surface-pile of sticks to a lodge, with its chamber above ground, and the previous burrow as its entrance from the pond. A burrow accidentally broken through at its upper end, and repaired with a covering of sticks and earth, would lead to a lodge above ground, and thus inaugurate a beaver lodge out of a broken burrow.

It is evidence of an important local variation of instinct, that in the Cascade Mountains the beavers live chiefly in burrows in the banks of streams, and rarely construct either lodges or dams. Dr. Newbury, in his report on the zoology of Oregon and California, says: 'We found the beavers in numbers, of which, when applied to beavers, I had no conception,' and yet 'we never saw their houses and seldom a dam.' Whether this local variation be due to a relapse from dam- and lodge-building instincts to the primitive burrowing instinct, or to a failure in the full development of the newer instinct, is immaterial. Probably, I think, looking to the high antiquity of the building instinct, and also to its being occasionally manifested by the Californian beavers, their case is to be regarded as one of relapsing instinct.

In selecting the site of their lodges beavers display much sagacity and forethought.

The severity of the climate in these high northern latitudes lays upon them the necessity of so locating their lodges as to be assured of water deep enough in their entrances, and also so protected in other respects, as not to freeze to the bottom;[1] otherwise they would perish with hunger, locked up in ice-bound habitations. To guard against this danger, the dam, also, must be sufficiently stable through the winter to maintain the water at a constant level; and this level, again, must be so adjusted with reference to the floor of the lodge as to enable them, at all times, to take in their cuttings from without as they are needed for food. When they leave their normal mode of life in the banks of the rivers, and undertake to live in dependence upon artificial ponds of their own formation, they are compelled to prevent the consequences of their acts at the peril of their lives.

On the upper Missouri, where the banks of the river are for miles together vertical, and rising from three to eight feet above its surface, the beavers resort to the device of making what are called 'beaver slides.' These are narrow inclined planes cut into the banks at intervals, the angle of inclination being 45° to 60°, so as to form a gradual descent from a point a few feet back from the edge of the bank to the level of the river. As Mr. Morgan observes, 'they furnish another conspicuous illustration of the fact that beavers possess a free intelligence, by means of which they are enabled to adapt themselves to the circumstances in which they are placed.'

Coming now to the habits of these animals in connection with the procuring and storing of food, it is first to be observed that 'the thick bark upon the trunks of large trees, and even upon those of medium size, is unsuitable

---

[1] To obviate this possibility, they often select as their site a place where a spring happens to rise in the bottom of the lake or pond.

for food; but the smaller limbs, the bark of which is tender and nutritious, afford the aliment which they prefer.' To obtain this food, the animals, as is well known, fell the trees by gnawing a ring round their base. Two or three nights' successive work by a pair of beavers is enough to bring down a half-grown tree, 'each family being left to the undisturbed enjoyment of the fruits of their own toil and industry.' 'When the tree begins to crackle they desist from cutting, which they afterwards continue with caution until it begins to fall, when they plunge into the pond usually, and wait concealed for a time, as if fearful that the cracking noise of the tree-fall might attract some enemy to the place.' It is of much interest that the beavers when thus felling trees know how to regulate the direction of the fall; by gnawing chiefly on the side of the trunk remote from the water, they make the tree fall towards the water, with the obvious purpose of saving as much as possible the labour of subsequent transport. For as soon as a tree is down, the next work is to cut off the branches, or such as are from two to six inches in diameter; and then, when they have been cleared of their twigs, to divide them into lengths sufficient to admit of the beavers transporting them to their lodges. The cutting into lengths is effected by making a number of semi-sections through the branch at more or less equal distances as it lies upon the ground, and then turning the branch half round and continuing the sections from the opposite side. 'To cut it (the branch) entirely through from the upper side would require an incision of such width as to involve a loss of labour.' The thicker the branch, the closer together are the sections made, and consequently the shorter are the resulting portions—the reason, of course, being that the strength of the animal would not be sufficient to transport a thick piece of timber of the same length as a thin piece which it is only just able to manage.

In moving cuttings of this description they are quite ingenious. They shove and roll them with their hips, using also their legs and tails as levers, moving sideways in the act. In this way they move the larger pieces from the more or less elevated ground on which the deciduous trees are found, over the uneven but generally descending surface to the pond. . . . After one of these cuttings has been transported to the water, a beaver, placing one end of it under his throat, pushes it before him to the place where it is to be sunk.

The sinking is no doubt partly effected by mere soaking; but there is also some evidence to show that the beavers have a method of anchoring down their supplies. Thus they have been observed towing pieces of brush to their lodges, and then, while holding the large end in their mouths, 'going down with it to the bottom, apparently to fix it in the mud bottom of the pond.' A brush-heap being thus formed, the cuttings from the felled trees are stuck through the brushwork, without which 'protection they would be liable to be floated off by the strong currents, and thus be lost to the beavers at the time when their lives might depend upon their safe custody.'

Lastly, as a method whereby the beavers can save themselves the trouble of cutting, transporting, and anchoring all at the same time, they are prone, when circumstances permit, to fell a tree growing near enough to their pond to admit of its branches being submerged in the water. The animals then well know that the branches and young shoots will remain preserved throughout the winter without any further trouble from them. But of course the supply of trees thus growing conveniently near a beaver-pond is too limited to last long.

We have next to consider the most wonderful, and I think the most psychologically puzzling structures that are presented as the works of any animal; I mean, of course, the dams and canals.

The object of the dam is that of forming an artificial pond, the use of which is to afford refuge to the animals as well as water connection with their lodges. Therefore the level of the pond must in all cases be higher than that of the lodge- and burrow-entrances, and it is usually maintained two or three feet above them.

As the dam is not an absolute necessity to the beaver for the maintenance of his life—his normal habitation being rather natural ponds and rivers, and the burrows in their banks— it is, in itself considered, a remarkable fact that he should have voluntarily transferred himself, by means of dams and ponds of his own construction, from a natural to an artificial mode of life.

In external appearance there are two distinct kinds of dams, although all are constructed on the same principle. One, the more common, is the 'stick dam,' which is composed of interlaced stick and pole work upon the lower face, with an embankment of earth mixed with the same materials on the upper face. The other is the 'solid-bank dam,' which differs from the former in having much more brush and mud worked into its construction, especially upon its surfaces; the result being that the whole formation looks like a solid bank of earth. In the first kind of dam the surplus water percolates through the structure along its entire length; but in the second kind the discharge takes place through a single furrow in the crest, which, remarkable though the fact unquestionably is, the beavers intentionally form for this purpose.

In the construction of the dam, stones are used here and there to give down-weight and solidity. These stones weigh from one to six pounds, and are carried by the beavers in the same way as they carry their mud—namely, by walking on their hind legs while holding their burden against the chest with their fore-paws. The solid dams are much firmer in their consistence than the stick dams; for while a horse might walk across the former, the weight of a man would be too great to be sustained by the latter. Each kind of dam is adapted to the locality in which it is built, the difference between the two kinds being due to the following cause. As a stream gains water and force in its descent, it develops banks, and also a broader and deeper channel. These banks assume a vertical form in the level areas where the soil is alluvial. Thus, an open stick-work dam could not in such places be led off from either bank; and even if it could, the force and depth of the stream would carry it away. Therefore in such places the beavers build their solid-bank dams, while in shallow and comparatively sluggish waters they content themselves with the smaller amount of labour involved in eth building of a stick dam.

To give some idea of the proportions of a dam, I shall epitomise a number of measurements given by Mr. Morgan:—

	Feet
Height of structure from base line	2 to 6
Difference in depth of water above and below dam	4 to 5
Width of base or section	6 to 18
Length of slope, lower face	6 to 13
Length of slope, upper face	4 to 8

The only other measurement is that of length, and this, of course, varies with the width of water to be spanned. Where this width is considerable the length of a dam may be prodigious, as the following quotation will show:—

Some of the dams in this region are not less remarkable for their prodigious length, a statement of which, in fact, would scarcely be credited unless verified by actual measurement. The largest one yet mentioned measures 260 feet, but there are dams 400 and even 500 feet long.

There is a dam in two sections, situated upon a tributary of the main branch of the Esconauba River, about a mile and a half north-west of the Washington Main. One section measures 110 and the other 400 feet, with an interval of natural bank, worked here and there, of 1,000 feet. A solid-bank dam, 20 feet in length, was first constructed across the channel of the stream, from bank to bank, with the usual opening for the surplus water, five feet wide. As the water rose and overflowed the bank on the left side, the dam was extended for 90 feet, until it reached ground high enough to confine the pond. This natural bank extended up the stream, and nearly parallel with it, for 1,000 feet, where the ground again subsided, and allowed the water in the upper part of the pond to flow out and around into the channel of the stream below the dam. To meet this emergency a second dam, 420 feet long, was constructed. For the greater part of its length it is low, but in some places it is two and a half and three feet high, and constructed of stick-work on the land, and with an earth embankment on its outer face. In effect, therefore, it is one structure 1,530 feet in length, of which 530 feet in two sections is artificial, and the remainder natural bank, but worked here and there where depressions in the ground required raising by artificial means.

It is truly an astonishing fact that animals should engage in such vast architectural labours with what appears to be the deliberate purpose of securing, by such very artificial means, the special benefits that arise from their high engineering skill. So astonishing, indeed, does this fact appear, that as sober-minded interpreters of fact we would fain look for some explanation which would not necessitate the inference that these actions are due to any intelligent appreciation, either of the benefits that arise from the labour, or of the hydrostatic principles to which this labour so clearly refers. Yet the more closely we look into the subject, the more impossible do we find it to account for the facts by any such easy method. Thus it seems perfectly certain that the beavers, properly and strictly speaking, understand the use of their dams in maintaining a certain level of water. For it is unquestionable that in the solid-bank dams, as already observed, a regular opening or trough is cut at one part of its crest to provide for the overflow; and now it has to be added that this opening is purposely widened or narrowed with reference to the amount of water in the stream at different times, so as to ensure the maintenance of a constant level in the pond. Similarly, though by different means, the same end is secured in the case of the stick dams. For 'in most of these dams the rapidity or slowness with which the surplus water is discharged is undoubtedly regulated by the beavers; otherwise the level of the pond would continually vary. There must be a constant tendency to enlarge the orifices through which the water passes,' when the stream is small, and *vice versâ*; otherwise the lodges would be either inundated or have their sub-aquatic entrances exposed.[1] Moreover, a very little consideration is enough to show that in stick dams the tendency to increased leakage from the effects of percolation, and to a settling down of the dam as its materials decay from underneath, must demand unceasing vigi-

---

[1] In times of considerable 'freshet' the former case sometimes occurs; the beavers not being able to provide for a very considerable overflow through their dams, the latter become then wholly submerged. When again exposed, the animals take great pains in repairing the injuries sustained.

lance and care to avert the consequences. And accordingly it is found that 'in the fall of the year a new supply of materials is placed upon the lower face of these dams to compensate this waste from decay.'

Now, it is obvious that we have here presented a continual variation of conditions, imposed by continual variations in the amount of water coming down; and it is a matter of observation that these variations are met by the beavers in the only way that they can be met—namely, by regulating the amount of flow taking place through the dams. It will therefore be seen that we have here to consider a totally different case from that of the operation of pure instinct, however wonderful such operation may be. For the adaptations of pure instinct only have reference to conditions that are unchanging; so that if in this case we suppose pure instinct to account for all the facts, we must greatly modify our ideas of what pure instinct is taken to mean. Thus we must suppose that when the beavers find the level of their ponds rising or falling, the discomfort which they experience acts as a stimulus to cause them, without intelligent purpose, either to widen or to narrow the orifices in their dams as the case may be. And not only so, but the conditions of stimulation and response must be so nicely balanced that the animals widen or narrow these orifices with a more or less precise *quantitative* reference to the degree of discomfort, actual or prospective, which they experience. Now it seems to me that even thus far it is an extremely difficult thing to believe that the mechanism of pure or wholly unintelligent instinct could admit of sufficient refinement to meet so complex a case of compensating adaptation; and, as we shall immediately see, this difficulty increases still more as we contemplate additional facts relating to these structures.

Thus it sometimes happens that in large dams the pressure of the water which they keep back is so considerable that their stability is endangered. In such cases it has been observed by Mr. Morgan that, at a short distance beneath the main dam, another and lower dam is thrown across the stream, with the result of forming a shallow pond between the two. This pond is—

Of no apparent use for beaver occupation, but yet subserving the important purpose of setting back water to the depth of twelve or fifteen inches; . . . . and the small dam, by maintaining the water a foot deep below the great dam, diminishes to this extent the difference in level above and below, and neutralises to the same extent the pressure of the water in the pond above against the main structure.

'Whether,' adds Mr. Morgan, with commendable caution, 'the lower dam was constructed with this motive and for this object, or is explainable on some other hypothesis, I shall not venture an opinion.' But as, he further adds, 'I have also found the same precise work repeated below other large dams,' we are led to conclude that their correlation cannot at least be accidental; and as it is of so definite a character, there really seems no 'other hypothesis' open to us than that of its having reference to the stability of the main dam. Yet, if this is the case, it becomes in my opinion simply impossible to attribute the fact to the operation of pure instinct.

Again, Mr. Morgan observed one case in which, higher up stream than the main dam, there was constructed another dam, ninety-three feet long, and two and a half feet high at the centre :—

A dam at this point is apparently of no conceivable use to improve the lake for beaver occupation. It has one feature, also, in which it differs from other dams except those upon lake outlets, and that consists in its elevation, at all points, of about two feet above the level of the lake at ordinary stages of the water. In all other dams, except those upon lake outlets, and in most of the latter, the water stands quite near their crests, while in the one under consideration it stood about two feet below it. This fact suggests at least the inference, although it may have but little of probability to sustain it, that it was constructed with special reference to sudden rises of the lake in times of freshet, and that it was designed to hold this surplus water until it could be gradually discharged through the dam into the great space below. It would at least subserve this purpose very efficiently, and thus protect the dam below it from the effects of freshets. To ascribe the origin of this dam to such motives of intelligence is to invest this animal with a higher degree of sagacity than we have probable reason to concede to him, and yet it is proper to mention the relation in which these dams stand to each other—whether that relation is regarded as accidental or intentional.

As before, we have here to commend the caution displayed by the closing sentence ; but, as useless dams are not found in other places, the inference clearly is that the dam in question, both as regards its exceptional position and exceptional height, can only be explained by supposing the structure to have been designed for the use which it unquestionably served. That is to say, if we do not entertain this explanation, there is no other to be suggested; and although in any ordinary or occasional instance of the display of animal intelligence in such a degree as this I should not hesitate to attribute the facts to accident, in the case of the beaver there are such a multitude of constantly recurring facts, all and only referable to a practical though not less extraordinary appreciation of hydrostatic principles, that the hypothesis of accident must here, I think, be laid aside. To substantiate this statement I shall detail the facts concerning the beaver-canals.

As Mr. Morgan, who first discovered and described these astonishing structures, observes,—

Remarkable as the dam may still be considered, from its structure and objects, it scarcely surpasses, if it may be said to equal, these water-ways, here called canals, which are excavated through the low lands bordering their ponds for the purpose of reaching the hard wood, and for affording a channel for its transportation to their lodges. To conceive and execute such a design presupposes a more complicated and extended process of reasoning than that required for the construction of a dam, and, although a much simpler work to perform when the thought was fully developed, it was far less to have been expected from a mute animal.

These canals are developed in this way. One of the principal objects served by a dam thrown across a small stream, is that of flooding the low ground so as to obtain water connection with the first high ground upon which hard wood is to be found, such connection being convenient, or even necessary, for the purposes of transport.

Where the pond fails to accomplish this fully, and also where the banks are defined and mark the limits of the pond, the deficiency is supplied by the canals in question. On descending surfaces, as has elsewhere been stated, beavers roll and drag their short cuttings down into the ponds. But where the ground is low it is generally so uneven and rough as to render it extremely difficult, if not impossible, for the beavers to move them for any considerable distance by physical force. Hence the canal for floating them across the intervening level ground to the pond. The necessity for it is so apparent as to diminish our astonishment at its construction; and yet that the beaver should devise a canal to surmount this difficulty is not the less remarkable.

The canals, which are made by excavation, are usually from three to five feet wide, three feet deep, and perhaps hundreds of feet long—the length of course depending on the distance between the lodge and the wood supply. They are cut in the form of trenches, having perpendicular sides and abrupt ends. All roots of trees, under-brush, &c., are cleared away in their course, so as to afford an unobstructed passage. These canals are of such frequent occurrence that it is impossible to attribute them to accident; they are evidently made, at the cost of much labour, with the deliberate purpose of putting them to the use for which they are designed. In executing this purpose there is sometimes displayed a depth of engineering forethought over details of structure required by the circumstances of special localities, which is even more astonishing than the execution of the general idea. Thus it not unfrequently happens that when a canal has been run for a certain distance, a rise in the level of the ground renders it impossible to continue the structure further from the water supply or lodge-pond, without either incurring a great amount of labour in digging the canal with progressively deepening sides, or leaving the trench empty of water, and so useless. In such cases the beavers resort to various expedients, according to the nature of the ground.

Mr. Morgan gives an interesting sketch of one such case, where the canal is excavated through low ground for a distance of 450 feet, when it reaches the first rise of ground, and throughout this distance, being level with the pond, it is supplied with water from this source. Where the rise begins a dam is made, and the canal is then continued for 25 feet at a level of one foot higher than before. This higher level reach is supplied with water collected from still higher levels by another dam, extending for 75 feet upon one side of the canal and 25 feet on the other, in the form of a crescent with its concavity directed towards the highlands, so as to collect all the drainage water, and concen-

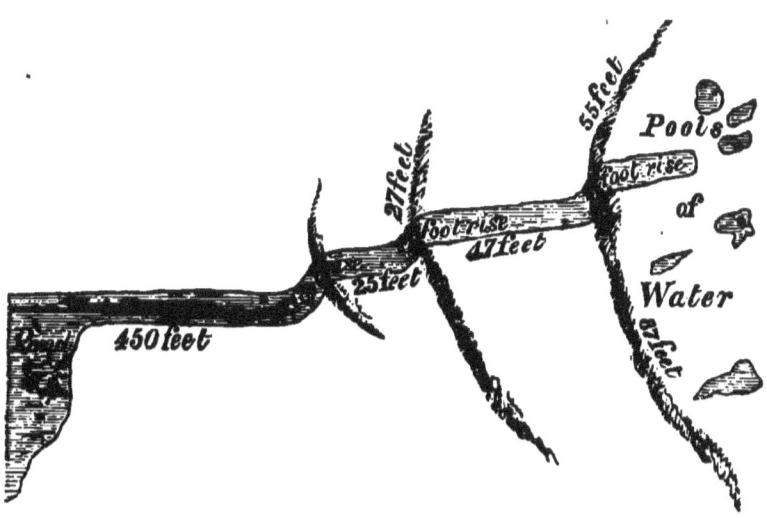

trate it into the second reach of the canal. Beyond this larger dam there is another abrupt rise of a foot, and the canal is there continued for 47 feet more, where a third dam is built resembling the second in construction, only having a still wider span on either side of the canal (142 feet), so as to catch a still larger quantity of drainage water to supply the third or uppermost reach of the canal. We have, therefore, here presented, not only a perfect application of the principle of 'locks,' which are used in canals of human construction, but also the principle of collecting water to supply the reaches situated on the slope by means of elaborately constructed dams of wide extent, and of the best form for the purpose. There is thus shown much too great a concurrence of engineering principles to the attainment of one object to admit of our attributing the facts to accident. On this structure Mr. Morgan observes:—

The crests of these dams where they cross the canals are depressed, or worn down, in the centre, by the constant passage of beavers over them while going to and fro and dragging their cuttings. This canal with its adjuncts of dams and its manifest objects is a remarkable work, transcending very much the ordinary estimates of the intelligence of the beaver. It served to bring the occupants of the pond into easy connection by water with the trees that supplied them with food, as well as to relieve them from the tedious and perhaps impossible task of transporting their cuttings 500 feet over uneven ground unassisted by any descent.

Again, in another case, also sketched by Mr. Morgan, another device is resorted to, and one which, having reference to the particular circumstances of the case, is the best that could have been adopted. Here the canal, proceeding from the pond to the woodland 150 feet distant, encounters at the woodland a rising slope covered with hard wood. Thereupon the canal bifurcates, and the two diverging branches or prongs are carried in opposite directions along the base of the woodland rise, one for a distance of 100 and the other for 115 feet. The level being throughout the same, the water from the pond supplies the two branch-canals as well as the trunk. Both branches end with abrupt vertical faces. Now the object of these branches is sufficiently apparent:—

After the rising ground, and with it the hard wood trees, were reached at the point where it branches, there was no very urgent necessity for the branches. But their construction along the base of the high ground gave them a frontage upon the canal of 215 feet of hard-wood lands, thus affording to them, along this extended line, the great advantages of water transportation for their cuttings.

One more proof of engineering purpose in the construction of canals will be sufficient to place beyond all question the fact that beavers form these canals, as they form their dams, with a far-seeing perception of the suitability of highly artificial means to the attainment of particular ends, under a variety of special circumstances. Mr. Morgan observed one or two instances where the land included in a wind or loop of a river was cut through by a beaver canal across the narrowest part, 'apparently to shorten the distance in going up and down by water.' Judging from the figures which he gives, drawn to measurement, there can be no question that such was the object; and as these structures may be one or two hundred feet in length, and represent the laborious excavation of some 1,500 cubic feet of soil, the animals must be actuated by the most vivid conception of the subsequent saving in labour that is to be effected by making an artificial communication across the chord of an arc, instead of always going round the natural curve of a stream.

Regarding now together all these facts relating to the psychology of the beaver, it must be confessed, as I said at the outset, that we have presented to us a problem perhaps the most difficult of any that we have to encounter in the whole range of animal intelligence. On the one hand, it seems incredible that the beaver should attain to such a level of abstract thought as would be implied by his forming his various structures with the calculated purpose of achieving the ends which they undoubtedly subserve. On the other hand, as we have seen, it seems little less than impossible that the formation of these structures can be due to instinct. Yet one or other hypothesis, either singly or in combination, must be resorted to. The case, it will be observed, thus differs from that of the more wonderful performances of instinct elsewhere, such as that of ants and bees, inasmuch as the performances here are so complex and varied, as well as having reference to physical principles of a much more recondite or less observable nature. The case from its theoretical side being thus one of much difficulty, I think it will be better to postpone its discussion till in 'Mental Evolution' I come to treat of the whole subject of instinct in relation to intelligence.

I must not, however, conclude this epitome of the facts without alluding to the only other publication on the habits of the beaver which is of distinctly scientific value. This is a short but interesting paper by Prof. Alexander Agassiz.[1] He says that the largest dam he has himself seen measured 650 feet in length, and $3\frac{1}{2}$ feet in height, with a small number of lodges in the vicinity of the pond. The number of lodges is always thus very small in proportion to the size of the dam, the greatest number of lodges that he has observed upon one pond being five. It is evident from this that beavers are not really gregarious in their habits, and that their dams and canals 'are the work of a comparatively small number of animals; but to make up for the numbers the work of succeeding inhabitants of any one pond must have been carried on for centuries to accomplish the gigantic results we find in some localities.'

In once case Prof. Agassiz obtained what may be termed geological evidence of the truth of an opinion advanced by Mr. Morgan, that beaver-works may be hundreds if not thousands of years in course of continuous formation. For the purpose of obtaining a secure foundation for a mill dam erected above a beaver dam, it was necessary to clear away the soil from the bottom of the beaver pond. This soil was found to be a peat bog. A trench was dug into the peat 12 feet wide by 1,200 feet long, and 9 feet deep; all the way along this trench old stumps of trees were found at various depths, some still bearing marks of having been gnawed by beavers' teeth. Agassiz calculated the growth of the bog as about a foot per century, so that here we have tolerably accurate evidence of an existing beaver dam being somewhere about a thousand years old.

The gradual growth of these enormous dams has the effect of greatly altering the configuration of the country where they occur. By taking levels from dams towards the sources of streams on which they occur, Agassiz was able ideally to reconstruct the original landscape before the growth of the dams, and he found that, 'from the

[1] Note on Beaver Dams (*Proc. Boston Soc. Nat. Hist.*, 1869, p. 101, *et seq.*).

nature of the surrounding country, the open spaces now joining the beaver ponds—the beaver meadows where the trees are scanty or small—must at one time have been all covered with forests.' At first the beavers 'began to clear the forest just in the immediate vicinity of the dams, extending in every direction, first up the stream as far as the nature of the creek would allow, and then laterally by means of their canals, as far as the level of the ground would allow, thus little by little clearing a larger area according to the time they have occupied any particular place.' In this way beavers may change the whole aspect of large tracts of country, covering with water a great extent of ground which was once thickly wooded.

CHAPTER XIII.

ELEPHANT.

THE intelligence of the elephant is no doubt considerable, although there is equally little doubt that it is generally exaggerated. Some of the most notorious instances of the display of remarkable sagacity by this animal are probably fabulous, or at least are not sufficiently corroborated to justify belief. Such, for instance, is the celebrated story told by Pliny with all the assurance of a '*certum est*,'[1] and repeated by Plutarch,[2] of the elephant, who having been beaten for not dancing properly, was afterwards found practising his steps alone in the light of the moon. Although this story cannot, in the absence of corroboration, be accepted as fact, we ought to remember, in connection with it, that many talking and piping birds unquestionably practise in solitude the accomplishments which they desire to learn.

Quitting, however, the enormous multitude of anecdotes, more or less doubtful, and which may or may not be true, I shall select a few well-authenticated instances of the display of elephant intelligence.

Memory.

As regards memory, several cases are on record of tamed elephants having become wild, and, on again being captured after many years, returning to all their old habits under domestication. Mr. Corse publishes in the 'Philosophical Transactions'[3] an instance which came under his own notice. He saw an elephant, which

---

[1] Plin., *Hist. Nat.*, viii. 1-13. [2] *De Solert. Anim.*, c. 12.
[3] *Philosophical Transactions*, 1799, p. 40.

was carrying baggage, take fright at the smell of a tiger and run off. Eighteen months afterwards this elephant was recognised by its keepers among a herd of wild companions, which had been captured and were confined in an enclosure. But when anyone approached the animal he struck out with his trunk, and seemed as fierce as any of the wild herd. An old hunter then mounted a tame elephant, went up to the feral one, seized his ear and ordered him to lie down. Immediately the force of old associations broke through all opposition, the word of command was obeyed, and the elephant while lying down gave a certain peculiar squeak which he had been known to utter in former days. The same author gives another and more interesting account of an elephant which, after having been for only two years tamed, ran wild for fifteen years, and on being then recaptured, remembered in all details the words of command. This, with several other well-authenticated facts of the same kind,[1] shows that the elephant certainly has an exceedingly tenacious memory, rendering credible the statement of Pliny, that in their more advanced age these animals recognise men who were their drivers when young.[2]

Emotions.

Concerning emotions, the elephant seems to be usually actuated by the most magnanimous of feelings. Even his proverbial vindictiveness appears only to be excited under a sense of remembered injustice. The universally known story of the tailor and the elephant doubtless had a foundation in fact, for there are several authentic cases on record of elephants resenting injuries in precisely the same way;[3] and Captain Shipp[4] personally tested the matter by giving to an elephant a sandwich of bread, butter, and cayenne pepper. He then waited for six

[1] See Bingley, *loc. cit.*, vol. i., pp. 148-51.
[2] *Hist. Nat.*, viii., 5.
[3] For these and other cases of vindictiveness, see Bingley, *loc. cit.*, vol. i., pp. 156-8.
[4] *Memoirs*, vol. i., p. 448.

weeks before again visiting the animal, when he went into the stable and began to fondle the elephant as he had previously been accustomed to do. For a time no resentment was shown, so that the Captain began to think that the experiment had failed; but at last, watching for an opportunity, the elephant filled his trunk with dirty water, and drenched the Captain from head to foot.

Griffiths says that at the siege of Bhurtpore, in 1805, the British army had been a long time before the city, and, owing to the hot dry winds, the ponds and tanks had dried up. There used therefore to be no little struggle for priority in procuring water at one of the large wells which still contained water:—

> On one occasion two elephant-drivers, each with his elephant, the one remarkably large and strong, and the other comparatively small and weak, were at the well together; the small elephant had been provided by his master with a bucket for the occasion, which he carried on the end of his proboscis, but the larger animal, being destitute of this necessary vessel, either spontaneously, or by the desire of his keeper, seized the bucket, and easily wrested it from his less powerful fellow-servant; the latter was too sensible of his inferiority openly to resent the insult, though it is obvious that he felt it; but great squabbling and abuse ensued between the keepers. At length the weaker animal, watching the opportunity when the other was standing with his side to the well, retired backwards a few paces in a very quiet and unsuspicious manner, and then, rushing forward with all his might, drove his head against the side of the other, and fairly pushed him into the well.

Great trouble was experienced in extricating this elephant from the well—a task which would, indeed, have been impossible but for the intelligence of the animal itself. For when a number of fascines, which had been employed by the army in conducting the siege, were thrown down the well, the elephant showed sagacity enough to arrange them with his trunk so as to construct a continuously rising platform, by which he gradually raised himself to a level with the ground.

Allied to vindictiveness for small injuries is revenge for large ones, and this is often shown in a terrible manner by wounded elephants. For instance, Sir E. Tennent writes:—

Some years ago an elephant which had been wounded by a native, near Hambangtotte, pursued the man into the town, followed him along the street, trampled him to death in the bazaar before a crowd of terrified spectators, and succeeded in making good its retreat to the jungle.

Many other cases of vindictiveness, more or less well authenticated, may be found mentioned by Broderip,[1] Bingley,[2] Mrs. Lee,[3] Swainson,[4] and Watson.[5] This trait of emotional character seems to be more generally present in the elephant than in any other animal, except perhaps the monkey.

Another emotion strongly developed in the elephant is sympathy. Numberless examples on this head might be adduced, but one or two may suffice. Bishop Huber saw an old elephant fall down from weakness, and another elephant was brought to assist the fallen one to rise. Huber says he was much struck with the almost human expression of surprise, alarm, and sympathy manifested by the second elephant on witnessing the condition of the first. A chain was fastened round the neck and body of the sick animal, which the other was directed to pull. For a minute or two the healthy elephant pulled strongly; but on the first groan given by its distressed companion it stopped abruptly, 'turned fiercely round with a loud roar, and with trunk and fore-feet began to loosen the chain from the neck.'

Again, Sir E. Tennent says:—

The devotion and loyalty which the herd evince to their leader are very remarkable. This is more readily seen in the case of a tusker than any other, because in a herd he is generally the object of the keenest pursuit by the hunters. On such occasions the others do their utmost to protect him from danger: when driven to extremity they place their leader in the centre

---

[1] *Zoological Recreations*, p. 315.
[2] *Animal Biography*, i., pp. 156-8.
[3] *Anecdotes of Animals*, p. 276.
[4] *Habits and Instincts of Animals*, p. 37.
[5] *Reasoning Power of Animals*, chap. iv.

and crowd so eagerly in front of him that the sportsmen have to shoot a number which they might otherwise have spared. In one instance a tusker, which was badly wounded by Major Rogers, was promptly surrounded by his companions, who supported him between their shoulders, and actually succeeded in covering his retreat to the forest.

Lastly, allusion may be made to the celebrated observation of M. le Baron de Lauriston, who was at Laknaor during an epidemic which stretched a number of natives sick and dying upon the road. The Nabob riding his elephant over the road was careless whether or not the animal crushed the men and women to death, but not so the elephant, which took great pains to pick his steps among the people so as not to injure them.

The following account of emotion and sagacity is quoted from the Rev. Julius Young's Memoirs of his father, Mr. Charles Young, the actor. The animal mentioned is the one that subsequently attained such widespread notoriety at Exeter Change, not only on account of his immense size, but still more because of his cruel death:—

In July 1810, the largest elephant ever seen in England was advertised as 'just arrived.' As soon as Henry Harris, the manager of Covent Garden Theatre, heard of it, he determined, if possible, to obtain it; for it struck him that if it were to be introduced into the new pantomime of 'Harlequin Padmenaba,' which he was about to produce at great cost, it would add greatly to its attraction. Under this impression, and before the proprietor of Exeter Change had seen it, he purchased it for the sum of 900 guineas. Mrs. Henry Johnston was to ride it, and Miss Parker, the columbine, was to play up to it. Young happened to be one morning at the box-office adjoining Covent Garden Theatre, when his ears were assailed by a strange and unusual uproar within the walls. On asking one of the carpenters the cause of it, he was told 'it was something going wrong with the elephant; he could not exactly tell what.' I am not aware what the usage may be nowadays, but then, whenever a new piece had been announced for presentation on a given night, and there was but scant time for its preparation, a rehearsal would take place after the night's regular performance was over, and the audience had been dismissed. One such there had been the night before my father's curiosity had been roused. As it had been arranged that Mrs. Henry Johnston, seated in a howdah on the elephant's back, should pass over a bridge in the centre of a numerous group of followers, it was thought expedient that the unwieldy monster's tractability should be tested. On stepping up to the bridge, which was slight and temporary, the sagacious brute drew back his fòre-feet and refused to budge. It is well known as a fact in natural history that the elephant, aware of his unusual bulk, will never trust its weight on any object which is unequal to its support. The stage-manager, seeing how resolutely the animal resisted every attempt made to compel or induce it to go over the bridge in question, proposed that they should stay proceedings till next day, when he might be in a better mood. It was during the repetition of the experiment that my father, having heard the extraordinary sounds, determined to go upon the stage, and see if he could ascertain the cause of them. The first sight that met his eyes kindled his indignation. There stood the high animal, with downcast eyes and flapping ears, meekly submitting to blow after blow from a sharp iron goad, which his keeper was driving ferociously into the fleshy part of his neck, at the root of the ear. The floor on which he stood was converted into a pool of blood. One of the proprietors, impatient at what he regarded as senseless obstinacy, kept urging the driver to proceed to still severer extremities, when Charles Young, who was a great lover of animals, expostulated with him, went up to the poor patient sufferer, and patted and caressed him; and when the driver was about to wield his instrument again, with even still more vigour, he caught him by the wrist as in a vice, and stayed his hand from further violence. While an angry altercation was going on between Young and the man of colour, who was the driver, Captain Hay, of the *Ashel*, who had brought over 'Chuny' in his ship, and had petted him greatly on the voyage, came in and begged to know what was the matter. Before a word of explanation could be given, the much-wronged creature spoke for himself; for, as soon as he perceived the entrance of his patron, he waddled up to him, and, with a look of gentle appeal, caught hold of his hand with his proboscis, plunged it into his bleeding wound, and then thrust it before his eyes. The gesture seemed to say, as plainly as if it had been enforced by speech, 'See how these cruel men treat Chuny. Can *you* approve of it?' The hearts of the hardest present were sensibly touched by what they saw, and among them that of the gentleman who had been so energetic in promoting its harsh treatment. It was under a far better impulse that he ran out into the street, purchased a few apples at a stall, and offered them to him. Chuny eyed him askance, took them, threw them beneath his feet, and when he had crushed them to pulp, spurned them from him. Young, who had gone into Covent Garden on the same errand as the gentleman who had preceded him, shortly after re-entered, and also held out to him some fruit, when, to the astonishment of the bystanders, the elephant ate every morsel, and after he had done so, twined his trunk with studied gentleness around Young's waist, marking by his action that, though he had resented a wrong, he did not forget a kindness.

It was in the year 1814 that Harris parted with Chuny to Cross, the proprietor of the menagerie at Exeter Change. One of the purchaser's first acts was to send Charles Young a life ticket of admission to his exhibition; and it was one of his little innocent vanities, when passing through the Strand with any friend, to drop in on Chuny, pay him a visit in his den, and show the intimate relations which existed between them. Some years after, when the elephant's theatrical career was run, and he was reduced to play the part of captive in one of the cages of Exeter Change, a thoughtless dandy one day amused himself by teasing him with the repeated offer of lettuces—a vegetable for which he was known to have an antipathy. At last he presented him with an apple, but, at the moment of his taking it, drove a large pin into his trunk, and then sprang out of his reach. The keeper seeing that the poor creature was getting angry, warned the silly fellow off, lest he should become dangerous. With a contemptuous shrug of the shoulder, he trudged off to the other end of the gallery, and there displayed his cruel ingenuity on other humbler beasts, till, after the absence of half-an-hour, he once more approached one of the cages opposite the elephant's. By this time he had forgotten his pranks with Chuny, but Chuny had not forgotten him; and as he was standing with his back towards him, he thrust his proboscis through the bars of his prison, twitched off the offender's hat, dragged it in to him, tore it to shreds, then threw it into the face of the offending gaby, consummating his revenge with a loud guffaw of exultation. All present proclaimed their approbation of this act of retributive justice, and the discomfited coxcomb had to retreat from the scene in confusion, jump into a hackney coach, and betake himself to the hatter's in quest of a new tile for his unroofed skull. The tragic end of poor Chuny must be within the recollection of many of my readers. From some cause unknown he went mad, and after poison had been tried in vain it took 152 shots, discharged by a detachment of the Guards, to despatch him.[1]

[1] Quoted in *Animal World*, March 1882.

ELEPHANT—EMOTIONS.

The elephant in many respects displays strange peculiarities of emotional temperament. Thus Mr. Corse says:—'If a wild elephant happens to be separated from its young for only two or three days, though giving suck, she never after recognises or acknowledges it;'[1] yet the young one knows its dam, and cries plaintively for her assistance.

Again, in the wild state, the spirit of exclusiveness shown by members of a herd (*i.e.* family) towards elephants of other herds is remarkable. Sir E. Tennent writes:—

If by any accident an elephant becomes hopelessly separated from his own herd, he is not permitted to attach himself to any other. He may browse in the vicinity, or frequent the same place to drink and to bathe; but the intercourse is only on a distant and conventional footing, and no familiarity or intimate association is under any circumstances permitted. To such a height is this exclusiveness carried, that even amidst the terror of an elephant corral, when an individual, detached from his own party in the *mêlée* and confusion, has been driven into the enclosure with an unbroken herd, I have seen him repulsed in every attempt to take refuge among them, and driven off by heavy blows with their trunks as often as he attempted to insinuate himself within the circle which they had formed for common security. There can be no reasonable doubt that this jealous and exclusive policy not only contributes to produce, but mainly serves to perpetuate, the class of solitary elephants which are known by the term *goondahs* in India, and which from their vicious propensities and predatory habits are called *Hora*, or *Rogues*, in Ceylon.[2]

The emotional temper, or rather transformation of emotional psychology, which is exhibited by the Rogues here mentioned, is as extraordinary as it is notorious. From being a peaceable, sympathetic, and magnanimous animal, the elephant, when excluded from the society of its kind, becomes savage, cruel, and morose to a degree unequalled in any other animal. The repulsive accounts of the bloodthirsty rage and wanton destructiveness of Rogues show that their actions are not due to sudden bursts of fury at the sight of man or his works, but rather to a

---

[1] *Philosophical Transactions*, 1873.
[2] *Natural History of Ceylon*, p. 114.

deliberate and brooding resolve to wage war on everything, so that the animal patiently lies in wait for travellers, rushing from his ambush only when he finds that the latter are within his power. As showing the cold-blooded determination of this murderous desire, I may quote the following case, as it was communicated to Sir E. Tennent:—

We had, says the writer, calculated to come up with the brute where it had been seen half an hour before; but no sooner had one of our men, who was walking foremost, seen the animal at the distance of some fifteen or twenty fathoms, than he exclaimed, 'There! there!' and immediately took to his heels, and we all followed his example. The elephant did not see us until we had run some fifteen or twenty paces from the spot where we turned, when he gave us chase, screaming frightfully as he came on. The Englishman managed to climb a tree, and the rest of my companions did the same; as for myself, I could not, although I made one or two superhuman efforts. But there was no time to be lost. The elephant was running at me with his trunk bent down in a curve towards the ground. At this critical moment Mr. Lindsay held out his foot to me, with the help of which and then of the branches of the tree, which were three or four feet above my head, I managed to scramble up to a branch. The elephant came directly to the tree and attempted to force it down, which he could not. He first coiled his trunk round the stem, and pulled it with all his might, but with no effect. He then applied his head to the tree, and pushed for several minutes, but with no better success. He then trampled with his feet all the projecting roots, moving, as he did so, several times round and round the tree. Lastly, failing in all this, and seeing a pile of timber, which I had lately cut, at a short distance from us, he removed it all (thirty-six pieces) one at a time to the root of the tree, and piled them up in a regular business-like manner; then placing his hind feet on this pile, he raised the fore part of his body, and reached out his trunk, but still he could not touch us, as we were too far above him. The Englishman then fired, and the ball took effect somewhere on the elephant's head, but did not kill him. It made him only the more furious. The next shot, however, levelled him to the ground. I afterwards brought the skull of the animal to Colombo, and it is still to be seen at the house of Mr. Armitage.[1]

[1] *Natural History of Ceylon*, p. 140.

Another highly curious trait in the emotional psychology of the elephant is the readiness with which the huge animal expires under the mere influence of what the natives call a 'broken heart.' The facts on this head are without a parallel in any other animal, and are the more remarkable from the fact that, so far as natural length of life is any token, the elephant may be said to have more vitality, or innate power of living, than any other terrestrial mammal. Again, to quote from Sir E. Tennent:—

Amongst the last of the elephants noosed was the *rogue*. Though far more savage than the others, he joined in none of their charges and assaults on the fences, as they uniformly drove him off, and would not permit him to enter their circle. When dragged past another of his companions in misfortune, who was lying exhausted on the ground, he flew upon him and attempted to fasten his teeth in his head; this was the only instance of viciousness which occurred during the progress of the corral. When tied up and overpowered, he was at first noisy and violent, but soon lay down peacefully, a sign, according to the hunters, that his death was at hand. Their prognostication was correct; he continued for about twelve hours to cover himself with dust like the others, and to moisten it with water from his trunk; but at length he lay exhausted, and died so calmly, that having been moving but a few moments before, his death was only perceived by the myriads of black flies by which his body was almost instantly covered, although not one was visible a moment before.[1]

But this peculiarity is not confined to rogue elephants. Thus Captain Yule, in his 'Narrative of an Embassy to Ava in 1855,' records an illustration of this tendency of the elephant to sudden death. One newly captured, the process of taming which was exhibited to the British Envoy, 'made vigorous resistance to the placing of a collar on its neck, and the people were proceeding to tighten it, when the elephant, which had lain down as if quite exhausted, reared suddenly on the hind quarters, and fell on its side—*dead!*'

Mr. Strachan noticed the same liability of the elephants to sudden death from very slight causes. 'Of the

[1] *Natural History of Ceylon*, p. 196.

fall,' he says, 'at any time, though on plain ground, they either die immediately, or languish till they die; their great weight occasioning them so much hurt by the fall.'[1]

And Sir E. Tennent observes that,—

In the process of taming, the presence of the tame ones can generally be dispensed with after two months, and the captive may then be ridden by the driver alone; and after three or four months he may be entrusted with labour, so far as regards docility; but it is undesirable, and even involves the risk of life, to work an elephant too soon; it has frequently happened that a valuable animal has lain down and died the first time it was tried in harness, from what the natives believed to be 'broken heart,' certainly without any cause inferable from injury or previous disease.[2]

Nor is this tendency to die under the influence of mere emotion restricted to the effect of a 'broken heart;' it seems also to occur under the power of strong emotional disturbances of other kinds. For instance, an elephant caught and trained by Mr. Cripps is thus alluded to by Sir E. Tennent:—

This was the largest elephant that had been tamed in Ceylon; he measured upwards of nine feet at the shoulders, and belonged to the caste so highly prized for the temples. He was gentle after his first capture, but his removal from the corral to the stables, though only a distance of six miles, was a matter of the extremest difficulty; his extraordinary strength rendering him more than a match for the attendant decoys. He on one occasion escaped, but was recaptured in the forest; and he afterwards became so docile as to perform a variety of tricks. He was at length ordered to be removed to Colombo; but such was his terror on approaching the fort, that on coaxing him to enter the gate he became paralysed in the extraordinary way elsewhere alluded to, and *died on the spot.*

General Intelligence.

The higher mental faculties of the elephant are more advanced in their development than in any other animal, except the dog and monkey. I shall, therefore, devote

---

[1] *Phil. Trans.*, A.D. 1701, vol. xxiii., p. 1052.
[2] *Loc. cit.*, p. 216.

some considerable space to the narration of instances of its display. The general fact that elephants are habitually employed in certain parts of India for the purposes of building, storing timber, &c., in itself shows a level of docile intelligence which only that of the dog can rival; but I shall here confine myself to stating special instances of the display of sagacity unusually high, even for the elephant.

Capt. Shipp, in his 'Memoirs,' gives the following incident, of which he was an eye-witness. During a march with guns in the mountainous districts of India, the force of which he was a member came to a steep ascent. A staircase of logs was prepared to enable the elephants to ascend the slope. When all was ready the first elephant was led to the bottom of the staircase:—

He looked up, shook his head, and when forced by his driver, roared piteously. There can be no question, in my opinion, but that this sagacious animal was competent instinctively to judge of the practicability of the artificial flight of steps thus constructed; for the moment some little alteration had been made, he seemed willing to approach. He then commenced his examination and scrutiny by pressing with his trunk the trees that had been thrown across; and after this he put his fore-leg on with great caution. . . . The next step for him to ascend by was a projecting rock, which he could not remove. Here the same sagacious examination took place, the elephant keeping his flat side close to the side of the trunk, and leaning against it. The next step was against a tree, but this, on the first pressure of his trunk, he did not like. Here the driver made use of the most endearing epithets, such as 'Wonderful,' 'My life,' 'Well done, my dear,' 'My dove,' 'My son,' 'My wife;' but all these endearing appellations, of which elephants are so fond, would not induce him to try again. Force was at length resorted to, and the elephant roared terrifically, but would not move.

Something was then altered, the elephant was satisfied, and at last succeeded in mounting to the top of the staircase:—

On reaching the top his delight was visible in a most eminent degree; he caressed his keepers, and threw dirt about in a most playful manner. Another elephant, a much younger animal, had now to follow. He had watched the ascent of the other with the utmost interest, making motions all the while as though he was assisting him by shouldering him up the acclivity, in such gestures as I have seen some men make when spectators of gymnastic exercises. When he saw his comrade up, he evinced his pleasure by giving a salute something like the sound of a trumpet. When called upon to take his turn, however, he seemed much alarmed, and would not act at all without force.

After a performance similar to that of the previous elephant, however, he too neared the top, when 'the other, who had already performed his task, extended his trunk to the assistance of his brother in distress, round which the younger animal entwined his, and thus reached the summit.' There was then a cordial greeting between the two animals, 'as if they had been long separated from each other, and had just escaped from some perilous achievement. They mutually embraced each other, and stood face to face for a considerable time, as if whispering congratulations.'[1]

Mr. Jesse says: 'I was one day feeding the poor elephant (who was so barbarously put to death at Exeter Change) with potatoes, which he took out of my hand. One of them, a round one, fell on the floor, just out of reach of his proboscis.' After several ineffectual attempts to reach it, 'he at length *blew* the potato against the opposite wall with sufficient force to make it rebound, and he then without difficulty secured it.'[2]

This remarkable observation has fortunately been corroborated by Mr. Darwin. He writes:—

I have seen, as I dare say have others, that when a small object is thrown on the ground beyond the reach of one of the elephants at the Zoological Gardens, he blows through his trunk on the ground beyond the object, so that the current reflected on all sides may drive the object within his reach.[3]

The observation has also been corroborated by other observers.[4]

[1] *Memoirs*, vol. ii., p. 64 *et seq.*
[2] Jesse, *Gleanings in Natural History*, vol. i, p. 19.
[3] *Descent of Man*, p. 96.
[4] See *Animal Kingdom*, vol. iii., p. 374.

The following is quoted from Mr. Watson's book : [1]—

Of the elephant's sense and judgment the following instance is given as a well-known fact in a letter of Dr. Daniel Wilson, Bishop of Calcutta, to his son in England, printed in a Life of the bishop, published a few years ago. An elephant belonging to an Engineer officer in his diocese had a disease in his eyes, and had for three days been completely blind. His owner asked Dr. Webb, a physician intimate with the bishop, if he could do anything for the relief of the animal. Dr. Webb replied that he was willing to try, on one of the eyes, the effect of nitrate of silver, which was a remedy commonly used for similar diseases in the human eye. The animal was accordingly made to lie down, and when the nitrate of silver was applied, uttered a terrific roar at the acute pain which it occasioned. But the effect of the application was wonderful, for the eye was in a great degree restored, and the elephant could partially see. The doctor was in consequence ready to operate similarly on the other eye on the following day; and the animal, when he was brought out and heard the doctor's voice, lay down of himself, placed his head quietly on one side, curled up his trunk, drew in his breath like a human being about to endure a painful operation, gave a sigh of relief when it was over, and then, by motions of his trunk and other gestures, gave evident signs of wishing to express his gratitude. Here we plainly see in the elephant memory, understanding, and reasoning from one thing to another. The animal remembered the benefit that he had felt from the application to one eye, and when he was brought to the same place on the following day and heard the operator's voice, he concluded that a like service was to be done to his other eye.

The fact that elephants exhibit this sagacious fortitude under surgical operations—thus resembling, as we shall afterwards observe, both dogs and monkeys—is corroborated by another instance given in Bingley's 'Animal Biography,'[2] and serves to render credible the following story given in the same work :—

In the last war in India a young elephant received a violent wound in its head, the pain of which rendered it so frantic and ungovernable that it was found impossible to persuade the animal to have the part dressed. Whenever any one approached

---

[1] *Reasoning Power of Animals*, pp. 54–5.
[2] Bingley, *Animal Biography*, vol. i., p. 155.

it ran off with fury, and would suffer no person to come within several yards of it. The man who had care of it at length hit upon a contrivance for securing it. By a few words and signs he gave the mother of the animal sufficient intelligence of what was wanted; the sensible creature immediately seized her young one with her trunk, and held it firmly down, though groaning with agony, while the surgeon completely dressed the wound; and she continued to perform this service every day till the animal was perfectly recovered.[1]

Again, as still further corroboration of this point, I may quote the following from Sir E. Tennent's 'Natural History of Ceylon:'—

Nothing can more strongly exhibit the impulse to obedience in the elephant than the patience with which, at the order of his keeper, he swallows the nauseous medicines of the native elephant-doctors; and it is impossible to witness the fortitude with which (without shrinking) he submits to excruciating surgical operations for the removal of tumours and ulcers to which he is subject, without conceiving a vivid impression of his gentleness and intelligence. Dr. Davy when in Ceylon was consulted about an elephant in the Government stud, which was suffering from a deep, burrowing sore in the back, just over the back-bone, which had long resisted the treatment ordinarily employed. He recommended the use of the knife, that issue might be given to the accumulated matter, but no one of the attendants was competent to undertake the operation. 'Being assured,' he continues, 'that the creature would behave well, I undertook it myself. The elephant was not bound, but was made to kneel down at his keeper's command; and with an amputating knife, using all my force, I made the incision required through the tough integuments. The elephant did not flinch, but rather inclined towards me when using the knife; and merely uttered a low, and as it were suppressed groan. In short, he behaved as like a human being as possible, as if conscious (as I believe he was) that the operation was for his good, and the pain unavoidable.

Major Skinner witnessed the following display of intelligent action by a large herd of wild elephants. During the hot season at Nenera Kalama the elephants have a difficulty in finding water, and are therefore

---

[1] Bingley, *Animal Biography*, vol. i., p. 155.

obliged to congregate in large numbers where water is to be obtained. Being stationed near a water supply, and knowing that a large herd of elephants were in the neighbourhood, Major Skinner resolved to watch their proceedings. On a moonlight night, therefore, he climbed a tree about four hundred yards from the water, and waited patiently for two hours before he heard or saw anything of the elephants. At length he saw a huge beast issue from the wood, and advance cautiously across the open ground to within a hundred yards of the tank, where he stood perfectly motionless; and the rest of the herd, meanwhile, were so quiet that not the least sound was to be heard from them. Gradually, at three successive advances, halting some minutes after each, he moved up to the water's edge, in which, however, he did not think proper to quench his thirst, but remained for several minutes listening in perfect stillness. He then returned cautiously and slowly to the point at which he had issued from the wood, from whence he came back with five other elephants, with which he proceeded, somewhat less slowly than before, to within a few yards of the tank, where he posted them as patrols. He then re-entered the wood and collected the whole herd, which must have amounted to between eighty and a hundred, and led them across the open ground with the most extraordinary composure and quiet till they came up to the five sentinels, when he left them for a moment, and again made a reconnaissance at the edge of the tank. At last, being apparently satisfied that all was safe, he turned back, and obviously gave the order to advance; 'for in a moment,' says Major Skinner, 'the whole herd rushed to the water with a degree of unreserved confidence so opposite to the caution and timidity which had marked their previous movements, that nothing will ever persuade me that there was not rational and preconcerted co-operation throughout the whole party, and a degree of responsible authority exercised by the patriarch-leader.'[1]

Mr. H. L. Jenkins writes to me:—

What I particularly wish to observe is that there are good reasons for supposing that elephants possess abstract ideas; for instance, I think it is impossible to doubt that they acquire through their own experience notions of hardness and weight, and the grounds on which I am led to think this are as follows.

[1] See his letter to Sir E. Tennent in *Nat. Hist. of Ceylon*, pp 118–20.

A captured elephant after he has been taught his ordinary duty, say about three months after he is taken, is taught to pick up things from the ground and give them to his mahout sitting on his shoulders. Now for the first few months it is dangerous to require him to pick up anything but soft articles, such as clothes, because the things are often handed up with considerable force. After a time, longer with some elephants than others, they appear to take in a knowledge of the nature of the things they are required to lift, and the bundle of clothes will be thrown up sharply as before, but heavy things, such as a crowbar or piece of iron chain, will be handed up in a gentle manner; a sharp knife will be picked up by its handle and placed on the elephant's head, so that the mahout can also take it by the handle. I have purposely given elephants things to lift which they could never have seen before, and they were all handled in such a manner as to convince me that they recognised such qualities as hardness, sharpness, and weight. You are quite at liberty to make any use of these remarks you please if they are of service.

Again, as Dr. Lindley Kemp observes,[1] 'the manner in which tame elephants assist in capturing wild ones affords us an instance of reasoning in an animal,' &c.; and similarly, Mr. Darwin observes: 'It is, I think, impossible to read the account given by Sir E. Tennent of the behaviour of the female elephants used as decoys, without admitting that they intentionally practise deceit.'[2]

The following is an extract from the more interesting of the observations to which Mr. Darwin here alludes, and I think it is impossible to read them without assenting to his judgment. Several herds of wild elephants having been driven into a corral, two tame decoys were ridden into it :—

One was of prodigious age, having been in the service of the Dutch and English Governments in succession for upwards of a century. The other, called by her keeper 'Siribeddi,' was about fifty years old, and distinguished for gentleness and docility. She was a most accomplished decoy, and evinced the utmost relish for the sport. Having entered the corral noiselessly, carrying a mahout on her shoulders with the headman of

[1] *Indications of Instinct*, p. 129.   [2] *Descent of Man*, p. 69.

the noosers seated behind him, she moved slowly along with a sly composure and an assumed air of easy indifference; sauntering leisurely in the direction of the captives, and halting now and then to pluck a bunch of grass or a few leaves as she passed. As she approached the herd they put themselves in motion to meet her, and the leader, having advanced in front and passed his trunk gently over her head, turned and paced slowly back to his dejected companions. Siribeddi followed with the same listless step, and drew herself up close behind him, thus affording the nooser an opportunity to stoop under her and slip the noose over the hind foot of the wild one. The latter instantly perceived his danger, shook off the rope, and turned to attack the man. He would have suffered for his temerity had not Siribeddi protected him by raising her trunk and driving the assailant into the midst of the herd, when the old man, being slightly wounded, was helped out of the corral, and his son, Ranghanie, took his place.

The herd again collected in a circle, with their heads towards the centre. The largest male was singled out, and two tame ones pushed boldly in, one on either side of him, till the three stood nearly abreast. He made no resistance, but betrayed his uneasiness by shifting restlessly from foot to foot. Ranghanie now crept up, and holding the rope open with both hands (its other extremity being made fast to Siribeddi's collar), and watching the instant when the wild elephant lifted its hind foot, succeeded in passing the noose over its leg, drew it close, and fled to the rear. The two tame elephants instantly fell back, Siribeddi stretched the rope to its full length, and whilst she dragged out the captive, her companion placed himself between her and the herd to prevent any interference.

In order to tie him to a tree he had to be drawn backwards some twenty or thirty yards, making furious resistance, bellowing in terror, plunging on all sides, and crushing the smaller timber, which bent like reeds beneath his clumsy struggles. Siribeddi drew him steadily after her, and wound the rope round the proper tree, holding it all the time at its full tension, and stepping cautiously across it when, in order to give it a second turn, it was necessary to pass between the tree and the elephant. With a coil round the stem, however, it was beyond her strength to haul the prisoner close up, which was, nevertheless, necessary in order to make him perfectly fast; but the second tame one, perceiving the difficulty, returned from the herd, confronted the struggling prisoner, pushed him shoulder to shoulder, and head to head, forcing him backwards, whilst at every step Siribeddi hauled in the slackened rope till she brought him fairly up to the foot of the tree, where he was made fast by the cooroowe people. A second noose was then passed over the other hind-leg, and secured like the first, both legs being afterwards hobbled together by ropes made from the fibre of the kitool or jaggery palm, which, being more flexible than that of the cocoa-nut, occasions less formidable ulcerations. The two decoys then ranged themselves, as before, abreast of the prisoner on either side, thus enabling Ranghanie to stoop under them and noose the two fore-feet as he had already done the hind; and these ropes being made fast to the tree in front, the capture was complete, and the tame elephants and keepers withdrew to repeat the operation on another of the herd.

The second victim singled out from the herd was secured in the same manner as the first. It was a female. The tame ones forced themselves in on either side as before, cutting her off from her companions, whilst Ranghanie stooped under them and attached the fatal noose, and Siribeddi dragged her out amidst unavailing struggles, when she was made fast by each leg to the nearest group of strong trees. When the noose was placed upon her fore-foot, she seized it with her trunk, and succeeded in carrying it to her mouth, where she would speedily have severed it had not a tame elephant interfered, and placing his foot on the rope pressed it downwards out of her jaws. . . . The conduct of the tame ones during all these proceedings was truly wonderful. They displayed the most perfect conception of every movement, both of the object to be attained and of the means to accomplish it. They manifested the utmost enjoyment in what was going on. There was no ill-humour, no malignity in the spirit displayed, in what was otherwise a heartless proceeding, but they set about it in a way that showed a thorough relish for it, as an agreeable pastime. Their caution was as remarkable as their sagacity; there was no hurrying, no confusion, they never ran foul of the ropes, were never in the way of the animals already noosed; and amidst the most violent struggles, when the tame ones had frequently to step across the captives, they in no instance trampled on them, or occasioned the slightest accident or annoyance. So far from this, they saw intuitively a difficulty or a danger, and addressed themselves unbidden to remove it. In tying up one of the larger elephants, he contrived, before he could be hauled close up to the tree, to walk once or twice round it, carrying the rope with him; the decoy, perceiving the advantage he had thus gained over the nooser, walked up of her own accord, and pushed him backwards with her head, till she made him unwind himself again; upon which the rope was hauled tight and made fast. More than once, when a wild one was extending his trunk, and would have intercepted the rope about to be placed over his leg, Siribeddi, by a sudden motion of her own trunk, pushed his aside, and prevented him; and on one occasion, when successive efforts had failed to put the noose over the fore-leg of an elephant which was already secured by one foot, but which wisely put the other to the ground as often as it was attempted to pass the noose under it, I saw the decoy watch her opportunity, and when his foot was again raised, suddenly push in her own leg beneath it, and hold it up till the noose was attached and drawn tight.

One could almost fancy there was a display of dry humour in the manner in which the decoys thus played with the fears of the wild herd, and made light of their efforts at resistance. When reluctant they shoved them forward, when violent they drove them back; when the wild ones threw themselves down, the tame ones butted them with head and shoulders, and forced them up again. And when it was necessary to keep them down, they knelt upon them, and prevented them from rising, till the ropes were secured.

At every moment of leisure they fanned themselves with a bunch of leaves, and the graceful ease with which an elephant uses his trunk on such occasions is very striking. It is doubtless owing to the combination of a circular with a horizontal movement in that flexible limb; but it is impossible to see an elephant fanning himself without being struck by the singular elegance of motion which he displays. The tame ones, too, indulged in the luxury of dusting themselves with sand, by flinging it from their trunks; but it was a curious illustration of their delicate sagacity, that so long as the mahout was on their necks, they confined themselves to flinging the dust along their sides and stomach, as if aware that to throw it over their heads and back would cause annoyance to their riders.[1]

Sir E. Tennent has also some observations on other uses to which tame elephants are put, which are well worth quoting. Thus, speaking of the labour of piling timber, he says that the elephant

---

[1] *Natural History of Ceylon*, pp. 181-94.

manifests an intelligence and dexterity which are surpricing to a stranger, because the sameness of the operation enables the animal to go on for hours disposing of log after log, almost without a hint or direction from his attendant. For example, two elephants employed in piling ebony and satinwood in the yards attached to the commissariat stores at Colombo, were so accustomed to their work, that they were able to accomplish it with equal precision and with greater rapidity than if it had been done by dock-labourers. When the pile attained a certain height, and they were no longer able by their conjoint efforts to raise one of the heavy logs of ebony to the summit, they had been taught to lean two pieces against the heap, up the inclined plane of which they gently rolled the remaining logs, and placed them trimly on the top.

It has been asserted that in their occupations 'elephants are to a surprising extent the creatures of habit,' that their movements are altogether mechanical, and that 'they are annoyed by any deviation from their accustomed practice, and resent any constrained departure from the regularity of their course.' So far as my own observation goes, this is incorrect; and I am assured by officers of experience, that in regard to changing his treatment, his hours or his occupation, an elephant evinces no more consideration than a horse, but exhibits the same pliancy and facility.

At one point, however, the utility of the elephant stops short. Such is the intelligence and earnestness he displays in work, which he seems to conduct almost without supervision, that it has been assumed that he would continue his labour, and accomplish his given task, as well in the absence of his keeper as during his presence. But here his innate love of ease displays itself, and if the eye of his attendant be withdrawn, the moment he has finished the thing immediately in hand, he will stroll away lazily, to browse or enjoy the luxury of fanning himself and blowing dust over his back.

The means of punishing so powerful an animal is a question of difficulty to his attendants. Force being almost inapplicable, they try to work on his passions and feelings, by such expedients as altering the nature of his food or withholding it altogether for a time. On such occasions the demeanour of the creature will sometimes evince a sense of humiliation as well as of discontent. In some parts of India it is customary, in dealing with offenders, to stop their allowance of sugar canes or of jaggery; or to restrain them from eating their own share of fodder and leaves till their companions shall have finished;

and in such cases the consciousness of degradation betrayed by the looks and attitudes of the culprit is quite sufficient to identify him, and to excite a feeling of sympathy and pity.

The elephant's obedience to his keeper is the result of affection, as well as of fear; and although his attachment becomes so strong that an elephant in Ceylon has been known to remain out all night, without food, rather than abandon his mahout, lying intoxicated in the jungle, yet he manifests little difficulty in yielding the same submission to a new driver in the event of a change of attendants.[1]

Lastly, Sir E. Tennent writes:—

One evening, whilst riding in the vicinity of Candy, towards the scene of the massacre of Major Davies' party in 1803, my horse evinced some excitement at a noise which approached us in the thick jungle, and which consisted of a repetition of the ejaculation *urmph! urmph!* in a hoarse and dissatisfied tone. A turn in the forest explained the mystery, by bringing us face to face with a tame elephant, unaccompanied by any attendant. He was labouring painfully to carry a heavy beam of timber, which he balanced across his tusks, but, the pathway being narrow, he was forced to bend his head to one side to permit it to pass endways; and the exertion and this inconvenience combined led him to utter the dissatisfied sounds which disturbed the composure of my horse. On seeing us halt, the elephant raised his head, reconnoitred us for a moment, then flung down the timber, and voluntarily forced himself backwards among the brushwood so as to leave a passage, of which he expected us to avail ourselves. My horse hesitated: the elephant observed it, and impatiently thrust himself deeper into the jungle, repeating his cry of *urmph!* but in a voice evidently meant to encourage us to advance. Still the horse trembled; and, anxious to observe the instinct of the two sagacious animals, I forebore any interference: again the elephant of his own accord wedged himself further in amongst the trees, and manifested some impatience that we did not pass him. At length the horse moved forward; and when we were fairly past, I saw the wise creature stoop and take up its heavy burden, trim and balance it on its tusks, and resume its route as before, hoarsely snorting its discontented remonstrance.

Dr. Erasmus Darwin records an observation which was communicated to him by a 'gentleman of undoubted

[1] *Natural History of Ceylon*, pp. 181–94.

veracity,' of an elephant in India which the keeper was in the habit of leaving to play the part of nurse to his child when he and his wife had occasion to go away from home. The elephant was chained up, and whenever the child in its creeping about came to the end of the elephant's tether, he used gently to draw it back again with his trunk.

In 'Nature,' vol. xix., p. 385, Mr. J. J. Furniss writes:—

In Central Park one very hot day my attention was drawn to the conduct of an elephant which had been placed in an enclosure in the open air. On the ground was a large heap of newly-mown grass, which the sagacious animal was taking up by the trunkful, and laying carefully upon his sun-heated back. He continued the operation until his back was *completely thatched*, when he remained quiet, apparently enjoying the result of his ingenuity.

Mr. Furniss in a later communication (vol. xx., p. 21) continues:—

Since the publication of my former letter (as above), I have received additional data bearing on the subject from Mr. W. A. Conklin, the superintendent of the Central Park Menagerie. I am informed by him that he has frequently observed elephants, when out of doors in the hot sunshine, thatch their backs with hay or grass; that they do so to a certain extent when under cover in the summer time, and when the flies which then attack the animals, often so fiercely as to draw blood, are particularly numerous; but that they never attempt to thatch their backs in winter. This seems to prove that they act intelligently for the attainment of a definite end. It would be interesting to learn whether elephants in their wild state are in the habit of so thatching their backs. It seems more probable to suppose that in their native wilds they would avail themselves of the natural shade afforded by the jungle, and that the habit is one which has been developed in consequence of their changed surroundings in captivity.

Mr. G. E. Peal writes to 'Nature' (vol. xxi., p. 34):—

One evening, soon after my arrival in Eastern Assam, and while the five elephants were as usual being fed opposite the bungalow, I observed a young and lately caught one step up to a bamboo-stake fence, and quietly pull one of the stakes up,

Placing it under foot, it broke a piece off with the trunk, and after lifting it to its mouth threw it away. It repeated this twice or thrice, and then drew another stake and began again. Seeing that the bamboo was old and dry I asked the reason of this, and was told to wait and see what it would do. At last it seemed to get a piece that suited, and holding it in the trunk firmly, and stepping the left fore-leg well forward, passed the piece of bamboo under the armpit, so to speak, and began to scratch with some force. My surprise reached its climax when I saw a large elephant leech fall on the ground, quite six inches long and thick as one's finger, and which, from its position, could not easily be detached without this scraper or scratcher which was deliberately made by the elephant. I subsequently found that it was a common occurrence. Such scrapers are used by every elephant daily.

On another occasion, when travelling at a time of the year when the large flies are so tormenting to an elephant, I noticed that the one I rode had no fan or wisp to beat them off with. The mahout, at my order, slackened pace and allowed her to go to the side of the road, when for some moments she moved along rummaging the smaller jungle on the bank; at last she came to a cluster of young shoots well branched, and after feeling among them and selecting one, raised her trunk and neatly stripped down the stem, taking off all the lower branches and leaving a fine bunch on top. She deliberately cleaned it down several times, and then laying hold at the lower end broke off a beautiful fan or switch about five feet long, handle included. With this she kept the flies at bay as we went along, flapping them off on each side.

Say what we may, these are both really *bonâ fide* implements, each intelligently made for a definite purpose.

My friend Mrs. A. S. H. Richardson sends me the following. The Rev. Mr. Townsend, who narrated the episode, is personally known to her:—

An elephant was chained to a tree in the compound opposite Mr. Townsend's house. Its driver made an oven at a short distance, in which he put his rice-cakes to bake, and then covered them with stones and grass and went away. When he was gone, the elephant with his trunk unfastened the chain round his foot, went to the oven and uncovered it, took out and ate the cakes, re-covered the oven with the stones and grass as before, and went back to his place. He could not fasten the chain again round his own foot, so he twisted it round and round it, in order to look the same, and when the driver returned the elephant was standing with his back to the oven. The driver went to his cakes, discovered the theft, and, looking round, caught the elephant's eye as he looked back over his shoulder out of the corner of it. Instantly he detected the culprit, and condign punishment followed. The whole occurrence was witnessed from the windows by the family

CHAPTER XIV.

THE CAT.

THE cat is unquestionably a highly intelligent animal, though when contrasted with its great domestic rival, the dog, its intelligence, from being cast in quite a different mould, is very frequently underrated. Comparatively unsocial in temperament, wanderingly predaceous in habits, and lacking in the affectionate docility of the canine nature, this animal has never in any considerable degree been subject to those psychologically transforming influences whereby a prolonged and intimate association with man has, as we shall subsequently see, so profoundly modified the psychology of the dog. Nevertheless, as we shall immediately find, the cat is not only by nature an animal remarkable for intelligence, but in spite of its naturally imposed disadvantages of temperament, has not altogether escaped those privileges of nurture which unnumbered centuries of domestication could scarcely fail to supply. Thus, as contrasted with most of the wild species of the genus when tamed from their youngest days, the domestic cat is conspicuously of less uncertain temper towards its masters—the uncertainty of temper displayed by nearly all the wild members of the feline tribe when tamed being, of course, an expression of the interference of individual with hereditary experience. And, as contrasted with all the wild species of the genus when tamed, the domestic cat is conspicuous in alone manifesting any exalted development of affection towards the human kind; for in many individual cases such affection, under favouring circumstances, reaches a level fully comparable to that which it attains in the dog. We do not know the wild stock from which the domestic cat originally sprang, and therefore cannot estimate the extent of the psychological results which human agency has here produced; but it is worth while in this connection to remember that the nearest ally of the domestic cat is the wild cat, and that this animal, while so closely resembling its congener in size and anatomical structure, differs so enormously from it in the branch of psychological structure which we are considering, that there is no animal on the face of the earth so obstinately untamable.

As regards the wild species of the tribe in general, it may be said that they all exhibit the same unsocial, fierce, and rapacious character. Bold when brought to bay, they do not court battle with dangerous antagonists, but prefer to seek safety in flight. Even the proverbial courage of the lion is now known, as a rule, to consist in 'the better part of valour;' and those exceptional individuals among tigers which adopt a 'man-eating' propensity, snatch their human victims by stealth. That the larger feline animals possess high intelligence would be shown, even in the absence of information concerning their ordinary habits, by the numerous tricks which they prove themselves capable of learning at the hands of menagerie-keepers; though in such cases the conflict of nature with nurture renders even the best-trained specimens highly uncertain in their behaviour, and therefore always more or less dangerous to the 'lion-kings.' The only wild species that is employed for any practical purpose—the cheetah—is so employed by utilising directly its natural instincts; it is shown the antelope, and runs it down after the manner of all its ancestors.

Returning now to the domestic cat, it is commonly remarked as a peculiar and distinctive trait in its emotional character that it shows a strongly rooted attachment to places as distinguished from persons. There can be no question that this peculiarity is a marked feature in the psychology of domestic cats considered as a class, although of course individual exceptions occur in abundance. Probably this feature is a survival of an instinctive attachment to dens or lairs bequeathed to our cats by their wild progenitors.

The only other feature in the emotional life of cats which calls for special notice is that which leads to their universal and proverbial treatment of helpless prey. The feelings that prompt a cat to torture a captured mouse can only, I think, be assigned to the category to which by common consent they are ascribed—delight in torturing for torture's sake. Speaking of man, John S. Mill somewhere observes that there is in some human beings a special faculty or instinct of cruelty, which is not merely a passive indifference to the sight of physical sufferings, but an active pleasure in witnessing or causing it. Now, so far as I have been able to discover, the only animals in which there is any evidence of a class of feelings in any way similar to these—if, indeed, in the case even of such animals the feelings which prompt actions of gratuitous cruelty really are similar to those which prompt it in man—are cats and monkeys. With regard to monkeys I shall adduce evidence on this point in the chapter which treats of these animals. With regard to cats it is needless to dwell further upon facts so universally known.

General Intelligence.

Coming now to the higher faculties, it is to be noted as a general feature of interest that all cats, however domesticated they may be, when circumstances require it, and often even quite spontaneously, throw off with the utmost ease the whole mental clothing of their artificial experience, and return in naked simplicity to the natural habits of their ancestors. This readiness of cats to become feral is a strong expression of the shallow psychological influence which prolonged domestication has here exerted, in comparison with that which it has produced in the case of the dog. A pet terrier lost in the haunts of his ancestors is almost as pitiable an object as a babe in the wood; a pet cat under similar circumstances soon finds itself quite at home. The reason of this difference is, of course, that the psychology of the cat, never having lent itself to the practical uses of, and intelligent dependency on, man, has never, :s in the case of the dog, been under the cumulative influence of human agency in becoming further and further bent away from its original and naturally imposed position of self-reliance; so that when now a severance takes place between a cat and its human protectors, the animal, inheriting unimpaired the transmitted experience of wild progenitors, knows very well how to take care of itself.

Having made these general remarks, I shall now pass on to quote a few instances showing the highest level of intelligence to which cats attain.

As to observation, Mrs. Hubbard tells me of a cat which she possessed, and which was in the habit of poaching young rabbits to 'eat privately in the seclusion of a disused pigsty.' One day this cat caught a small black rabbit, and instead of eating it, as she always did the brown ones, brought it into the house unhurt, and laid it at the feet of her mistress. 'She clearly recognised the black rabbit as an unusual specimen, and apparently thought it right to show it to her mistress.' Such was 'not the only instance this cat showed of zoological discrimination,' for on another occasion, 'having caught another unusual animal—viz., a stoat—she also brought this alive into the house for the purpose of exhibiting it.'

Mr. A. Percy Smith informs me of a cat which he possesses, and which, to test her intelligence, he used to punish whenever her kittens misbehaved. Very soon this had the effect of causing the cat herself to train the kittens, for whenever they misbehaved 'she swore at them and boxed their ears, until she taught the kittens to be clean.'

Mr. Blackman, writing from the London Institution, tells me of a cat which he has, and which without tuition began to 'beg' for food, in imitation of a terrier in the same house whose begging gesture it must have observed to be successful in the obtaining of tit-bits. The cat, however, would never beg unless it was hungry;—

And no coaxing could persuade it to do so unless it felt so inclined. The same cat also, whenever it wanted to go out, would come into the sitting-room, and make a peculiar noise to attract attention: failing that mode being successful, it would pull one's dress with its claw and then having succeeded in attracting the desired attention, it would walk to the street door and stop there, making the same cry until let out.

Coming now to cases indicative of reason in cats, Mr. John Martin, writing from St. Clement's, Oxford, informs me: 'I have a cat which a short time ago had kittens, and from some cause or other her milk failed. My housekeeper saw her carrying a piece of bread to them.' The process of reasoning here is obvious.

Mr. Bidie, writing from the Government Museum of Madras to 'Nature' (vol. xx., p. 96), relates this instance of reasoning in a cat:—

In 1877 I was absent from Madras for two months, and left in my quarters three cats, one of which, an English tabby, was a very gentle and affectionate creature. During my absence the quarters were occupied by two young gentlemen, who delighted in teasing and frightening the cats. About a week before my return the English cat had kittens, which she carefully concealed behind bookshelves in the library. On the morning of my return I saw the cat, and patted her as usual, and then left the house for about an hour. On returning to dress I found that the kittens were located in a corner of my dressing-room, where previous broods had been deposited and nursed. On questioning the servant as to how they came there, he at once replied, 'Sir, the old cat taking one by one in her mouth, brought them here.' In other words, the mother had carried them one by one in her mouth from the library to the dressing-room, where they lay quite exposed. I do not think I have heard of a more remarkable instance of reasoning and affectionate confidence in an animal, and I need hardly say that the latter manifestation gave me great pleasure. The train of reasoning seems to have been as follows: 'Now that my master has returned there is no risk of the kittens being injured by the two young savages in the house, so I will take them out for my protector to see and admire, and keep them in the corner in which all my former pets have been nursed in safety.'

Dr. Bannister writes me from Chicago, of a cat belonging to his friend the late Mr. Meek, the palæontologist, who drew my correspondent's attention to the fact:—

He had fixed upright on his table a small looking-glass, from which he used to draw objects from nature, reversed on wood. The cat seeing her image in this glass made several attempts to investigate it, striking at it, &c. Then coming apparently to the conclusion that there was something between her and the other animal, she very slily and cautiously approached it, keeping her eye on it all the while, and struck her paw around behind the mirror, becoming seemingly much surprised at finding nothing there. This was done repeatedly, until she was at last convinced that it was beyond her comprehension, or she lost interest in the matter.

Mr. T. B. Groves communicates an almost precisely similar observation to 'Nature' (vol. xx., p. 291), of a cat which, on first seeing his own reflection in a mirror, tried to fight it. Meeting with resistance from the glass, the cat next ran behind the mirror. Not finding the object of his search, he again came to the front, and while keeping his eyes deliberately fixed on the image, felt round the edge of the glass with one paw, whilst with his head twisted round to the front he assured himself of the persistence of the reflection. He never afterwards condescended to notice a mirror.

The following is communicated to me by a correspondent whose name I cannot obtain permission to publish. I am sure, however, that it is communicated in good faith, and the incident can scarcely be supposed to have been due to accident. After describing the cat and the parrot in their amiable relationship, my correspondent proceeds:—

One evening there was no one in the kitchen. Cook had gone upstairs, and left a bowl full of dough to rise by the fire. Shortly after, the cat rushed up after her, mewing, and making what signs she could for her to go down; then she jumped up and seized her apron, and tried to drag her down. As she was in such a state of excitement cook went, and found 'Polly' shrieking, calling out, flapping her wings and struggling violently, 'up to her knees' in dough, and stuck quite fast.

No doubt if she had not been rescued she would have sunk in the morass and been smothered.

I shall here introduce two or three cases to show the ingenious devices to which clever cats will resort for the purpose of capturing prey.

Mr. James Hutchings writes in 'Nature' (vol. xii., p. 330) an account of an old tom cat using a young bird, which had fallen out of its nest, as a decoy for the old birds. The cat touched the young bird with his paw when it ceased to flutter and cry, in order that, by thus making it display its terror, the old cock bird, which was all the while flying about in great consternation, might be induced to approach near enough to be caught. Many times the cock bird did so, and the cat made numerous attempts to catch it, but without success. All the while a kitten had to be kept from killing the young bird. As this scene continued for a long time—in fact, till terminated by Mr. Hutchings—and as there does not appear to have been any opportunity for errors of observation, I think the case worth recording.

The following case is communicated to me by Mr. James G. Stevens, of St. Stephen, New Brunswick:—

Looking out on the garden in front of my residence, I observed a robin alight on a small tree : it was midwinter, the ground covered with about a foot of *light* snow. A cat came stealthily along, with difficulty making her way through the snow until within about three feet of the tree where the bird was ; the robin was sluggishly resting on a twig distant three feet from the ground or surface of snow ; the cat could not well, owing to the softness of the snow, venture to make a spring. She crouched down and at first gently stirred herself, evidently with the purpose of causing the bird to move. The first attempt failed. She again more actively stirred herself by a shaking motion. She again failed, when she stirred herself vigorously again and started the bird, which flew about fifty feet away, and alighted on a small low bush on the *northern* side of a *close-boarded* fence. The cat keenly watched the flight and the alighting of the bird ; as quickly as she could cross through the snow, she then *took a circuit of about one hundred feet*, watching the place where the bird was all the while, and covering her march by making *available every bush to hide her.* When out of range of vision of the bird she more actively made for the fence, leaped over it, came up on the *southern* side of it, and jumped on it, calculating her distance so accurately that she came within a foot of the bush where the bird was, and at once sprung. She missed her prey, but I thought she proved herself a cunning hunter. If this case is worth relating you may use the name of Judge Stevens, of St. Stephen, New Brunswick, as a witness to the same.

Again, I quote the following case communicated to 'Nature' by Dr. Frost, because, although it shows an almost incredible amount of far-sighted stratagem, I cannot on the one hand see much room for mal-observation, and on the other hand it is, as I shall show, to some extent corroborated by an independent observation of my friend Dr. Klein, and another correspondent:—

> Our servants have been accustomed during the late frost to throw the crumbs remaining from the breakfast-table to the birds, and I have several times noticed that our cat used to wait there in ambush in the expectation of obtaining a hearty meal from one or two of the assembled birds. Now, so far, this circumstance in itself is not an 'example of abstract reasoning.' But to continue. For the last few days this practice of feeding the birds has been left off. The cat, however, with an almost incredible amount of forethought, was observed by myself, together with two other members of the household, to scatter crumbs on the grass with the obvious intention of enticing the birds.[1]

Although this account, as I have said, borders on the incredible, I have allowed it to pass, because up to a certain point it is, as I have also said, corroborated by an observation communicated to me by my friend Dr. Klein, F.R.S.

Dr. Klein satisfied himself that the cat he observed had established a definite association between crumbs already sprinkled on the garden walk, and sparrows coming to eat them; for as soon as the crumbs were sprinkled on the walk, the cat used to conceal himself from the walk in a neighbouring shrubbery, there to await in ambush the coming of the birds. The latter, however, showed themselves more wide awake than the cat, for there was a wall running behind the shrubbery, from the top of which the birds could see the cat in his supposed concealment, and then a long line of sparrows used to wait watching the cat and the crumbs at the same time, but never venturing to fly down to the latter until the former, wearied with waiting, went away. In this case the reasoning observation

---

[1] *Nature*, vol. xix., p. 519.

of the cat—'crumbs attract birds, therefore I will wait for birds when crumbs are scattered'—was as complete as in the case of Dr. Frost's cat, but the reasoning in the latter case seems to have proceeded a stage further—'therefore I will scatter crumbs to attract birds.'

Now, in the face of the definite statement made by Dr. Frost, that his cat did advance to this further stage of reasoning, I have not felt justified in suppressing his remarkable observation. And, as lending still further credence to the account, I may quote the corroborative observation of another correspondent in 'Nature,' which is of value because forming an intermediate step between the intelligence displayed by Dr. Klein's cat and that displayed by Dr. Frost's. This correspondent says:—

A case somewhat similar to that mentioned by Dr. Frost, of a cat scattering crumbs, occurred here within my own knowledge. During the recent severe winter a friend was in the habit of throwing crumbs outside his bedroom window. The family have a fine black cat, which, seeing that the crumbs brought birds, would occasionally hide herself behind some shrubs, and when the birds came for their breakfast, would pounce out upon them with varying success. The crumbs had been laid out as usual one afternoon, but left untouched, and during the night a slight fall of snow occurred. On looking out next morning my friend observed puss busily engaged scratching away the snow. Curious to learn what she sought, he waited, and saw her take the crumbs up from the cleared space and lay them one by one after another on the snow. After doing this she retired behind the shrubs to wait further developments. This was repeated on two other occasions.[1]

Taking, then, these three cases together, we have an ascending series in the grades of intelligence from that displayed by Dr. Klein's cat, which merely observed that crumbs attracted birds, through that of the cat which exposed the concealed crumbs for the purpose of attracting birds, to that of Dr. Frost's cat, which actually sprinkled the crumbs. Therefore, although, if the last-mentioned or most remarkable case had stood alone, I should not have felt justified in quoting it, as we find it thus led up to by other and independent observations, I do not feel that I

[1] *Nature*, vol. xx., p. 197.

should be justified in suppressing it. And, after all, regarded as an act of reason, the sprinkling of crumbs to attract birds does not involve ideas or inferences very much more abstruse or remote than those which are concerned in some of the other and better corroborated instances of the display of feline intelligence, which I shall now proceed to state.

In the understanding of mechanical appliances, cats attain to a higher level of intelligence than any other animals, except monkeys, and perhaps elephants. Doubtless it is not accidental that these three kinds of animals fall to be associated in this particular. The monkey in its hands, the elephant in its trunk, and the cat in its agile limbs provided with mobile claws, all possess instruments adapted to manipulation, with which no other organs in the brute creation can properly be compared, except the beak and toes of the parrot, where, as we have already seen, a similar correlation with intelligence may be traced. Probably, therefore, the higher aptitude which these animals display in their understanding of mechanical appliances is due to the reaction exerted upon their intelgence by these organs of manipulation. But, be this as it may, I am quite sure that, excepting only the monkey and elephant, the cat shows a higher intelligence of the special kind in question than any other animal, not forgetting even the dog. Thus, for instance, while I have only heard of one solitary case (communicated to me by a correspondent) of a dog which, without tuition, divined the use of a thumb-latch, so as to open a closed door by jumping upon the handle and depressing the thumb-piece, I have received some half-dozen instances of this display of intelligence on the part of cats. These instances are all such precise repetitions of one another, that I conclude the fact to be one of tolerably ordinary occurrence among cats, while it is certainly very rare among dogs. I may add that my own coachman once had a cat which, certainly without tuition, learnt thus to open a door that led into the stables from a yard into which looked some of the windows of the house. Standing at these windows when the cat did not see me, I have many times witnessed her *modus operandi.* Walking up to the door with a most matter-of-course kind of air, she used to spring at the half-hoop handle just below the thumb-latch. Holding on to the bottom of this half-hoop with one fore-paw, she then raised the other to the thumb-piece, and while depressing the latter, finally with her hind legs scratched and pushed the doorposts so as to open the door. Precisely similar movements are described by my correspondents as having been witnessed by them.

Of course in all such cases the cats must have previously observed that the doors are opened by persons placing their hands upon the handles, and, having observed this, the animals forthwith act by what may be strictly termed rational imitation. But it should be observed that the process as a whole is something more than imitative. For not only would observation alone be scarcely enough (within any limits of thoughtful reflection that it would be reasonable to ascribe to an animal) to enable a cat upon the ground to distinguish that the essential part of the process as performed by the human hand consists, not in grasping the handle, but in depressing the latch; but the cat certainly never saw any one, after having depressed the latch, pushing the doorposts with his legs; and that this pushing action is due to an originally deliberate intention of opening the door, and not to having accidentally found this action to assist the process, is shown by one of the cases communicated to me (by Mr. Henry A. Gaphaus); for in this case, my correspondent says, 'the door was not a loose-fitting one by any means, and I was surprised that by the force of one hind leg she should have been able to push it open after unlatching it.' Hence we can only conclude that the cats in such cases have a very definite idea as to the mechanical properties of a door; they know that to make it open, even when unlatched, it requires to be *pushed*—a very different thing from trying to imitate any particular action which they may see to be performed for the same purpose by man. The whole psychological process, therefore, implied by the fact of a cat opening a door in this way is really most complex. First the animal must have observed that the door is opened by the hand grasping the handle and moving the latch. Next she must reason, by 'the logic of feelings'—If a hand can do it, why not a paw? Then, strongly moved by this idea, she makes the first trial. The steps which follow have not been observed, so we cannot certainly say whether she learns by a succession of trials that depression of the thumb-piece constitutes the essential part of the process, or, perhaps more probably, that her initial observations supplied her with the idea of clicking the thumb-piece. But, however this may be, it is certain that the pushing with the hind feet after depressing the latch must be due to adaptive reasoning unassisted by observation; and only by the concerted action of all her limbs in the performance of a highly complex and most unnatural movement is her final purpose attained.

Again, several very similar cases are communicated to me of cats spontaneously, or without tuition, learning to knock knockers and ring bells. Of course in both cases the animals must have observed the use to which knockers and bells are put, and when desiring a door to be opened, employ these signals for the purpose. It betokens no small amount of observation and reasoning in a cat to jump at a knocker with the expectation of thereby summoning a servant to open the door—especially as in some of the cases the jump is not a random jump at the knocker, but a deliberate and complex action, having for its purposes the raising and letting fall of the knocker. For instance, Mr. Belshaw, writing to 'Nature' (vol. xix., p. 659), says :—

I was sitting in one of the rooms, the first evening there, and hearing a loud knock at the front door was told not to heed it, as it was only this kitten asking admittance. Not believing it, I watched for myself, and very soon saw the kitten jump on to the door, hang on by one leg, and put the other fore-paw right through the knocker and rap twice.

In such cases the action closely resembles that of opening thumb-latches, but clearly is performed with the purpose of summoning some one else to open the door. Wonderful, however, as these cases of summoning by knockers undoubtedly are, I think they are surpassed by other cases in which the instrument used is the bell. For here it is not merely that cats perfectly well understand the use of bells as calls,[1] but I have one or two cases of cats jumping at bell-*wires* passing from outside into

[1] Some of my correspondents tell me of pet or drawing-room cats jumping on chairs and looking at bells when they want milk—this being their sign that they want the bell pulled to call the servant who brings the milk; and Mr. Lawson Tait tells me that one of his cats—of course without tuition—has gone a step further, in that she places her paws upon the bell as a still more emphatic sign that she desires it pulled. But Dr. Creighton Browne tells me of a cat which he has that goes a step further than this, and herself rings the bell. This is corroborative of Archbishop Whately's anecdote. 'This cat lived many years in my mother's family, and its feats of sagacity were witnessed by her, my sisters, and myself. It was known, not merely once or twice, but habitually, to ring the parlour bell whenever it wished the door to be opened. Some alarm was excited on the first occasion that it turned bell-ringer. The family had retired to rest, and in the middle of the night the parlour bell was rung violently; the sleepers were startled from their repose, and proceeded downstairs with poker and tongs, to intercept, as they thought, the predatory movements of some burglar; but they were equally surprised to find that the bell had been rung by pussy, who frequently repeated the act whenever she wished to get out of the parlour.' The cases, however, mentioned in the text are more remarkable than any of these, which, nevertheless, all tend to lead up to them as by a series of steps. Dogs attain to the level of asking by gesture their masters to ring bells. One instance will be sufficient to quote. Mr. Rae says in 'Nature' (vol. xix., p. 459): 'A small English terrier belonging to a friend has been taught to ring for the servant. To test if the dog knew *why* it rang the bell he was told to do so while the girl was in the room. The little fellow looked up in the most intelligent manner at the person giving the order (his master or mistress, I forget which), then at the servant, and refused to obey, although the order was repeated more than once. The servant left the room, and a few minutes afterwards the dog rang the bell immediately on being told to do so.'

It must also be added that dogs sometimes attain to the level of knocking knockers—though I should think this must be very rare with these animals, as I have only met with one case of it. This, however, is a remarkably good case, not only because it rests upon the authority of a famous observer, but also because it is so very definite as proving an act of reason. Dureau de la Malle had a terrier born in his house. It had never seen a knocker in its native home, and when grown up it was taken by its master to Paris. Getting fatigued by a walk in the streets, the animal returned to the house, but found the door shut, and it endeavoured vainly to attract the attention of those within by barking. At length a visitor called, knocked at the knocker, and gained admittance. The dog observed what had been done, and went in together with the visitor. The same afternoon he went in and out

houses the doors of which the cats desired to be opened.[1] My informants tell me that they do not know how these cats, from any process of observation, can have surmised that pulling the wire in an exposed part of its length would have the effect of ringing the bell; for they can never have observed any one pulling the wires. I can only suggest that in these cases the animals must have observed that when the bells were rung the wires moved, and that the doors were afterwards opened; then a process of inference must have led them to try whether jumping at the wires would produce the same effects. But even this, which is the simplest explanation possible, implies powers of observation scarcely less remarkable than the process of reasoning to which they gave rise.

As further instances corroborating the fact that both these faculties are developed in cats to a wonderful degree, I may add the following. Couch ('Illustrations of Instinct,' p. 196) gives a case within his own knowledge of a cat which, in order to get at milk kept in a locked cupboard, used to unlock the door by seating herself on an adjoining table, and 'repeatedly patting on the bow of the key with her paw, when with a slight pull on the door' she was able to open it; the lock was old, and the key turned in it ' on a very slight impulse.'

As a still further instance of the high appreciation of mechanical appliances to which cats attain, I shall quote an extract from a paper by Mr. Otto, which will have been read at the Linnean Society before this work is pub-

half a dozen times, gaining admittance on each occasion by springing at the knocker.

Lastly, Dr. W. H. Kesteven writes to 'Nature' (xx., p. 428) of a cat which used to knock at a knocker to gain admittance, in the way already described of so many other cats; but as showing how much more readily cats acquire this practice than dogs, it is interesting to note that Dr. Kesteven adds that a dog which lived in the same house ascertained that the cat was able to gain admittance by knocking, and yet did not imitate the action, but 'was in the habit of searching for her when he wanted to come in, and either waiting till she was ready to knock at the door, or inducing her to do it to please him.'

[1] Consul E. L. Layard gives in *Nature* (xx., p. 339) a precisely similar case of a cat habitually and without tuition ringing a bell by pulling at an exposed wire.

lished. After describing the case of a cat opening a thumb-latch in the same way as those already mentioned, this writer proceeds:—

At Parara, the residence of Parker Bowman, Esq., a full-grown cat was one day accidentally locked up in a room without any other outlet than a small window, moving on hinges, and kept shut by means of a swivel. Not long afterwards the window was found open and the cat gone. This having happened several times, it was at last found that the cat jumped upon the window-sill, placed her fore-paws as high as she could reach against the side, deliberately reached with one over to the swivel, moved it from its horizontal to a perpendicular position, and then, leaning with her whole weight against the window, swung it open and escaped.

To give only one other instance of high reasoning power in this animal, Mr. W. Brown, writing from Greenock to 'Nature' (vol. xxi., p. 397), gives a remarkable story of a cat, the facts in which do not seem to have admitted of mal-observation. While a paraffine lamp was being trimmed, some of the oil fell upon the back of the cat, and was afterwards ignited by a cinder falling upon it from the fire. The cat with her back 'in a blaze, in an instant made for the door (which happened to be open) and sped up the street about 100 yards,' where she plunged into the village watering-trough, and extinguished the flame. 'The trough had eight or nine inches of water, and puss was in the habit of seeing the fire put out with water every night.' The latter point is important, as it shows the data of observation on which the animal reasoned.

CHAPTER XV.

FOXES, WOLVES, JACKALS, ETC.

The general psychology of these animals is, of course, very much the same as that of the dog; but, from never having been submitted to the influences of domestication, their mental qualities present a sufficient number of differences from those of the dog to require another chapter for their consideration.

If we could subtract from the domestic dog all the emotions arising from his prolonged companionship with man, and at the same time intensify the emotions of self-reliance, rapacity, &c., we should get the emotional character now presented by the wolves and jackals. It is interesting to note that this genetic similarity of emotional character extends to what may be termed idiosyncratic details in cases where it has not been interfered with by human agency. Thus the peculiar, weird, and unaccountable class of emotions which cause wolves to bay at the moon has been propagated unchanged to our domestic dogs.

The intelligence of the fox is proverbial; but as I have not received many original observations on this head, I shall merely refer to some of the best authenticated observations already published, and shall begin with the instance narrated by Mr. St. John in his 'Wild Sports of the Highlands':—

When living in Ross-shire I went out one morning in July, before daybreak, to endeavour to shoot a stag, which had been complained of very much by an adjoining farmer, as having done great damage to his crops. Just after it was daylight I saw a large fox coming quietly along the edge of the plantation in which I was concealed; he looked with great care over the turf wall into the field, and seemed to long to get hold of some hares that were feeding in it, but apparently knew that he had no chance of catching one by dint of running; after considering a short time he seemed to have formed his plans, and having examined the different gaps in the wall by which the hares might be supposed to go in and out, he fixed upon the one that seemed the most frequented, and laid himself down close to it in an attitude like a cat watching a mouse. Cunning as he was, he was too intent on his own hunting to be aware that I was within twenty yards of him with a loaded rifle, and able to watch every movement that he made. I was much amazed to see the fellow so completely outwitted, and kept my rifle ready to shoot him if he found me out and attempted to escape. In the meantime I watched all his plans. He first with great silence and care scraped a small hollow in the ground, throwing up the sand as a kind of screen between his hiding-place and the hares' mews; every now and then, however, he stopped to listen, and sometimes to take a most cautious look into the field; when he had done this he laid himself down in a convenient position for springing upon his prey, and remained perfectly motionless with the exception of an occasional reconnoitre of the feeding hares. When the sun began to rise, they came one by one from the field to the cover of the plantation; three had already come in without passing by his ambush; one of them came within twenty yards of him, but he made no movement beyond crouching still more closely to the ground. Presently two came directly towards him; though he did not venture to look up, I saw by an involuntary motion of his ears that those quick organs had already warned him of their approach: the two hares came through the gap together, and the fox, springing with the quickness of lightning, caught one and killed her immediately; he then lifted up his booty and was carrying it off like a retriever, when my rifle-ball stopped his course by passing through his back-bone, and I went up and despatched him.

Numberless instances are on record showing the remarkable cunning of foxes in procuring bait from traps without allowing themselves to be caught. These cases are so numerous, and all display so much the same quality of intelligence, that it is impossible to doubt so great a concurrence of testimony. I shall only give two or three specific cases, to show the kind of intelligence that is in question. It will be observed that it is much the same as that which is displayed under similar circumstances by rats and wolverines, in which animals we have already considered it. In all these cases the intelligence displayed must justly be deemed to be of a very remarkable order. For, inasmuch as traps are not things to be met with in nature, hereditary experience cannot be supposed to have played any part in the formation of special instincts to avoid the dangers arising from traps, and therefore the astonishing devices by which these dangers are avoided can only be attributed to observation, coupled with intelligent investigation of a remarkably high character.

I extract the following from Couch's 'Illustrations of Instinct' (p. 175):—

Whenever a cat is tempted by the bait, and caught in a fox-trap, Reynard is at hand to devour the bait and the cat too, and fearlessly approaches an instrument which the fox must know cannot *then* do it any harm. Let us compare with this boldness the incredible caution with which the animal proceeds when tempted by the bait in a *set* trap. Dietrich aus dem Winkell had once the good fortune of observing, on a winter evening, a fox which for many preceding days had been allured with loop baits, and as often as it ate one it sat comfortably down, wagging its brush. The nearer it approached the trap, the longer did it hesitate to take the baits, and the oftener did it make the tour round the catching-place. When arrived near the trap it squatted down, and eyed the bait for ten minutes at least; whereupon it ran three or four times round the trap, then it stretched out one of its fore-paws after the bait, but did not touch it; again a pause, during which the fox stared immovably at the bait. At last, as if in despair, the animal made a rush and was caught by the neck. (Mag. Nat. Hist., N. S., vol. i., p. 512.)

In 'Nature,' vol. xxi., p. 132, Mr. Crehore, writing from Boston, says:—

Some years since, while hunting in Northern Michigan, I tried with the aid of a professional trapper to entrap a fox who made nightly visits to a spot where the entrails of a deer had been thrown. Although we tried every expedient that suggested itself to us we were unsuccessful, and, what seemed very singular, we always found the trap sprung. My companion insisted that the animal dug beneath it, and putting his paw beneath the jaw, pushed down the pan with safety to himself; but though the appearance seemed to confirm it, I could hardly credit his explanation. This year, in another locality of the same region, an old and experienced trapper assured me of its correctness, and said in confirmation that he had several times caught them, after they had made two or three successful attempts to spring the trap, by the simple expedient of setting it upside down, when of course the act of undermining and touching the pan would bring the paw within the grasp of the jaws.

In connection with traps, my friend Dr. Rae has communicated to me a highly remarkable instance of the display of reason on the part of the Arctic foxes. I have previously published the facts in my lecture before the British Association in 1879, and therefore shall here quote them from it :—

Desiring to obtain some Arctic foxes, Dr. Rae set various kinds of traps; but as the foxes knew these traps from previous experience, he was unsuccessful. Accordingly he set a kind of trap with which the foxes in that part of the country were not acquainted. This consisted of a loaded gun set upon a stand pointing at the bait. A string connected the trigger of the gun with the bait, so that when the fox seized the bait he discharged the gun, and thus committed suicide. In this arrangement the gun was separated from the bait by a distance of about 30 yards, and the string which connected the trigger with the bait was concealed throughout nearly its whole distance in the snow. The gun-trap thus set was successful in killing one fox, but never in killing a second; for the foxes afterwards adopted either of two devices whereby to secure the bait without injuring themselves. One of these devices was to bite through the string at its exposed part near the trigger, and the other device was to burrow up to the bait through the snow at right angles to the line of fire, so that, although in this way they discharged the gun, they escaped with perhaps only a pellet or two in the nose. Now both of these devices exhibited a wonderful degree of what I think must fairly be called power of reasoning. I have carefully interrogated Dr. Rae on all the circumstances of the case, and he tells me that in that part of the world traps are never set with strings; so that there can have been no special association in the foxes' minds between strings and traps. Moreover, after the death of fox No. 1, the track on the snow showed that fox No. 2, notwithstanding the temptation offered by the bait, had expended a great deal of scientific observation on the gun before he undertook to sever the cord. Lastly, with regard to burrowing at right angles to the line of fire, Dr. Rae justly deemed this so extraordinary a circumstance, that he repeated the experiment a number of times, in order to satisfy himself that the direction of the burrowing was really to be attributed to thought, and not to chance.[1]

[1] I have requested Dr. Rae to write out all the particulars of these remarkable observations, and the following is the response which he has kindly made:—' When trapping foxes in Hudson's Bay it sometimes happens that certain of these acute animals, probably from having seen their companions caught, studiously avoid the ordinary steel and wooden traps, however carefully set. The trapper then sets one or more guns in a peculiar manner, having a line 15 or 20 yards long uniting the trigger with a bait, on taking hold of which the fox sets the gun off, and commits suicide. The double object of the bait being placed so near the gun is that the fox may be certainly killed—not wounded only—and that the head alone should be hit, and the body not riddled all over with shot, which would spoil the skin. It is also necessary to mention that four or five inches of slack line must be allowed for contraction of the line by change from a dry to a moist atmosphere, which otherwise would cause so great a strain on the trigger that the gun would be discharged without the bait being touched. So as to conceal as far as possible all connection between bait and gun, that part of the line next the bait is carefully hid under the snow.

' When the fox takes the bait, he will have lifted it five inches (the length of the slack line) from its normal position before the gun goes off; consequently, instead of pointing the gun at the bait, it is aimed fully eight or nine inches higher, at the probable position of the brain of the animal when the gun is discharged.

' For reasons which scarcely require explanation, foxes very generally go about in pairs (long before the snow disappears), not necessarily always close together, because they have a better chance of finding food if separated some distance from each other.

' After one or more foxes have been shot, the trapper on visiting his guns perhaps finds that a fox has first cut the line connecting the bait with the gun, and then gone up and eaten the bait; or, if the gun has been set on a drift bank of snow, he or she has scraped a trench ten or twelve inches deep up to the bait, taken hold of it whilst lying in the trench, set the gun off, and then trotted coolly away with the food (taken, one may say, from the gun's mouth) safe and uninjured, as is clearly evinced by there being no mark of blood on the tracks.

' In pulling the bait whilst in the trench, the fox would drag it five inches, or the length of the slack line, *downwards*, and therefore his *head* and *nose* would be completely out of harm's way, both because of the snow protection, and also these parts of his body being twelve or thirteen inches below the line of aim.

' In the cases seen by myself, and by a friend of greater experience, the trench was always scraped at right angles, or nearly so, to the line of fire of the gun. This at first sight may appear erroneous, but on

Dr. Rae also informs me with regard to wolves, that 'they have been frequently known to take the bait from a gun without injury to themselves, by first cutting the line of communication between the two.'[1] He adds:—

I may also mention what I have been told, although I have never had an opportunity of seeing it, that wolves watch the fishermen who set lines in deep water for trout, through holes in the ice on Lake Superior, and very soon after the man has left, the wolf goes up to the place, takes hold of the stick which is placed across the hole and attached to the line, trots off with it along the ice until the bait is brought to the surface, then returns and eats the bait and the fish, if any happens to be on the hook. The trout of Lake Superior are very large, and the baits are of a size in proportion.

Mr. Murray Browne, Inspector of the Local Government Board, writes to me from Whitehall as follows:—

I once, at the Devil's Glen, Wicklow, found a fox fast in a trap by the foot. We did not like to touch him, but got sticks and poked at the trap till we got it open. The process took ten minutes or a quarter-hour. When first we came up the fox strained to get free, and looked frightfully savage; but we had not poked at the trap more than a very short time before the whole expression of his face changed, he lay perfectly quiet (though we must at times have hurt him); and when at last we had got the trap completely off his foot, he still lay quiet,

reflection it really is not so, for if the trench is to be a shelter one— thinking, as the fox must have done, that the gun or something coming from it was the danger to be protected from or guarded against—it must be made across the line of fire, for if scratched in the direction of fire it would afford little or no protection or concealment, and the reasoning power or intelligence of the fox would be at fault.

'My belief is that one of these knowing foxes had seen his or her companion shot, or found it dead shortly after it had been killed, and not unnaturally attributed the cause of the mishap to the only strange thing it saw near, namely, the gun.

'It was evident that in all cases they had studied the situation carefully, as was sufficiently shown by their tracks in the snow, which indicated their extremely cautious approach when either the string-cutting or trench-making dodge was resorted to, in attempting to obtain the coveted bait without injury to themselves.'

[1] It will be remembered that, from evidence previously detailed, both the wolverine or glutton and certain deer have been shown capable of similarly obviating the danger of gun-traps.

and looked calmly at us, as if he knew we were friends. In fact, we had some little difficulty in getting him to move away, which he did readily enough when he chose. Was not this a case of reason and good sense *overpowering* natural instinct?

Couch says ('Illustrations of Instinct,' p. 178): 'Derham quotes Olaus in his account of Norway as having himself witnessed the fact of a fox dropping his tail among the rocks on the sea-shore to catch the crabs below, and hauling up and devouring such as laid hold of it.'

Under the present heading I must not omit to refer to an interesting class of instincts which are manifested by those species of the genus *Canis*, whose custom it is to hunt in packs. The instincts to which I refer are those which lead to a combination among different members of the same pack for the capture of prey by stratagem. These instincts, which no doubt arose and are now maintained by intelligent adaptation to the requirements of the chase, I shall call 'collective instincts.' Thus Sir E. Tennent writes:—

At dusk, and after nightfall, a pack of jackals, having watched a hare or a small deer take refuge in one of these retreats, immediately surrounded it on all sides; and having stationed a few to watch the path by which the game entered, the leader commences the attack by raising the cry peculiar to their race, and which resembles the sound 'okkay' loudly and rapidly repeated. The whole party then rush into the jungle and drive out the victim, which generally falls into the ambush previously laid to entrap it.

A native gentleman, who had favourable opportunities of observing the movements of these animals, informed me that when a jackal has brought down his game and killed it, his first impulse is to hide it in the nearest jungle, whence he issues with an air of easy indifference to observe whether anything more powerful than himself may be at hand, from which he might encounter the risk of being despoiled of his capture. If the coast be clear he returns to the concealed carcass and carries it away, followed by his companions. But if a man be in sight, or any other animal to be avoided, my informant has seen the jackal seize a cocoa-nut husk in his mouth, or any similar substance, and fly at full speed, as if eager to carry off his pretended prize, returning for the real booty at some more convenient season.[1]

Again, Jesse records the following display of the same instinct by the fox, as having been communicated to him by a friend on whose veracity he could rely:—

Part of this rocky ground was on the side of a very high hill, which was not accessible for a sportsman, and from which both hares and foxes took their way in the evening to the plain below. There were two channels or gullies made by the rains, leading from these rocks to the lower ground. Near one of these channels, the sportsman in question, and his attendant, stationed themselves one evening in hopes of being able to shoot some hares. They had not been there long, when they observed a fox coming down the gully, and followed by another. After playing together for a little time, one of the foxes concealed himself under a large stone or rock, which was at the bottom of the channel, and the other returned to the rocks. He soon, however, came back, chasing a hare before him. As the hare was passing the stone where the first fox had concealed himself, he tried to seize her by a sudden spring, but missed his aim. The chasing fox then came up, and finding that his expected prey had escaped, through the want of skill in his associate, he fell upon him, and they both fought with so much animosity, that the parties who had been watching their proceedings came up and destroyed them both.

Similarly, Mr. E. C. Buck records ('Nature,' viii., 303) the following interesting observation made by his friend Mr. Elliot, B.C.S., Secretary to Government, N.W.P.:—

He saw two wolves standing together, and shortly after noticing them was surprised to see one of them lie down in a ditch, and the other walk away over the open plain. He watched the latter, which deliberately went to the far side of a herd of antelopes standing in the plain, and drove them, as a sheep-dog would a flock of sheep, to the very spot where his companion lay in ambush. As the antelopes crossed the ditch, the concealed wolf jumped up as in the former case, seized a doe, and was joined by his colleague.

Mr. Buck draws attention to another closely similar display of collective instinct of wolves in the same district observed by a 'writer of one of the books on Indian sport.'

[1] *Nat. Hist. of Ceylon*, p. 35.

With reference to this case I wrote to 'Nature' as follows. The friend to whom I allude was the late Dr. Brydon, C.B. (the 'last man' of the Afghan expedition of 1841), whom I knew intimately for several years, and always found his observations on animals to be trustworthy:—

In response to the appeal which closes Mr. Buck's interesting letter ('Nature,' vol. viii., p. 302), the following instance of 'collective instinct' exhibited by an animal closely allied to the wolf, viz., the Indian jackal, deserves to be recorded. It was communicated to me by a gentleman (since deceased) on whose veracity I can depend. This gentleman was waiting in a tree to shoot tigers as they came to drink at a large lake (I forget the district), skirted by a dense jungle, when about midnight a large axis deer emerged from the latter and went to the water's edge. Then it stopped and sniffed the air in the direction of the jungle, as if suspecting the presence of an enemy; apparently satisfied, however, it began to drink, and continued to do so for a most inordinate length of time. When literally swollen with water it turned to go into the jungle, but was met on its extreme margin by a jackal, which, with a sharp yelp, turned it again into the open. The deer seemed much startled, and ran along the shore for some distance, when it again attempted to enter the jungle, but was again met and driven back in the same manner. The night being calm, my friend could hear this process being repeated time after time—the yelps becoming successively fainter and fainter in the distance, until they became wholly inaudible. The stratagem thus employed was sufficiently evident. The lake having a long narrow shore intervening between it and the jungle, the jackals formed themselves in line along it while concealed within the extreme edge of the cover, and waited until the deer was waterlogged. Their prey, being thus rendered heavy and shortwinded, would fall an easy victim if induced to run sufficiently far, *i.e.*, if prevented from entering the jungle. It was, of course, impossible to estimate the number of jackals engaged in this hunt, for it is not impossible that as soon as one had done duty at one place, it outran the deer to await it in another.

A native servant who accompanied my friend told him that this was a stratagem habitually employed by the jackals in that place, and that they hunted in sufficient numbers 'to leave nothing but the bones.' As it is a stratagem which could only be effectual under the peculiar local conditions described, it must appear that this example of collective instinct is due to 'separate expression,' and not to 'inherited habit.'

Cases of collective instinct are not of unfrequent occurrence among dogs. For the accuracy of the two following I can vouch. A small Skye and a large mongrel were in the habit of hunting hares and rabbits upon their own account, the small dog having a good nose, and the larger one great fleetness. These qualities they combined in the most advantageous manner, the terrier driving the cover towards his fleet-footed companion which was waiting for it outside.

The second case is remarkable for a display of sly sagacity. A friend of mine in Ross-shire had a small terrier and a large Newfoundland. One day a shepherd called upon him to say that his dogs had been worrying sheep the night before. The gentleman said there must be some mistake, as the Newfoundland had not been unchained. A few days afterwards the shepherd again called with the same complaint, vehemently asserting that he was positive as to the identity of the dogs. Consequently the owner set one watch upon the kennel and another outside the sheep enclosure, directing them (in consequence of what the shepherd had told him) not to interfere with the action of the dogs. After this had been done several nights in succession, the small dog was observed to come at daydawn to the place where the large one was chained; the latter immediately slipped his collar, and the two animals made straight for the sheep. Upon arriving at the enclosure the Newfoundland concealed himself behind a hedge, while the terrier drove the sheep towards his ambush, and the fate of one of them was quickly sealed. When their breakfast was finished the dogs returned home, and the larger one, thrusting his head into his collar, lay down again as though nothing had happened. Why this animal should have chosen to hunt by stratagem prey which it could easily run down, I cannot suggest; but there can be little doubt that so wise a dog must have had some good reason.

A similar instance of the display of collective instinct is thus narrated by M. Dureau de la Malle:—

I had at one time two sporting dogs, the one an excellent pointer with a very smooth skin, and of remarkable beauty and intelligence; the other was a spaniel with long and thick hair, but which had not been taught to point, but only coursed in the woods like a harrier. My *château* is situated on a level spot of ground, opposite to copse wood filled with hares and rabbits. When sitting at my window, I have observed these two dogs, which were at large in the yard, approach and make signs to each other, and first glancing at me, as if to see if I offered any obstacle to their wishes, step away very gently, then quicken their pace when they were at a little distance from my sight, and finally dart off at full speed when they thought I could neither see them nor order them back. Surprised at this mysterious manœuvre, I followed them, and witnessed a singular sight. The pointer, who seemed to be the leader of the enterprise, had sent the spaniel out to beat the bushes, and give tongue at the opposite extremity of the bushwood. As to himself, he made with slow steps the circuit of the wood by following it along the border, and I observed him stop before a passage much frequented by rabbits, and there point. I continued at a distance to observe how the intrigue was going to end. At length I heard the spaniel, which had started a hare, drive it with much tongue towards the place where its companion was lying in ambush, and the moment that the hare came out of the passage to gain the fields, the latter darted upon it and brought it to me with an air of triumph. I have seen these two dogs repeat this same manœuvre more than a hundred times; and this conformity has convinced me that it was not accidental, but the result of a concerted agreement and combined plan of operations understood beforehand.

Again, among Mr. Darwin's MSS., I find a letter from Mr. H. Reeks (1871), which says that the wolves of Newfoundland adopt exactly the same stratagem for the capture of deer in winter as that which is adopted by the hunters. That is to say, some of the pack secrete themselves in one or more of the *leeward* deer-paths in the forest or 'belting,' while one or two wolves make a circuit round the herd of deer to windward. The herd invariably retreats by one of its accustomed runs, and 'it rarely happens . . . . that the wolves do not manage by this stratagem to secure a doe or young stag.' And Leroy, in his book on Animal Intelligence, narrates closely similar facts of the wolves of Europe as having fallen within his own observation.

CHAPTER XVI.

THE DOG.

THE intelligence of the dog is of special, and indeed of unique interest from an evolutionary point of view, in that from time out of record this animal has been domesticated on account of the high level of its natural intelligence; and by persistent contact with man, coupled with training and breeding, its natural intelligence has been greatly changed. In the result we see, not only a general modification in the way of dependent companionship and docility, so unlike the fierce and self-reliant disposition of all wild species of the genus; but also a number of special modifications, peculiar to certain breeds, which all have obvious reference to the requirements of man. The whole psychological character of the dog may therefore be said to have been moulded by human agency with reference to human requirements, so that now it is not more true that man has in a sense created the structure of the bull-dog and greyhound, than that he has implanted the instincts of the watch-dog and pointer. The definite proof which we thus have afforded of the transforming and creating influence exerted upon the mental character and instincts of species by long and persistent training, coupled with artificial selection, furnishes the strongest possible corroboration of the theory which assigns psychological development in general to the joint operation of individual experience coupled with natural selection. For thousands of years man has here been virtually, though unconsciously, performing what evolutionists may regard as a gigantic experiment upon the potency of individual experience accumulated by heredity; and now there stands before us this most wonderful monument of his labours—the culmination of his experiment in the transformed psychology of the dog.

In my next work I shall treat of this subject with the fulness that it deserves—especially in its relation to the origin of instincts and the development of the moral sense; but to enter upon this topic at present would demand more space than can be allowed.

To do full justice to the psychology of the dog a separate treatise would be required. Here I can only trace a sketch.

Memory.

As regards memory, one or two instances will suffice. Mr. Darwin writes: 'I had a dog who was savage and averse to all strangers, and I purposely tried his memory after an absence of five years and two days. I went near the stable where he lived, and shouted to him in my old manner; he showed no joy, but instantly followed me out walking, and obeyed me, as if I had parted with him only an hour before.'[1]

It is not only persons or places that dogs remember for long periods. I had a setter in the country, which one year I took up with me to town for a few months. While in town he was never allowed to go out without a collar on which was engraved my address. A ring upon this collar made a clinking sound, and the setter soon learnt to associate the approach of this sound with the prospect of a walk. Three years afterwards I again took this setter up to town. He remembered every nook and corner of my house in town, and also his way about the streets, and the first time that I brought his collar, slightly clinking as before, he showed by his demonstrations of joy that he well remembered the sound with all its old associations, although he had not heard this sound for three years.

Emotions.

The emotional life of the dog is highly developed—more highly, indeed, than that of any other animal. His

[1] *Descent of Man*, p. 74.

gregarious instincts, united with his high intelligence and constant companionship with man, give to this animal a psychological basis for the construction of emotional character, having a more massive as well as more complex consistency than that which is presented even in the case of the monkey, which, as we shall afterwards see, attains to a remarkably high level in this respect.

Pride, sense of dignity, and self-respect are very conspicuously exhibited by well-treated dogs. As with man, so with the friend of man, it is only those whose lines of fortune have fallen in pleasant places, and whose feelings may therefore be said to have profited by the refining influences of culture, that display in any conspicuous measure the emotions in question. 'Curs of low degree,' and even many dogs of better social position, have never enjoyed those conditions essential to moral refinement, which alone can engender a true sense of self-respect and dignity. A 'low-life' dog may not like to have his tail pulled, any more than a gutter child may like to have his ears boxed; but here it is physical pain rather than wounded pride that causes the smart. Among 'high-life' dogs, however, the case is different. Here wounded sensibilities and loss of esteem are capable of producing much keener suffering than is mere physical pain; so that among such dogs a whipping produces quite a different and a much more lasting effect than in the case of their rougher brethren, who, as soon as it is over, give themselves a shake and think no more about it. As evidence of the delicacy of feeling to which dogs of aristocratic estate may attain, I shall give one or two among many instances that I could render.

A reproachful word or look from any of his friends would make a Skye terrier that I owned miserable for a whole day. If we had ever ventured to strike him I do not know what would have happened, for his sentiments were quite abreast of the age with respect to moral repugnance to the use of the lash. Thus, for instance, at one time when all his own friends were out of town, he was taken for a walk every day in the park by my brother, to whose care he had been entrusted. He enjoyed his walks very much, and was wholly dependent upon my brother for obtaining them. Nevertheless, one day while he was amusing himself with another dog in the park, my brother, in order to persuade him to follow, struck him with a glove. The terrier looked up at his face with an astonished and indignant gaze, deliberately turned round, and trotted home. Next day he went out with my brother as before, but after he had gone a short distance he looked up at his face significantly, and again trotted home with a dignified air. After thus making his protest in the strongest way he could, the dog ever afterwards refused to accompany him.

This terrier habitually exhibited a strong repugnance to corporal punishment, even when inflicted upon others. Thus, whenever or wherever he saw a man striking a dog, whether in the house or outside, near at hand or at a distance, he used to rush in to interfere, snarling and snapping in a most threatening way. Again, when driving with me in a dog-cart, he always used to hold the sleeve of my coat every time I touched the horse with the whip. As bearing upon this sensitiveness of feeling produced in dogs by habitually kind treatment, I shall here give an extract from the letter of one of my correspondents (Mrs. E. Picton). It relates to a Skye terrier which had a strong aversion to being washed:—

In process of time this aversion increased so much that all the servants I had refused to perform the ablutions, being in terror of doing so from the ferocity the animal evinced on such occasions. I myself did not choose to undertake the office, for though the animal was passionately attached to me, such was his horror of the operation, that even I was not safe. Threats, beating, and starving were all of no avail; he still persisted in his obstinacy. At length I hit upon a new device. Leaving him perfectly free, and not curtailing his liberty in any way, I let him know, by taking no notice of him, that he had offended me. He was usually the companion of my walks, but now I refused to let him accompany me. When I returned home I took no notice of his demonstrative welcome, and when he came looking up at me for caresses when I was engaged either in reading or needlework, I deliberately turned my head aside. This state of things continued for about a week or ten days, and the poor animal looked wretched and forlorn. There was evidently a conflict going on within him, which told visibly on his outward appearance. At length one morning he crept quietly up to me, and gave me a look which said as plainly as any spoken words could have done, 'I can stand it no longer; I submit.' And submit he did quite quietly and patiently to one of the roughest ablutions it had ever been his lot to experience; for by this time he sorely needed it. After it was over he bounded to me with a joyous bark and wag of his tail, saying unmistakably, 'I know all is right now.' He took his place by my side as his right when I went for my walk, and retained from that time his usually glad and joyous expression of countenance. When the period for the next ablution came round the old spirit of obstinacy resumed its sway for a while, but a single look at my averted countenance was sufficient for him, and he again submitted without a murmur. Must there not have been something akin to the reasoning faculty in the breast of an animal who could thus for ten days carry on such a struggle?

This strong effect of silent coldness shows that the loss of affectionate regard caused the terrier more suffering than beating, starving, or even the hated bath; and as many analogous cases might be quoted, I have no hesitation in adducing this one as typical of the craving for affectionate regard, which is manifested by sensitive dogs.

In this connection I may point out the remarkable change which has been produced in the domestic dog as compared with wild dogs, with reference to the enduring of pain. A wolf or a fox will sustain the severest kinds of physical suffering without giving utterance to a sound, while a dog will scream when any one accidentally treads upon its toes. This contrast is strikingly analogous to that which obtains between savage and civilised man: the North American Indian, and even the Hindoo, will endure without a moan an amount of physical pain—or at least bodily injury—which would produce vehement expressions of suffering from a European. And doubtless the explanation is in both cases the same—namely, that refinement of life engenders refinement of nervous organisation, which renders nervous lesions more intolerable.

As evidence of the idea of caste in a dog, I shall quote only one instance, although many others might be given: this also may be taken as typical. I extract it from St. John's 'Wild Sports of the Highlands,' where, speaking of his retriever, this very good observer states: 'He struck up an acquaintance with a ratcatcher and his cur, thoroughly entering into their way of business; but the moment he saw me he instantly cut his humble friends, and denied all acquaintance with them in the most comical manner.'[1]

Dogs likewise display in a high degree the feelings of emulation and jealousy. I once had a terrier which took great pains, and manifested paternal delight, in teaching his puppy to hunt rabbits. But in time the puppy outgrew his father in strength and fleetness, so that in the chase, in spite of straining every nerve, the father used to be gradually distanced. His whole demeanour then changed, and every time that he found his son drawing away from him he used in desperation to seize the receding tail of the youngster. Although the son was now much stronger than the father, he never used to resent this exercise of paternal authority, even though the rabbit were close under his nose.

Of jealousy in dogs innumerable instances might be given, but I shall merely quote one from my bulky correspondence on this head. It is sent me by Mr. A. Oldham:—

He had grown old, and having some affection in his legs which made walking difficult to him, he had sunk into a very stagnant sort of life, when a Scotch terrier was brought to live with us, and treated with much favour. All Charlie's old vigour revived upon the advent of this rival. He exhibited agonies of jealousy, and has since spent his life in following, watching, and imitating him. He insists on doing everything that Jack does. Although he had previously given up walking, he now makes a point of going out whenever Jack does so. Several times he has started with us, but finding that Jack was not of the party, has turned back and quietly gone home. In the same way,

---

[1] So many cases are on record of large dogs (especially of the Newfoundland breed) throwing troublesome curs into the water, and again rescuing them if they show danger of drowning, that we can scarcely fail to accept them as true. Such cases exhibit a wonderful play of human-like emotions.

although before he ate nothing but meat, he now eats any food that is also given to Jack; and if Jack is caressed he watches for some time, and then bursts out whining and barking. I have seen the same rage manifested by a fine cockatoo at the sight of his mistress carrying on her wrist and stroking affectionately a little green parrot. Such jealousy seems to me a very advanced emotion, as it has passed beyond the stage when it may be supposed to be caused by a fear of other animals monopolising *material* benefits which they desire for themselves; it is excited solely by seeing *affection* or *attention* bestowed by those they love upon other animals. The actions in which Charlie tries to participate—such as walking far, plunging into cold water after sticks, &c.—are in themselves extremely disagreeable to him, and he performs them only that he may obtain a share in the companionship and notice bestowed upon Jack.

Akin to jealousy is the sense of justice. If a master is not equal in his ways towards his dogs, the dogs are very apt to discover the injustice and to resent it accordingly. The well-known observation of the great Arago may be taken as a typical one in this connection. Having been detained by a storm at a country inn, and having ordered a chicken for his dinner, Arago was warming himself by the kitchen fire, when he saw the innkeeper put the fowl on the spit and attempt to seize a turnspit dog lying in the kitchen. The brute, however, refused to enter the wheel, got under a table, and showed fight. On Arago asking what could be the meaning of such conduct, the host replied that the dog had some excuse, that it was not his turn but his comrade's, who did not happen to be in the kitchen. Accordingly, the other turnspit was sent for, and he entered the spit very willingly, and turned away. When the fowl was half roasted Arago took him out, and the other dog, no longer smarting under the sense of injustice, now took his turn without any opposition, and completed the roasting of the fowl.

Deceitfulness is another trait in canine character of which numberless instances might be given; but here, again, it seems unnecessary to quote more than one or two cases as illustrative of the general fact. Another of my correspondents, after giving several examples of the display of hypocrisy of a King Charles spaniel, proceeds:—

He showed the same deliberate design of deceiving on other occasions. Having hurt his foot he became lame for a time, during which he received more pity and attention than usual. For months after he had recovered, whenever he was harshly spoken to, he commenced hobbling about the room as if lame and suffering pain from his foot. He only gave up the practice when he gradually perceived that it was unsuccessful.

The following instance, which I observed myself, I regard as more remarkable. It has already been published in 'Nature' (vol. xii., p. 66), from which I quote it:

The terrier used to be very fond of catching flies upon the window-panes, and if ridiculed when unsuccessful was evidently much annoyed. On one occasion, in order to see what he would do, I purposely laughed immoderately every time he failed. It so happened that he did so several times in succession—partly, I believe, in consequence of my laughing—and eventually he became so distressed that he positively *pretended* to catch the fly, going through all the appropriate actions with his lips and tongue, and afterwards rubbing the ground with his neck as if to kill the victim: he then looked up at me with a triumphant air of success. So well was the whole process simulated that I should have been quite deceived, had I not seen that the fly was still upon the window. Accordingly I drew his attention to this fact, as well as to the absence of anything upon the floor; and when he saw that his hypocrisy had been detected he slunk away under some furniture, evidently very much ashamed of himself.

This allusion to the marked effects of ridicule upon a dog leads to a consideration of the next emotion with which I feel certain that some dogs are to be accredited. I mean the emotion of the ludicrous. This same terrier used, when in good humour, to perform several tricks, which I know to have been self-taught, and which clearly had the object of exciting laughter. For instance, while lying on his side and violently grinning, he would hold one leg in his mouth. Under such circumstances, nothing pleased him so much as having his joke duly appreciated, while if no notice was taken of him he would become sulky. On the other hand, nothing displeased him so much as being laughed at when he did not intend to be ridiculous, as could not be more con clusively proved than by the fact of his behaviour in pretending to catch the fly. Mr. Darwin observes: 'Dogs show what may be fairly called a sense of humour, as distinct from mere play; if a bit of stick or other such object be thrown to one, he will often carry it away for a short distance; and then squatting down with it on the ground close before him, will wait until his master comes close to take it away. The dog will seize it and rush away in triumph, repeating the same manœuvre, and evidently enjoying the practical joke.'[1]

General Intelligence.

I have very definite evidence of the fact that dogs are able to communicate to one another simple ideas. The communication is always effected by gesture or tones of barking, and the ideas are always of such a simple nature as that of a mere 'follow me.' According to my own observations, the dogs must be above the average of canine intelligence, and the gesture they invariably employ is a contact of heads, with a motion between a rub and a butt. It is quite different from anything that occurs in play, and is always followed by a definite course of action. I must add, however, that although the information thus conveyed is always definite, I have never known a case in which it was complex—anything like asking or telling the way, which several writers have said that dogs can do, being, I believe, quite out of the question. One example will suffice. A Skye terrier (not quite pure) was asleep in the room where I was, while his son lay upon a wall which separates the lawn from the high road. The young dog, when alone, would never attack a strange one, but was a keen fighter when in company with his father. Upon the present occasion a large mongrel passed along the road, and shortly afterwards the old dog awoke and went sleepily downstairs. When he arrived upon the door-step his son ran up to him and

[1] *Descent of Man*, p. 71.

made the sign just described. His whole manner immediately altered to that of high animation. Clearing the wall together, the two animals ran down the road as terriers only can when pursuing an enemy. I watched them for a mile and a half, within which distance their speed never abated, although the object of their pursuit had not from the first been in sight.

It is almost superfluous to give cases illustrating the well-known fact that dogs communicate their desires and ideas to man; but as the subject of the communication by signs will afterwards be found of importance in connection with the philosophy of communication by words, I shall here give a few examples of dogs communicating by signs with man, which for my purpose will be the more valuable the less they are recognised as unusual.

Lieutenant-Gen. Sir John H. Lefroy, C.B., K.C.M.G., F.R.S., writes me that he has a terrier which it is the duty of his wife's maid to wash and feed. 'It was her habit after calling her mistress in the morning to go out and milk a goat which was tethered near the house, and give "Button" the milk. One morning, being rather earlier than usual, instead of going out at once she took up some needlework and began to occupy herself. The dog endeavoured in every possible way to attract her attention and draw her forth, and at last pushed aside the curtain of a closet, and never having been taught to fetch or carry, took between his teeth the cup she habitually used, and brought it to her feet. I inquired into every circumstance strictly on the spot, and was shown where he found the cup.'

Similarly I select the following case from a great number of others that I might quote, because it is so closely analogous to the above. It is communicated to me by Mr. A. H. Baines:—

There is a drinking-trough for him in my sitting-room: if at any time it happens to be without water when he goes to drink, he scratches the dish with his fore-paws in order to call attention to his wants, and this is done in an authoritative way, which generally has the desired effect. Another Pomeranian—a member of the same family—when quite young used to soak hard biscuits in water till soft enough to eat. She would carry the biscuit in her mouth to the drinking-trough, drop it in and leave it there for a few minutes, and then fish it out with her paw.

One more instance of the communication of ideas by gestures will no doubt be deemed sufficient. It is one of a kind which has many analogies in the literature of canine intelligence.

Dr. Beattie relates this case of canine sagacity, of which the scene was a place near Aberdeen. The Dee being frozen, a gentleman named Irvine was crossing the ice, which gave way with him about the middle of the river. Having a gun, he was able to keep himself from sinking by placing it across the opening. 'The dog made many fruitless efforts to save his master, and then ran to a neighbouring village, where he saw a man, and with the most significant gestures pulled him by the coat, and prevailed on him to follow. The man arrived on the spot in time to save the gentleman's life.'

Numberless other instances of the same kind might be given, and they display a high degree of intelligence. Even the idea of saving life implies in itself no small amount of intelligence; but in such cases as these we have added the idea of going for help, communicating news of a disaster, and leading the way to its occurrence.

Having thus as briefly as possible considered the emotional and the more ordinary intellectual faculties of the dog, I shall pass on to the statement of cases showing the higher and more exceptional developments of canine sagacity.

Were the purpose of this work that of accumulating anecdotes of animal intelligence, this would be the place to let loose a flood of facts, which might all be well attested, relating to the high intelligence of dogs. But as my aim is rather that of suppressing anecdotes, except in so far as facts are required to prove the presence in animals of the sundry psychological faculties which I believe the different classes to present, I shall here, as elsewhere, follow the method of not multiplying anecdotes further than seems necessary fully to demonstrate the highest level of intelligence to which the animal under consideration can certainly be said to attain. But in order that any who read these pages for the sake of the anecdotes which they necessarily present may not be disappointed by meeting with cases already known to them, I shall draw my material mainly from the facts communicated to me by private correspondents, alluding to previously published facts only as supplementary to those now published for the first time. It may be well to explain to my numerous correspondents that I select the following cases for quoting, not because they are the most sensational that I have received, but rather because they either contain nothing sufficiently exceptional to excite the criticism of incredulity, or because they happen to have been corroborated by the more or less similar cases which I quote from other correspondents.

As showing the high general intelligence of the dog, I shall first begin with the collie. It is certain that many of these dogs can be trusted to gather and drive sheep without supervision. It is enough on this head to refer to the well-known anecdotes of the poet Hogg in his 'Shepherd's Calendar,' concerning his dog 'Sirrah.'

Williams, in his book on 'Dogs and their Ways,' says (p. 124) that a friend of his had a collie which, whenever his master said the words 'Cast, cast,' would run off to seek any sheep that might be cast, and on finding it would at once assist it to rise. He also knew of another dog (p. 102), which would perform the same office even in the absence of his master, going the round of the fields and pastures by himself to right all the sheep that he found to be cast.[1]

One of my correspondents (Mr. Laurie Gentles) sends me an account of a sheep-dog belonging to a friend of his (Mr. Mitchell, of Inverness-shire) which strayed to a neighbouring farm, and took up his residence with the farmer. On the second night after the dog arrived at the farm the farmer 'took the dog down to the meadow to see if the cattle were all right. To his dismay he found that

[1] For many other instances of sheep-dog sagacity, see Watson, *Reasoning Power of Animals*, under 'Shepherd's Dog.'

the fence between his meadow and his neighbour's had got broken down, and that the whole of his neighbour's cattle had got mixed up with his. By the help of the dog the strange cattle were driven back into their proper meadow, and the fence put into temporary repair. The *next* night, at the same hour, the gentleman started off to look after the cattle. The dog, however, was not to be seen. On arriving at the meadow, what was the gentleman's astonishment to find that the dog had preceded him! His astonishment soon changed into delighted approbation when he found the dog sitting on the broken fence between the two meadows, and daring the cattle from either side to cross. The cattle had during the interval between the first and second visits broken down the fence, and had got mixed up with each other. The dog had quietly gone off on his own account to see if all was right, and finding a similar accident to the one the previous evening, had *alone* and *unaided* driven back the *strange* cattle to their proper meadow, and had mounted guard over the broken fence as I have already indicated.'

Colonel Hamilton Smith says that the cattle-dogs of Cuba and Terra Firma are very wise in managing cattle, but require to display different tactics from the cattle-dogs of Europe:—

When vessels with live stock arrive at any of the West India harbours, these animals, some of which are nearly as large as mastiffs, are wonderfully efficient in assisting to land the cargo. The oxen are hoisted out with a sling passing round the base of their horns; and when an ox, thus suspended by the head, is lowered, and allowed to fall into the water, so that it may swim to land, men sometimes swim by the side of it and guide it, but they have often dogs of this breed which will perform the service equally well; for, catching the perplexed animal by the ears, one on each side, they will force it to swim in the direction of the landing-place, and instantly let go their hold when they feel it touch the ground, as the ox will then naturally walk out of the water by itself.[1]

That this sagacity need not be due to special tuition, may be inferred from a closely similar display sponta-

---

[1] *Naturalist's Library*, vol. x, p. 154 (quoted by Watson).

neously shown in the following case. It is communicated to me by a correspondent, Mr. A. H. Browning. This gentleman was looking at a litter of young pigs in their sty, and when he went away the door of the sty was inadvertently left unfastened. The pigs all escaped into his garden. My correspondent then proceeds:—

My attention was called to my dog appearing in a great state of excitement, *not* barking (he seldom barks), but whining and performing all sorts of antics (in a human subject I should have said 'gesticulating'). The herdmen and myself returned to the sty; we caught but one pig, and put him back; no sooner had we done so than the dog ran after each pig in succession, brought him back to the sty by the ear, and then went after another, until the whole number were again housed.

In Lord Brougham's 'Dialogues on Instinct' (iii.) there is narrated the story told to the author by Lord Truro of a dog that used to worry sheep at night. The animal quietly submitted to be tied up in the evening, but when everybody was asleep he used to slip his collar, worry the sheep, and, returning before dawn, again get into his collar to avoid suspicion. I allude to this remarkable display of sagacity because I am myself able fully to corroborate it by precisely similar cases. A friend of mine (the late Mr. Sutherland Murray) had a dog which was always kept tied up at night, but nevertheless the neighbouring farmers complained of having detected him as the culprit when watching to find what dog it was that committed nightly slaughter among their sheep. My friend, therefore, set a watch upon his dog, and found that when all was still he slipped his collar, and after being absent for some hours, returned and slipped his head in again.

A precisely similar case is given further back, and others are communicated to me by two correspondents (Mr. Goodbehere, of Birmingham, and Mr. Richard Williams, of Buffalo). The latter says:—

And here let me ask if you are aware of the cunning and sagacity of these sheep-killing dogs, that they never kill sheep on the farm to which they belong, or in the immediate vicinity, but often go miles away; that they always return before daylight, and before doing so wash themselves in some stream to get rid of the blood.

In Germany I knew a large dog that was very fond of grapes, and at night used to slip his collar in order to satisfy his propensity; and it was not for some time that the thief was suspected, owing to his returning before daylight and appearing innocently chained up in his kennel.

A closely similar case is recorded in Mr. Duncan's book on 'Instinct' of a dog belonging to the Rev. Mr. Taylor, of Colton. The only difference is that the delinquent dog slipped and afterwards readjusted a muzzle instead of a collar.

In connection with sly sagacity I may also give another story contained in my correspondence, although in this case I am specially requested by my correspondent not to publish his name. I can, therefore, only say that he occupies a high position in the Church, and that the dog (a retriever) was his own property :—

The dog was lying one evening before the kitchen fire where the cook had prepared a turkey for roasting. She left the kitchen for a few moments, when the dog immediately carried away the turkey and placed it in the cleft of a tree close to the house, but which was well concealed by the surrounding laurels. So rapid were his movements that he returned to his post before the cook had come back, and stretching himself before the fire, looked 'as innocent as a child unborn.' Unfortunately for him, however, a man who was in the habit of taking him to shoot, saw him carrying away his prize and watched his progress. On coming into the kitchen the man found the dog in his old place pretending to be asleep. Diver's conduct was all along dictated by a desire to conceal his theft, and if he were a man I should have said that he intended, in case of inquiry, to prove an alibi.

Mr. W. H. Bodley writes me of a retriever dog that belonged to him :—

Before he came to me he lived where another dog of similar size was kept, and on one occasion they fought. Having been chastised for this, on future occasions when they quarrelled they used to swim over a river of some breadth, where they could not be interfered with, and fight out their quarrel on the other side. What seems to me noteworthy in this conduct is the *self-restraint* manifested under the influence of *passion*, and the mutual understanding to defer the fight till they could prosecute it unmolested; like two duellists crossing the Channel to fight in France.

It is, of course, a well-known thing that dogs may easily be taught the use of coin for buying buns, &c. In the 'Scottish Naturalist' for April, 1881, Mr. Japp vouches for the fact that a collie which he knew was in the habit of purchasing cakes with coppers without ever having been taught the use of coin for such purposes. This fact, however, of a dog spontaneously divining the use of money requires corroboration, although it is certain that many dogs have an instinctive idea of giving peace-offerings, and the step from this to the idea of barter may not be large. Thus, to give only two illustrations, Mr. Badcock writes to me that a friend of his had a dog which one day had a quarrel with a companion dog, so that they parted at variance. 'On the next day the friend appeared with a biscuit, which he presented as a peace-offering.' Again, Mr. Thomas D. Smeaton writes to me of his dog that he 'has an amusing practice when he is restored to favour after some slight offence, of immediately picking up and carrying anything that is handiest, stone, stick, paper: it is a deliberate effort to please, a sort of good-will offering, a shaking hands over the past.'

I am indebted for the following to Mr. Goodbehere, of Birmingham; it may be taken as typical of many similar cases:—

My friend (Mr. James Canning, of Birmingham) was acquainted with a small mongrel dog who on being presented with a penny or a halfpenny would run with it in his mouth to a baker's, jump on to the top of the half-door leading into the shop, and ring the bell behind the door until the baker came forward and gave him a bun or a biscuit in exchange for the coin. The dog would accept any small biscuit for a halfpenny, but nothing less than a bun would satisfy him for a penny. On one occasion the baker (being annoyed at the dog's too frequent visits), after receiving the coin, refused to give the dog any thing in exchange, and on every future occasion the latter (who declined being *taken in* a second time) would put the coin on the floor, and not permit the baker to pick it up until he had received its equivalent.

Mr. R. O. Backhouse writes to me :—

My dog is a broken-haired rabbit-coursing dog, and is very intelligent. I took him one day to an exhibition of pictures and objects of interest, among which were statues and a bust of Sir Walter Scott. It was a local exhibition, and as there was jewellery, some one had to sit up all night with it as guard. I volunteered, and as we were looking about and sitting on a stand of flowers, my dog suddenly began to bark, and made as if he had found some one hiding. On looking round I found that it was the bust of Sir Walter Scott standing among the flowers, and in which he evidently recognised sufficient likeness to a human being to think the supposed man had no business there at so late an hour.

I adduce this instance because it serves as a sort of introduction to the more remarkable faculty which I cannot have the least doubt is manifested by some dogs—the faculty, namely, of recognising portraits as representing persons, or possibly of mistaking portraits for persons.

Mr. Crchore, writing to 'Nature' (vol. xxi., p. 132), says :—

A Dandie-Dinmont terrier, after the death of his mistress, was playing with some children in a room into which was brought a photograph (large) of her that he had never previously seen. It was placed upon the floor leaning against the wall. In the words of my informant, who witnessed it, the dog, when he suddenly caught sight of the picture, crouched and trembled all over, his whole body quivering. Then he crept along the floor till he reached it, and, seating himself before it, began to bark loudly, as if he would say, 'Why don't you speak to me?' The picture was moved to other parts of the room, and he followed, seating himself before it and repeating his barking.

Mr. Charles W. Peach also gives an account in 'Nature' (vol. xx., p. 196) of a large dog recognising his portrait :—

When it (the portrait) was brought to my house, my old dog was present with the family at the unveiling ; nothing was said to him, nor invitation given to him to notice it. We saw that his gaze was steadily fixed on it, and he soon became excited and whined, and tried to lick and scratch it, and was so much taken up with it that we—although so well knowing his intelligence—were all quite surprised—in fact, could scarcely believe that he should know it was my likeness. We, however, had sufficient proof after it was hung up in our parlour. The room was rather low, and under the picture stood a chair : the door was left open, without any thought about the dog; he, however, soon found it out, when a low whining and scratching was heard by the family, and on search being made, he was in the chair trying to get at the picture. After this I put it up higher, so as to prevent its being injured by him. This did not prevent him from paying attention to it, for whenever I was away from home, whether for a short or a long time—sometimes for several days—he spent most of his time gazing on it, and as it appeared to give him comfort the door was always left open for him. When I was long away he made a low whining, as if to draw attention to it. This lasted for years—in fact, as long as he lived.

From this account it appears that when in the first instance the dog's attention was drawn to the picture it was on the floor in the line of the dog's sight; the behaviour of the animal then and subsequently was too marked and peculiar to admit of mistake.

Another correspondent in 'Nature' (vol. xx., p. 220), alluding to the previous letter, writes :—

Having read Mr. Peach's letter on 'Intellect in Brutes,' as shown by the sagacity he witnessed in his dog, I have been asked to send a similar anecdote, which I have often told to friends. Many years ago my husband had his portrait taken by J. Phillips, R.A., and subsequently went to India, leaving the portrait in London to be finished and framed. When it was sent home, about two years after it was taken, it was placed on the floor against the sofa, preparatory to being hung on the wall. We had then a very handsome black-and-tan setter, which was a great pet in the house. As soon as the dog came into the room he recognised his master, though he had not seen him for two years, and went up to the picture and licked the face. When this anecdote was told to Phillips, he said it was the highest compliment that had ever been paid him.

Similarly, in the same periodical (vol. xx., p. 220), Mr. Henry Clark writes :—

Some years ago a fine arts exhibition was held at Derby. A portrait of a Derby artist (Wright) was thus signalised :—'The artist's pet dog distinguished this from a lot of pictures upon the floor of the studio by licking the face of the portrait.'

Again, I learn from Dr. Samuel Wilks, F.R.S , that a friend of his, whom I shall call Mrs. E., has a terrier which recognised her portrait. 'The portrait is now (1881) hanging in the Royal Academy. When it first arrived home the dog barked at it, as it did at strangers; but after a day or two, when Mrs. E. opened the door to show the portrait to some friends, the dog went straight to the picture and licked the hand. The picture is a three-quarter length portrait of a lady with the hand at the bottom of the picture.'

Lastly, my sister, who is a very conscientious and accurate observer, witnessed a most unmistakable recognition of portraits as representative of persons on the part of a small but intelligent terrier of her own. At my request she committed the facts to writing shortly after they occurred. The following is her statement of them:—

I have a small terrier who attained the age of eight months without ever having seen a large picture. One day three nearly life-sized portraits were placed in my room during his absence. Two were hung up, and one left standing against the wall on the floor awaiting the arrival of a picture-rod. When the dog entered the room he appeared much alarmed by the sight of the pictures, barking in a terrified manner first at one and then at another. That is to say, instead of attacking them in an aggressive way with tail erect, as he would have done on thus encountering a strange person, he barked violently and incessantly at some distance from the paintings, with tail down and body elongated, sometimes bolting under the chairs and sofas in the extremity of his fear, and continuing barking from there. Thinking it might be merely the presence of strange objects in the room which excited him, I covered the faces of the portraits with cloths and turned the face of the one on the floor to the wall. The dog soon after emerged from his hiding-place, and having looked intently at the covered pictures and examined the back of the frame on the ground, became quite quiet and contented. I then uncovered one of the pictures, when he immediately flew at it, barking in the same frightened manner as before. I then re-covered that one and took the cover off another. The dog left the covered one and rushed at the one which was exposed. I then turned the face of the one on the floor to the room, and he flew at that with increased fierceness. This I did many times, covering and uncovering each picture alternately, always with the same result. It was only when all three paintings were uncovered at the same time, and he saw one looking at him in whatever direction he turned, that he became utterly terrified. He continued in this state for nearly an hour, at the end of which time, although evidently very nervous and apt to start, he ceased to bark. After that day he never took any more notice of the pictures during the three months he remained in the house. He was then absent from the house for seven months. On his return he went with me into the room where the portraits were hung, immediately on his arrival. He was evidently again much startled on first seeing them, for he rushed at one, barking as he had done on the first occasion, but he only gave three or four barks when he ran back to me with the same apologetic manner as he has when he has barked at a well-known friend by mistake.

It will have been observed that in all these cases the portraits, when first recognised as bearing resemblance to human beings, were placed on the floor, or in the ordinary line of the dog's sight. This is probably an important condition to the success of the recognition. That it certainly was so in the case of my sister's terrier was strikingly proved on a subsequent occasion, when she took the animal into a picture-shop where there were a number of portraits hanging round the walls, and also one of Carlyle standing on the floor. The terrier did not heed those upon the walls, but barked excitedly at the one upon the floor. This case was further interesting from the fact that there were a number of purchasers in the shop who were, of course, strangers to the terrier; yet he took no notice of them, although so much excited by the picture. This shows that the pictorial illusion was not so complete as to make the animal suppose the portrait to be a real person; it was only sufficiently so to make it feel a sense of bewildered uncertainty at the kind of life-in-death appearance of the motionless representation.

If, notwithstanding all this body of mutually corroborative cases, it is still thought incredible that dogs should be able to recognise pictorial representations,[1] we should

[1] Since my MS. went to press I have myself met with a striking

do well to remember that this grade of mental evolution is reached very early in the psychical development of the human child. In my next work I shall adduce evidence to show that children of one year, or even less, are able to distinguish pictures as representations of particular objects, and will point at the proper pictures when asked to show these objects.

Coming now to cases more distinctly indicative of reason in the strict sense of the word, numberless ordinary acts performed by dogs indisputably show that they possess this faculty. Thus, for instance, Livingstone gives the following observation.[1] A dog tracking his master along a road came to a place where three roads diverged. Scenting along two of the roads and not finding the trail, he ran off on the third without waiting to smell. Here, therefore, is a true act of inference. If the track is not on A or B, it must be on C, there being no other alternative.

Again, it is not an unusual thing for intelligent dogs, who know that their masters do not wish to take them out, to leave the house and run a long distance in the direction in which they suppose their masters are about to go, in order that when they are there found the distance may be too great for their masters to return home for the purpose of shutting them up. I have myself known several terriers that would do this, and one of the instances I shall give *in extenso* (quoted from an account which I published at the time in 'Nature'); for I think it displays remarkably complex processes of far-seeing calculation:—

The terrier in question followed a conveyance from the house in which I resided in the country, to a town ten miles distant. *He only did this on one occasion*, and about five months afterwards was taken *by train* to the same town as a present to some friends there. Shortly afterwards I called upon these friends in a different conveyance from the one which the dog had previously followed; but the latter may have known that the two conveyances belonged to the same

display of the recognition of a portrait by a dog. The portrait was one of myself, and the dog a half-bred setter and retreiver of my own

[1] *Missionary Travels*, chap. i.

house. Anyhow, after I had put up the horses at an inn, I spent the morning with the terrier and his new masters, and in the afternoon was accompanied by them to the inn. I should have mentioned that the inn was the same as that at which the conveyance had been put up on the previous occasion, five months before. Now, the dog evidently remembered this, and, reasoning from analogy, inferred that I was about to return. This is shown by the fact that he stole away from our party—although at what precise moment he did so I cannot say, but it was certainly *after* we had arrived at the inn, for subsequently we all remembered his having entered the coffee-room with us. Now, not only did he infer from a single precedent that I was going home, and make up his mind to go with me, but he also further reasoned thus:—'As my previous master lately sent me to town, it is probable that he does not want me to return to the country; therefore, if I am to seize this opportunity of resuming my poaching life, I must now steal a march upon the conveyance. But not only so, my former master may possibly pick me up and return with me to my proper owners; therefore I must take care only to intercept the conveyance at a point sufficiently far without the town to make sure that he will not think it worth his while to go back with me.'

Complicated as this train of reasoning is, it is the simplest one I can devise to account for the fact that slightly beyond the *third* milestone the terrier was awaiting me, lying right in the middle of the road with his face towards the town. I should add that the second two miles of the road were quite straight, so that I could easily have seen the dog if he had been merely running a comparatively short distance in front of the horses. Why this animal should never have returned to his former home on his own account I cannot suggest, but I think it was merely due to an excessive caution which he also manifested in other things. However, be the explanation of this what it may, as a fact he never did venture to come back upon his own account, although there never was a subsequent occasion upon which any of his former friends went to the town but the terrier was seen to return with them, having always found some way of escape from his intended imprisonment.

The Rev. J. C. Atkinson gives an account ('Zoologist,' vol. vii., p. 2338) of his terrier, which, on starting a water-rat out of reeds into the running stream, would not plunge directly after it, knowing that the rat would beat him at swimming. But the moment the rat plunged, the dog ran four or five yards down the bank, and there waited till the water-rat, being carried down stream, appeared upon the surface, when he pounced upon it successfully.

Cases of this kind might be multiplied indefinitely, and they appear to show a true faculty of reason or inferring.

Professor W. W. Bailey, writing from Broun University to 'Nature' (xxii., p. 607), says:—

A friend of mine, a naturalist, and a very conscientious man, whose word can be implicitly trusted, gives the following, to which he was an eye-witness. His grandfather, then a very old but hale and hearty man, had a splendid Newfoundland. There was a narrow and precipitous road leading from the fields to the house. It was regarded as a very dangerous place. One day when the old gentleman was doing some work about the farm his horse became alarmed, and started off with the waggon along this causeway. The chances were that he would dash himself and the empty waggon to pieces. At once the dog seemed to take in the situation, although until that time he had been impassive. He started after the horse at full speed, overtook him, caught the bridle, and by his strength arrested the frightened creature until help could reach him. My friend gives many other stories of this fine dog, and thinks he had a decided sense of humour. I will repeat that both of these tales come to me well authenticated, and I could, by seeking permission, give names and places.

Couch gives the following, which is worth quoting, as showing the intelligence of dogs in attacking unusual prey:—

On the first discovery of the prey (crabs) a terrier runs in to seize it, and is immediately and severely bitten in the nose. But a sedate Newfoundland dog of my acquaintance proceeds more soberly in his work. He lays his paw on it to arrest it in its escape; then tumbling it over he bares his teeth, and, seizing it with the mouth, throws the crab aloft. It falls upon the stones; the shell is cracked beyond redemption, and then the dainty dish is devoured at his leisure.[1]

[1] *Illustrations of Instinct*, p. 179.

I myself know a large dog in Germany which used to kill snakes by dexterously tossing them in the air a great number of times, too quickly to admit of the snake biting. When the snake was thus quite confused, the dog would tear it in pieces. This dog can never have been poisoned by the bite of a snake; but he seems to have had an instinctive idea that the snake might be more harmful in its bite than other animals; for while he was bold in fighting with dogs, and did not then object to receiving his fair share of laceration, he was extremely careful never to begin to tear a snake till he had thoroughly bewildered it by tossing it as described.

The reasoning displayed by dogs may not always be of a high order, but little incidents, from being of constant occurrence among all dogs, are the more important as showing the reasoning faculty to be general to these animals. I shall therefore give a few cases to show the kind of reasoning that is of constant occurrence.

Mr. Stone writes to me from Norbury Park concerning two of his dogs, one large and the other small. Both being in a room at the same time,

one of them, the larger, had a bone, and when he had left it the smaller dog went to take it, the larger one growled, and the other retired to a corner. Shortly afterwards the larger dog went out, but the other did not appear to notice this, and at any rate did not move. A few minutes later the large dog was heard to bark out of doors; the little dog then, without a moment's hesitation, went straight to the bone and took it. It thus appears quite evident that she reasoned—'That dog is barking out of doors, therefore he is not in this room, therefore it is safe for me to take the bone.' The action was so rapid as to be clearly a consequence of the other dog's barking.

Again, Mr. John Le Conte, writing from the University of California, tells me of a dog which used to hunt rabbits in an extensive pasture-ground where there was a hollow tree, which frequently served as a place of refuge for the rabbits when they were pressed:—

On one occasion a rabbit was 'started,' and all of the dogs, with the exception of 'Bonus,' dashed off in full pursuit. We were astonished to observe that the sedate 'Bonus,' foregoing the intense excitement of the chase, deliberately trotted by a short cut to a hollow oak trunk, and crouching at its base calmly awaited the advent of the fleeing rabbit. And he was not disappointed (they frequently escaped without being reduced to this extremity), for the pursuing dogs pressed the rabbit so hard that, after making a long detour, it made for the place of refuge. As it was about entering the hollow trunk, the crouching 'Bonus' captured the astonished rodent.

Similarly, Dr. Andrew Wilson, F.R.S.E., writes me as follows :—

There is a shrubbery near the house, about 200 or 300 yards long, and running in the shape of a horseshoe. A small terrier used to start a rabbit nearly every morning, at the end of the shrubbery next the house, and hunt him through the whole length of it to the other end, where the rabbit escaped into an old drain. The dog then appears to have come to the conclusion that the chord of a circle is shorter than its arc, for he raised the rabbit again, and instead of following him through the shrubbery as usual, he took the short cut to the drain, and was ready and in waiting on the rabbit when he arrived, and caught him.

A somewhat similar instance is communicated to me by Mr. William Cairns, of Argyll House, N.B. :—

I was watching the operations of a little Skye terrier on a wheatstack which was in the course of being thrashed, when suddenly a very large rat bounced off, just from under Fan's nose. It darted into a pit of water about a dozen yards from the stack, and tried to escape. Fan, however, plunged after, and swam for some distance, but found she was being left behind. So she turned to the shore again *and ran round to the other side of the pit, and was ready and caught it just on landing.*

I never saw anything more remarkable. If it was not reason, I do not know how it is possible that it could come much more closely to the exercise of that faculty.

Dr. Bannister, editor of the 'Journal of Nervous and Mental Diseases,' writes me from Chicago, that having spent a winter in Alaska, he 'had a good opportunity to study animal intelligence in the Eskimo dogs,' and he reports it as 'a fact of common occurrence,' when the dogs are drawing sledges on the ice near the coast, that on coming to sinuosities in the coast-line, they spontaneously leave the beaten track and strike out so as to 'cut across the windings by going straight from point to point' of land. This is frequently done even when the leading dog 'could not see the whole winding of the beaten track; he seemed to reason that the route must lead around the headlands, and that he could economise travel by cutting across.'

It will be remembered in connection with these dogs, that Mr. Darwin in the 'Descent of Man' (p. 75) quotes Dr. Hayes, who, in his work on 'The Open Polar Sea,' 'repeatedly remarks that his dogs, instead of continuing to draw the sledges in a compact body, diverged and separated when they came to thin ice, so that their weight might be more evenly [and widely] distributed. This was often the first warning which the travellers received that the ice was becoming thin and dangerous.' Mr. Darwin remarks, 'This instinct may possibly have arisen since the time, long ago, when dogs were first employed by the natives in drawing their sledges; or the Arctic wolves, the parent stock of the Esquimaux dog, may have acquired an instinct, impelling them not to attack their prey in a close pack when on thin ice.'

Mrs. Horn writes me:—

One morning, soon after his usual time for starting, I saw the dog looking anxiously about, evidently afraid that my brother had gone without him. He looked into the room where we had breakfasted, but my brother was not there. He went up two or three stairs, and listened attentively. Then, to my astonishment, he came down, and going to the hat-stand in the hall, stood on his hind legs and sniffed at the great-coats hanging there, undoubtedly trying to ascertain whether my brother's coat was there or not.

Another correspondent (Mr. Westlecombe) writes:—

My cat had kittens, of which two were preserved, the rest being drowned. The dog tolerated the two kittens, but did not care about them with any friendship. When the kittens were a few weeks old, I—finding that I could get but one of them off my hands—determined to kill the other, and, as the quickest mode of death, to shoot it by a pistol close behind its head. The dog saw me do this in my garden, and in a few minutes afterwards she appeared with the other kitten dead in her mouth; she had killed it. If that was not reasoning I do not know what is.

Mr. W. F. Hooper writes me of a Newfoundland dog that was in the habit of accompanying the nursemaid and baby belonging to its mistress. On one occasion a keen wind began to blow, and the nursemaid drew her shawl over the child:—

The nursemaid had not taken many steps towards home before her progress was barred by the dog, who placed himself in the centre of the path and growled whenever she advanced. She was much alarmed, and tried to coax the dog to move, but Leo would not, and abated nothing of the hostile display. Half an hour passed, and the girl became nearly distracted. What could be the matter with the dog? Was she to be a prisoner all day? Would the animal fly at her throat? Was Leo suffering from hydrophobia? These and similar questions crossed the girl's mind. At length a suggestion of despair—it was nothing more—occurred to her. She thought it might win the dog round to good humour if she showed it the baby; so she removed the folds of her shawl and presented it at arm's length to the dog. The result was magical, and far in excess of all expectation, for not only did the dog cease to growl, but he began to gambol and caress, and removed himself from the path altogether, so that there was now a free course, and home was soon reached. The explanation of the whole affair is, when the nursemaid turned on her path thinking she had gone sufficiently far, the dog missed sight of the baby, and believed it was gone. Under this impression the dog converted himself into a sentinel, with the resolve that not one step should be taken towards home without the baby; and faithfully did the animal keep watch and ward until the demonstration was given that the child had not been left behind, but was still in the nurse's arms alive and well. I think this is an exhibition of intelligence worthy of being known to you.

I extract the following instance from Col. Hutchinson's 'Dog-breaking.' It is briefly alluded to in the 'Descent of Man.' The observer and narrator is Mr. Colquhoun:—

I may mention a proof of his sagacity. Having a couple of long shots across a pretty broad stream, I stopped a mallard with each barrel, but both were only wounded. I sent him across for the birds. He first attempted to bring them both, but one always struggled out of his mouth: he then laid down one intending to bring the other; but whenever he attempted to cross to me, the bird left fluttered into the water; he immediately returned again, laid down the first on the shore and recovered the other. The first now fluttered away, but he instantly secured it, and, standing over them both, seemed to cogitate for a moment; then, although on any other occasion he never ruffles a feather, deliberately killed one, brought over the other, and then returned for the dead bird.

The following, communicated to me by Mr. Blood, is a closely analogous, and therefore confirmatory case. He was out shooting with a companion, and three wild ducks were simultaneously dropped into a lake—one falling dead and the other two winged. Mr. Blood sent in his spaniel to retrieve,

and of course when the wounded birds saw her coming they swam out, so that she first reached the dead duck. She swam up to it, paused for a moment, and passing it went after the nearest wounded bird. Having caught this, she again hesitated, and apparently after consideration she gave it a chop and let it go, quieted for the present. She then caught and brought to land the other wounded duck, and going back she again reached the dead bird; but looking at the other and seeing that it was again moving, she went out and brought it in, and last of all brought the dead bird. The dog was a first-rate retriever and never injured game, so that it was an entirely new thing for her to kill a bird.

Again, Mr. Arthur Nicols, in 'Nature,' vol. xix., page 496, says:—

Can we conceive any human being reasoning more correctly than a dog did in the following instance? Towards the evening of a long day's snipe-shooting on Dartmoor, the party was walking down the bank of the Dart, when my retriever flushed a widgeon which fell to my gun in the river, and of course instantly dived. I said no word to the dog. He did not plunge into the river *then*, but galloped *down* stream some fifty or sixty yards, and then entered and dashed from side to side— it was about twenty or thirty feet wide—working up stream, and making a great commotion in the water until he came to the place where we stood. Then he landed and shook himself, and carefully hunted the near bank a considerable distance down, crossed to the opposite side, and diligently explored that bank. Two or three minutes elapsed, and the party was for moving on, when I called their attention to a sudden change in the dog's demeanour. His 'flag' was now up and going from side to side in that energetic manner which, as every sportsman knows, betokens a hot scent. I then knew that the bird was as safe as if it was already in my bag. Away through the heather went the waving tail, until twenty or thirty yards from the bank opposite to that on which we were standing there was a momentary scuffle; the bird just rose from the ground above the heather, the dog sprang into the air, caught it, came away at full gallop, dashed across the stream, and delivered it into my hands. Need I interpret all this for the experienced sportsman? The dog had learned from long experience in Australia and the narrow cañadas in the La Plata that a wounded duck goes down stream; if winged, his maimed wing sticks out and renders it impossible for him to go up, so he will invariably land and try to hide away from the bank. But if the dog enters at the place where the bird fell, the latter will go on with the stream for an indefinite distance, rising now and then for breath, and give infinite trouble. My dog had found out all this long since, and had proved the correctness of his knowledge times out of number, and by his actions had *taught me* the whole art and mystery of retrieving duck. His object, I say without a doubt, because I had numberless opportunities of observing it, was to fling the bird and force it to land by cutting it off lower down the stream. Then assuming, as his experience justified him, that the bird had landed, he hunted each bank in succession for the tra l, which he knew must betray the fugitive.

As showing in a higher, and therefore rarer degree, the ratiocinative faculty in dogs, I may quote a brief extract from my British Association lecture:—

My friend Dr. Rae, the well-known traveller and naturalist, knew a dog in Orkney which used to accompany his master to church on alternate Sundays. To do so he had to swim a channel about a mile wide; and before taking to the water he used to run about a mile to the north when the tide was flowing, and a nearly equal distance to the south when the tide was ebbing, 'almost invariably calculating his distance so well that he landed at the nearest point to the church.' In his letter to me Dr. Rae continues: 'How the dog managed to calculate the strength of the spring and neap tides at their various rates of speed, and always to swim at the proper angle, is most surprising.'

As a confirmatory case, I may also quote an extract from a letter sent me by Mr. Percival Fothergill. Writing of a retriever which he has, he says:—

I have seen her spring overboard from our gangway 16 feet from the water-line. The tides ran more than 5 knots, and she invariably came down to a little wharf abreast the ship, and gazed intently for small pieces of stick or straw, and having thus ascertained the drift of the tide (did as you mention of another dog), ran up tide and swam off. The sentry on the forecastle always kept a look-out for the dog, and threw over a line with a bowling knot, and she was hauled on board.

But one day she was observed to wait an unusual time on the wharf; no wood or straw gave her the required information. After waiting some time, she lay down on the planks, and dropped one paw into the water, and found by the feel which way the tide ran, got up, and ran up stream as usual.

Mr. George Cook writes me that he recently had a pointer, which one morning, when the grass was covered with frost, dragged a mat out of his kennel, from which he had got loose, to the lawn beneath the house windows, where he was found lying upon the mat, which thus served to protect him from the frost. The distance over which he had dragged the mat for this purpose was about 100 yards. Mr. Cook adds: 'I have since frequently seen him bring this mat out of his kennel and lay it in the sunshine, shifting it if a shadow came upon the place where he had laid it.'

The following is sent me by the Rev. F. J. Penky. He gives me the name of his friend the canon, but does not give me express permission to publish it. In quoting his account, therefore, I leave this name blank. He says:—

The following is an instance of sagacity—indeed, amounting to reason—in a dog, a French poodle that belonged to Colonel Pearson (not the lately beleaguered colonel at Ekowe, but a Colonel Pearson living some years ago at Lichfield). The circumstance happened to a friend of mine, Canon ——, rector of ——. I have the story from his own lips, but I have no permission for his name to be used in any publication, should the story be thought worthy of it. My friend the canon, I may say, has no leanings. Being a guest at luncheon with the dog's master, my friend fed the dog with pieces of beef. After luncheon the beef was taken into the larder. The dog did not think he had his fair share. What did he do? Now he had been taught to stand on his hind legs, put his paw on a lady's wrist, and hand her into the dining-room. He adopted the same tactics with my friend the canon, stood on his hind legs, put his paw on his arm, and made for the door. To see what would follow, Canon —— suffered himself to be led; but the sagacious dog, instead of steering for the dining-room, led him in the direction of the larder, along a passage, down steps, &c., and did not halt till he brought him to the larder, and close to the shelf where the beef had been put. The dog had a small bit given him for his sagacity, and Canon —— returned to the drawing-room. But the dog was still not satisfied. He tried the same trick again, but this time fruitlessly. The canon was not going again with him to the larder. What was Mori to do? And here comes the instance of reason in the poodle. Finding he could not prevail on the visitor to make a second excursion to the larder, he went out into the hall, took in his teeth Canon ——'s hat from off the hall table, and carried it under the shelf in the larder, where the coveted beef lay out of his reach. There he was found with the hat, waiting for the owner of the hat, and expecting another savoury bit when he should come for his hat.

Many anecdotes might be adduced of the cleverness which some dogs show in finding their way by train; but I shall give only three, and I select these, not only because they all mutually corroborate one another, but likewise because they all display such high intelligence on the part of the dogs.

Mr. Horsfall, in 'Nature,' vol. xx., p. 505, says:—

Last year we spent our holidays at L!an Bedr, Merionethshire. Our host has a house in the above village, and another at Harlech, a town three miles distant. His favourite dog, Nero, is of Norwegian birth, and a highly intelligent animal. He is at liberty to pass his time at either of the houses owned by his master, and he occasionally walks from one to the other. More frequently, however, he goes to the railway station at Llan Bedr, gets into the train, and jumps out at Harlech. Being most probably unable to get out of the carriage, he was on one occasion taken to Salsernau, the station beyond Harlech, when he left the carriage and waited on the platform for the return train to Harlech. If Nero did not make use of 'abstract reasoning' we may as well give up the use of the term.

Miss M. C. Young writes to me:—

You may perhaps think the following worthy of notice, as illustrating the comparative failure of *instinct* in an animal which has begun to *reason*. A friend of mine has a mongrel fox-terrier of remarkable intelligence, though undeveloped by any training. This dog has always shown a great fondness for accompanying any of the family on a railway journey, often having to be taken out of the train by force. One morning in the summer of 1877 the groom came, in great distress, to say that Spot had followed him to the station, and jumped into the train after a visitor's maid who was going to see her friends, and he (the groom) felt sure the dog would be stolen. The railway is a short single line, with three trains down and up each day, and my friend is well known to all the officials, so she sent to meet the next train, when the guard said the dog (apparently finding no *friend* in the train) had jumped out at a little roadside station about five miles distant. Most dogs would have found their way home easily, though the place itself was strange, but Spot did not appear till late in the evening, after ten hours' absence, and *dead tired*. On inquiry we found that the guard had seen nothing of her at 9 A.M., at 12 A.M., at 1 P.M., nor at 4 P.M.; but when he reached the little station on his return at 5.30, 'she was walking up and down the platform like a Christian,' jumped into his box, and jumped out again of her own accord at the right station for her home. She had evidently spent the interval in trying to find her way home on foot, and not succeeding, had resolved on returning the way she came.

Lastly, for the following very remarkable case I am indebted to my friend Mrs. A. S. H. Richardson:—

The Rev. Mr. Townsend, incumbent of Lucan, was formerly an engineer on the Dundalk line of railway. He had a very intelligent Scotch retriever dog, which used to have a habit of jumping into any carriage in which Mr. Townsend travelled; but this had been discontinued for a year when the following incident happened. Mr. Townsend and the dog were on the platform at Dundalk station; Mr. Townsend went to get a ticket for a lady, and during his absence the dog jumped into a carriage, and when the train started, was carried down to

Clones. There he found himself alone when he jumped out; he went into the station-master's office and looked about, then into the ticket-collector's and searched there, and then ran off to the town of Clones, a mile distant. There he searched the resident engineer's office, and not finding his master, returned to the station and went to the *up* platform. When the up train arrived, he jumped in, but was driven out by the guard. A ballast train then drew up, going on to a branch line which was being constructed to Caran, but which was not finished yet. The dog travelled on the engine as far as the line went, and then ran the remaining five miles to Caran, where Mr. Townsend's sister lived. He visited her house, and not finding his master, ran back to the station, and took a return train to Clones, where he slept, and was fed by the station-master. At four in the morning he took a goods train down to Dundalk, where he found Mr. Townsend.

It would be easy to continue multiplying anecdotes of canine intelligence; but I think a sufficient number of instances have now been given for the only purpose that I have in view—namely, that of exhibiting in a connected manner the various psychological faculties which are presented by dogs, and the level of development to which they severally attain. I may again remark that I have selected these instances for publication from among many others that I could have given, only because they conform to one or other of the general principles to which I everywhere adhere in the quoting of facts. That is to say, these facts are either matters of ordinary observation, and so intrinsically credible; or they stand upon the authority of observers well known to me as competent; or they are of a kind which do not admit of mal-observation; or, lastly, they are well corroborated by similar accounts received from independent observers. I think, therefore, that this sketch of the psychology of the dog is as accurate as the nature of the materials admits of my drawing it. If it is fairly open to criticism on any one side, I believe it is from the side of the dog-lovers, who may perhaps with justice complain that I have ignored a number of published facts, standing on more or less good authority, and appearing more wonderful than any of the facts that I have rendered. To this criticism I have only to answer that it is better to err on the safe side, and that if the facts which I have rendered are sufficient to prove the existence of all the psychological faculties which the dog can fairly be said to possess, it is of less moment that partly doubtful cases should be suppressed, where the only object of introducing them would be to show that some particular faculties were in some particular instances more highly developed than was the case in the instances here recorded.

CHAPTER XVII.

MONKEYS, APES, AND BABOONS.

WE now come to the last group of animals which we shall have occasion to consider, and these, from an evolutionary point of view, are the most interesting. Unfortunately, however, the intelligence of apes, monkeys, and baboons has not presented material for nearly so many observations as that of other intelligent mammals. Useless for all purposes of labour or art, mischievous as domestic pets, and in all cases troublesome to keep, these animals have never enjoyed the improving influences of hereditary domestication, while for the same reasons observation of the intelligence of captured individuals has been comparatively scant. Still more unfortunately, these remarks apply most of all to the most man-like of the group, and the nearest existing prototypes of the human race: our knowledge of the psychology of the anthropoid apes is less than our knowledge of the psychology of any other animal. But notwithstanding the scarcity of the material which I have to present, I think there is enough to show that the mental life of the *Simiadæ* is of a distinctly different type from any that we have hitherto considered, and that in their psychology, as in their anatomy, these animals approach most nearly to *Homo sapiens*.

Emotions.

Affection and sympathy are strongly marked—the latter indeed more so than in any other animal, not even excepting the dog. A few instances from many that might be quoted will be sufficient to show this.

Mr. Darwin writes:—

Rengger observed an American monkey (a Cebus) carefully driving away the flies which plagued her infant; and Duvancel saw a Hylobates washing the faces of her young ones in a stream. So intense is the grief of female monkeys for the loss of their young, that it invariably caused the death of certain kinds kept under confinement by Brehm in North Africa. Orphan monkeys were always adopted and carefully guarded by the other monkeys, both male and female.[1]

Again, Jobson says that whenever his party shot an orang-outang from their boat, the body was carried off by others before the men could reach the shore.

So, again, James Forbes, F.R.S., in his 'Oriental Memoirs,' narrates the following remarkable instance of the display of solicitude and care for a dead companion exhibited by a monkey:—

One of a shooting-party under a banian tree killed a female monkey, and carried it to his tent, which was soon surrounded by forty or fifty of the tribe, who made a great noise and seemed disposed to attack their aggressor. They retreated when he presented his fowling-piece, the dreadful effect of which they had witnessed and appeared perfectly to understand. The head of the troop, however, stood his ground, chattering furiously; the sportsman, who perhaps felt some little degree of compunction for having killed one of the family, did not like to fire at the creature, and nothing short of firing would suffice to drive him off. At length he came to the door of the tent, and, finding threats of no avail, began a lamentable moaning, and by the most expressive gesture seemed to beg for the dead body. It was given him; he took it sorrowfully in his arms and bore it away to his expecting companions. They who were witnesses of this extraordinary scene resolved never again to fire at one of the monkey race.

Of course it is not to be supposed from this instance that all, or even most monkeys display any care for their dead. A writer in 'Nature' (vol. ix. p. 243), for instance, says expressly that such is not the case with Gibbons (*Hylobates agilis*), which he has observed to be highly sympathetic to injured companions, but 'take no notice whatever' of dead ones.

[1] *Descent of Man*, p. 70.

Regarding their sympathy for injured companions this writer says:—

I keep in my garden a number of Gibbon apes (*Hylobates agilis*); they live quite free from all restraint in the trees, merely coming when called to be fed. One of them, a young male, on one occasion fell from a tree and dislocated his wrist; it received the greatest attention from the others, especially from an old female, who, however, was no relation; she used before eating her own plantains to take up the first that were offered to her every day, and give them to the cripple, who was living in the eaves of a wooden house; and I have frequently noticed that a cry of fright, pain, or distress from one would bring all the others at once to the complainer, and they would then condole with him and fold him in their arms.

Captain Hugh Crow, in his 'Narrative of my Life,' relates an interesting tale of the conduct of some monkeys on board his ship. He says:—

We had several monkeys on board; they were of different species and sizes, and amongst them was a beautiful little creature, the body of which was about ten inches or a foot in length, and about the circumference of a common drinking glass. This interesting little animal, which, when I received it from the Governor of the Island of St. Thomas, diverted me by its innocent gambols, became afflicted by the malady which unfortunately prevailed in the ship. It had always been a favourite with the other monkeys, who seemed to regard it as the last born and the pet of the family; and they granted it many indulgences which they seldom conceded to one another. It was very tractable and gentle in its temper, and never took advantage of the partiality shown to it. From the moment it was taken ill their attention and care of it redoubled; and it was truly affecting and interesting to see with what anxiety and tenderness they tended and nursed the little creature. A struggle often ensued among them for priority in those offices of affection; and some would steal one thing and some another, which they would carry to it untasted, however tempting it might be to their own palates. Then they would take it up gently in their fore-paws, hug it to their breasts, and cry over it as a fond mother would over her suffering child. The little creature seemed sensible of their assiduities, but it was wofully overpowered by sickness. It would sometimes come to me and look me pitifully in the face, and moan and cry like an infant, as if it besought me to give it relief; and we did everything we could think of to restore it to health: but, in spite of the united attention of its kindred tribes and ourselves, the interesting little creature did not survive long.

Here is a case which I myself witnessed at the Zoological Gardens, and published in the 'Quarterly Journal of Science,' from which I now quote:—

A year or two ago there was an Arabian baboon and an Anubis baboon confined in one cage, adjoining that which contained a dog-headed baboon. The Anubis baboon passed its hand through the wires of the partition, in order to purloin a nut which the large dog-headed baboon had left within reach—expressly, I believe, that it might act as a bait. The Anubis baboon very well knew the danger he ran, for he waited until his bulky neighbour had turned his back upon the nut with the appearance of having forgotten all about it. The dog-headed baboon, however, was all the time slyly looking round with the corner of his eye, and no sooner was the arm of his victim well within his cage than he sprang with astonishing rapidity and caught the retreating hand in his mouth. The cries of the Anubis baboon quickly brought the keeper to the rescue, when, by dint of a good deal of physical persuasion, the dog-headed baboon was induced to leave go his hold. The Anubis baboon then retired to the middle of his cage, moaning piteously, and holding the injured hand against his chest while he rubbed it with the other one. The Arabian baboon now approached him from the top part of the cage, and, while making a soothing sound very expressive of sympathy, folded the sufferer in its arms—exactly as a mother would her child under similar circumstances. It must be stated, also, that this expression of sympathy had a decidedly quieting effect upon the sufferer, his moans becoming less piteous so soon as he was enfolded in the arms of his comforter; and the manner in which he laid his cheek upon the bosom of his friend was as expressive as anything could be of sympathy appreciated. This really affecting spectacle lasted a considerable time, and while watching it I felt that, even had it stood alone, it would in itself have been sufficient to prove the essential identity of some of the noblest among human emotions with those of the lower animals.

As a beautiful instance of the display of sympathy, I may narrate an occurrence which was witnessed by my friend Sir James Malcolm—a gentleman on the accuracy of whose observation I can rely. He was on board a steamer where there were two common East India monkeys, one of which was older and larger than the other, though they were not mother and child. The smaller monkey one day fell overboard amidships. The larger one became frantically excited, and running over the bulwarks down to a part of the ship which is called 'the bend,' it held on to the side of the vessel with one hand, while with the other it extended to her drowning companion a cord with which she had been tied up, and one end of which was fastened round her waist. The incident astonished everyone on board, but unfortunately for the romance of the story the little monkey was not near enough to grasp the floating end of the cord. The animal, however, was eventually saved by a sailor throwing out a longer rope to the little swimmer, who had sense enough to grasp it, and so to be hauled on board.

The following account of the behaviour of a wounded monkey seems to suggest the presence of a class of emotions similar to those which we know as feelings of reproach. The observer was Capt. Johnson:—

I was one of a party of Jeekary in the Bahar district; our tents were pitched in a large mango garden, and our horses were picquetted in the same garden a little distance off. When we were at dinner a Syer came to us, complaining that some of the horses had broken loose in consequence of being frightened by monkeys (i.e. Macacus Orhesus) on the trees. As soon as dinner was over I went out with my gun to drive them off, and I fired with small shot at one of them, which instantly ran down to the lowest branch of the tree, as if he were going to fly at me, stopped suddenly, and coolly put his paw to the part wounded, covered with blood, and held it out for me to see. I was so much hurt at the time that it has left an impression never to be effaced, and I have never since fired a gun at any of the tribe. Almost immediately on my return to the party, before I had fully described what had passed, a Syer came to inform us that the monkey was dead. We ordered the Syer to bring it to us, but by the time he returned the other monkeys had carried the dead one off, and none of them could anywhere be seen.

This case is strikingly corroborated by the following

allusion to Sir W. Hoste's Memoirs, given by Jesse as follows:—

One of his officers, coming home after a long day's shooting, saw a female monkey running along the rocks, with her young one in her arms. He immediately fired, and the animal fell. On his coming up, she grasped her little one close to her breast, and with her other hand pointed to the wound which the ball had made, and which had entered above her breast. Dipping her finger in the blood, and then holding it up, she seemed to reproach him with being the cause of her death, and consequently that of the young one, to which she frequently pointed. 'I never,' says Sir William, 'felt so much as when I heard the story, and I determined never to shoot one of these animals as long as I lived.'[1]

Mr. Darwin says that most persons who have observed monkeys have seen them show a sense of the ludicrous. Here is an instance which I have myself observed, and now quote from my article in the 'Quarterly Journal of Science:'—

Several years ago I used to watch carefully the young orang-outang in the Zoological Gardens, and I am quite sure that she manifested a sense of the ludicrous. One example will suffice. Her feeding tin was of a somewhat peculiar shape, and when it was empty she used sometimes to invert it upon her head. The tin then presented a comical resemblance to a bonnet, and as its wearer would generally favour the spectators with a broad grin at the time of putting it on, she never failed to raise a laugh from them. Her success in this respect was evidently attended with no small gratification on her part.

But perhaps the strongest evidence of monkeys having an appreciation of the ludicrous is the same as that which we have seen to be presented in the case of certain dogs—namely, in the animals disliking ridicule. Abundant evidence on this head in the case of monkeys will be given further on.

That monkeys enjoy play no one can question who spends on hour or two in the monkey-house at the Zoological Gardens. According to Savage, chimpanzees congregate together for the sole purpose of play, when

---

[1] *Gleanings*, vol. iii. pp. 86-7.

they beat or drum with pieces of stick on sonorous pieces of wood.[1]

Curiosity is more strongly pronounced in monkeys than in any other animals. We all know the interesting illustration on this head furnished by the experiment of Mr. Darwin, who, in order to test the statement of Brehm that monkeys have an instinctive dread of snakes, and yet cannot 'desist from occasionally satiating their curiosity in a most human fashion, by lifting up the lid of the box in which the snakes were kept,' took a stuffed snake to the monkey-house at the Zoological Gardens. Mr. Darwin says:—

The excitement thus caused was one of the most curious spectacles I ever beheld. . . . I then placed a live snake in a paper bag, with the mouth loosely closed, in one of the larger compartments. One of the monkeys immediately approached, cautiously opened the bag, peeped in, and instantly dashed away. Then I witnessed what Brehm has described, for monkey after monkey, with head raised high and turned on one side, could not resist taking a momentary peep into the upright bag, at the dreadful object lying quietly at the bottom.[2]

Allied, perhaps, to curiosity, and so connected with the emotions, is what Mr. Darwin calls 'the principle of imitation.' It is proverbial that monkeys carry this principle to ludicrous lengths, and they are the only animals which imitate for the mere sake of imitating, as has been observed by Desor, though an exception ought to be made in favour of talking birds. The psychology of imitation is difficult of analysis, but it is remarkable as well as suggestive that it should be confined in its manifestations to monkeys and certain birds among animals, and to the lower mental levels among men. As Mr. Darwin says:—

The principle of imitation is strong in man, and especially, as I have myself observed, with savages. In certain morbid states of the brain, this tendency is exaggerated to an extraordinary degree; some hemiplegic patients and others, at the

---

[1] *Boston Journal of Nat. Hist.*, iv. p. 324.
[2] *Descent of Man*, p. 72.

commencement of inflammatory softening of the brain, unconsciously imitate every word that is uttered, whether in their own or in a foreign language, and every gesture or action which is performed near them.

The same sort of tendency is often observable in young children, so that it seems to be frequently distinctive of a certain stage or grade of mental evolution, and particularly in the branch *Primates*. Other animals, however, certainly imitate each other's actions to a certain extent, as I shall have occasion fully to notice in my next work.

As for the sterner emotions, rage may be so pronounced as to make a monkey exhaust itself with beating about its cage, or a baboon bite its own limbs till the blood flows.[1] Jealousy occurs in a correspondingly high degree, while retaliation and revenge are shown by all the higher monkeys when injury has been done to them, as any one may find by offering an insult to a baboon. The following is a good case of this, as it shows what may be called brooding resentment deliberately preparing a satisfactory revenge. Mr. Darwin writes:—

Sir Andrew Smith, a zoologist whose scrupulous accuracy was known to many persons, told me the following story of which he was himself an eye-witness. At the Cape of Good Hope, an officer had often plagued a certain baboon, and the animal, seeing him approaching one Sunday for parade, poured water into a hole and hastily made some thick mud, which he skilfully dashed over the officer as he passed by, to the amusement of many bystanders. For long afterwards the baboon rejoiced and triumphed whenever he saw his victim.[2]

General Intelligence.

Coming now to the higher powers, I shall give a few cases to show that monkeys certainly surpass all other animals in the scope of their rational faculty. Professor Croom Robertson writes me:—

I witnessed the following incident in the Jardin des Plantes, now many years ago; but it struck me greatly at the time, and I have narrated it repeatedly in the interval. A large ape—I

---

[1] *Descent of Man*, 71.   [2] *Ibid.*, p. 69.

believe anthropoid, but cannot tell the species—was in the great iron cage with a number of smaller monkeys, and was lording it over them with many wild gambols, to the amusement of a crowd of spectators. Many things—fruits and the like—had been thrown between the bars into the cage, which the ape was always forward to seize. At last some one threw in a small hand looking-glass, with a strongly made frame of wood. This the ape at once laid hold of, and began to brandish like a hammer. Suddenly he was arrested by the reflection of himself in the glass, and looked puzzled for a moment; then he darted his head behind the glass to find the other of his kind that he evidently supposed to be there. Astonished to find nothing, he apparently bethought himself that he had not been quick enough with his movement. He now proceeded to raise and draw the glass nearer to him with great caution, and then with a swifter dart looked behind. Again finding nothing, he repeated the attempt once more. He now passed from astonishment to anger, and began to beat with the frame violently on the floor of the cage. Soon the glass was shattered, and pieces fell out. Continuing to beat, he was in the course of one blow again arrested by his image in the piece of glass still remaining in the frame. Then, as it seemed, he determined to make one trial more. More circumspectly than ever the whole first part of the process was gone through with; more violently than ever the final dart made. His fury over this last failure knew no bounds. He crunched the frame and glass together with his teeth, he beat on the floor, he crunched again, till nothing but splinters was left.

Mr. Darwin writes: 'Rengger, a most careful observer, states that when first he gave eggs to his monkeys in Paraguay, they smashed them, and thus lost much of their contents; afterwards they generally hit one end against some hard body, and picked off the bits with their fingers. After cutting themselves only *once* with any sharp tool, they would not touch it again, or would handle it with the greatest caution. Lumps of sugar were often given them wrapped up in paper; and Rengger sometimes put a live wasp in the paper, so that in hastily unfolding it they got stung; after this had *once* happened, they always first held the packet to their ears to detect any movement within.'[1]

[1] *Descent of Man*, pp. 77-8.

The powers of observation and readiness to establish new associations thus rendered apparent, display a high level of general intelligence. Mr. Darwin further observes that Mr. Belt 'likewise describes various actions of a tamed cebus, which, I think, clearly show that this animal possessed some reasoning power.' The following is the account to which Mr. Darwin here refers, and I quote it *in extenso*, because, as I shall presently show, I have myself been able to confirm most of the observations on another monkey of the same genus:—

It would sometimes entangle itself round a pole to which it was fastened, and then unwind the coils again with the greatest discernment. Its chain allowed it to swing down below the verandah, but it could not reach to the ground. Sometimes, when there were broods of young ducks about, it would hold out a piece of bread in one hand, and when it had tempted a duckling within reach, seize it by the other, and kill it with a bite in the breast. There was such an uproar amongst the fowls on these occasions, that we soon knew what was the matter, and would rush out and punish Mickey (as we called him) with a switch; so that he was ultimately cured of his poultry-killing propensities. One day, when whipping him, I held up the dead duckling in front of him, and at each blow of the light switch told him to take hold of it, and at last, much to my surprise, he did so, taking it and holding it tremblingly in one hand. He would draw things towards him with a stick, and even used a swing for the same purpose. It had been put up for the children, and could be reached by Mickey, who now and then indulged himself in a swing on it. One day I had put down some bird-skins on a chair to dry, far beyond, as I thought, Mickey's reach; but, fertile in expedients, he took the swing and launched it towards the chair, and actually managed to knock the skins off in the return of the swing, so as to bring them within his reach. He also procured some jelly that was set out to cool in the same way. Mickey's actions were very human-like. When any one came near to fondle him, he never neglected the opportunity of pocket-picking. He would pull out letters, and quickly take them from their envelopes.[1]

I shall now proceed to state some further facts, showing the high level of intelligence to which monkeys of various kinds attain.

[1] *Naturalist in Nicaragua*, p. 119.

The orang which Cuvier had used to draw a chair from one end to the other of a room, in order to stand upon it so as to reach a latch which it desired to open; and in this we have a display of rationally adaptive action which no dog has equalled, although, as in the case before given of the dog dragging the mat, it has been closely approached. Again, Rengger describes a monkey employing a stick wherewith to prise up the lid of a chest, which was too heavy for the animal to raise otherwise. This use of a lever as a mechanical instrument is an action to which no animal other than a monkey has ever been known to attain; and, as we shall subsequently see, my own observation has fully corroborated that of Rengger in this respect. More remarkable still, as we shall also subsequently see, the monkey to which I allude as having myself observed, succeeded also by methodical investigation, and without any assistance, in discovering for himself the mechanical principle of the screw; and that monkeys well understand how to use stones as hammers is a matter of common observation since Dampier and Wafer first described this action as practised by these animals in the breaking open of oyster-shells. The additional observation of Gernelli Carreri of monkeys thrusting stones into the open valves of oysters so as to save themselves the trouble of smashing the shells, though not incredible, requires confirmation. But Mr. Haden, of Dundee, has communicated to me the following very remarkable appreciation of mechanical principles which he himself observed in a monkey (species not noted), and which would certainly be beyond the mental powers of any other animal:—

'A large monkey, confined alone in a large cage, had its sleeping-place in the form of a kind of hut in the centre of the cage. Springing near the hut was a tree, or imitation tree, the main branch of which ascended over the top of the hut, and then came forwards away from it. Whether the roof of the hut enabled this animal to gain any part of this branch, I did not observe, but only remarked its method at the time of gaining the part of the branch which led frontwards, and away from the hut. This could be done by means of the hut door, which, when opened, swung beneath this part of the branch. The door, either by accident or by the design of its construction, *swung to* each time the animal opened it to mount upon its top edge. After one or two efforts to mount by it in spite of its immediate swinging to, the creature procured a thick blanket which lay in the cage, and threw it over the door, having opened the same, so that its complete swinging to was prevented sufficiently for the creature to mount upon its free edge, and so gain that part of the branch which ran above it.'

The following, which I quote from 'Nature' (vol. xxiii., p. 533), also displays high intelligence:—

One of the large monkeys at the Alexandra Palace had been for some time suffering from the decay of the right lower canine, and an abscess, forming a large protuberance on the jaw, had resulted. The pain seemed so great, it was decided to consult a dentist as to what should be done; and, as the poor creature was at times very savage, it was thought that if the tooth had to be extracted, gas should be used for the safety of the operation. Preparations were made accordingly, but the behaviour of the monkey was quite a surprise to all who were concerned. He showed great fight on being taken out of the cage, and not only struggled against being put into a sack prepared with a hole cut for his head, but forced one of his hands out, and snapped and screamed, and gave promise of being very troublesome. Directly, however, Mr. Lewin Moseley, who had undertaken the operation, managed to get his hand on the abscess and gave relief, the monkey's demeanour changed entirely. He laid his head down quietly for examination, and, without the use of the gas, submitted to the removal of a stump of a tooth as quietly as possible.

According to D'Osbonville, certain monkeys that he observed in the wild state were in the habit of administering corporal chastisement to their young. After suckling and cleansing them, the mothers used to sit down and watch the youngsters play. These would wrestle, throw and chase each other, &c.; but if any of them grew malicious, the dams would spring up, and, seizing their offspring by the tail with one hand, correct them severely with the other.

We have already seen that dogs and cats display the idea of maintaining discipline among their progeny.

According to Houzeau the sacred monkey of India (*Semnopithecus entellus*) is very clever in catching snakes, and in the case of poisonous species destroy the fangs by breaking them against stones.[1]

Of the fact that monkeys act in co-operation, many proofs might be given, but one will suffice.

Lieutenant Schipp, in his Memoirs, says:—

A Cape baboon having taken off some clothes from the barracks, I formed a party to recover them. With twenty men I made a circuit to cut them off from the caverns, to which they always fled for shelter. They observed my movements, and detaching about fifty to guard the entrance, the others kept their post. We could see them collecting large stones and other missiles. One old grey-headed one, who had often paid us a visit at the barracks, was seen distributing his orders, as if a general. We rushed on to the attack, when, at a scream from him, they rolled down enormous stones on us, so that we were forced to give up the contest.

I shall here bring to a close my selections from the literature of monkey psychology, because I wish to devote a good deal of space to detailing a number of observations which have not yet been published. Thinking it desirable for the purposes of this work that an intelligent monkey should be subjected to close observation for some length of time, I applied to Mr. Sclater for the loan of one from the collection of the Zoological Society. He kindly consented to my proposal, and I selected a specimen of *Cebus fatuellus*, which appeared to me to be the most intelligent monkey in the collection. Not having facilities for keeping the animal in my own house, I consigned him to the charge of my sister (who lives close by), with the request that she should carefully note all points of interest connected with his intelligence. Therefore, from the day of his arrival till that of his departure she kept a diary,

Loc. cit., vol. i., p. 305.

or note-book, in which all the observations that she made when I was absent were entered. It was originally my intention to make an abstract of this note-book; but on afterwards reading it through for this purpose, it seemed to me that I should rather spoil matters by attempting a condensation. There is a certain graphic effect incidental to the diary form and spontaneous style of diction—the notes, of course, not having been written with a view to verbatim publication; and besides, as the psychology of monkeys has been so little studied, I think it is well to give all the details of a continuous series of observations. It is desirable to add that on occasions subsequent to the taking of this or that particular note, I generally had the opportunity of verifying the observation myself; but I may state that I attach no more importance to this circumstance than I should to verifying an observation of my own; for as a careful observer of animals I have quite as much confidence in my sister as in myself. It only remains to explain that my mother, being an invalid, is confined most of the time to her bedroom; and that the monkey was kept there for the first six weeks of his stay at her house, partly in order that he might be under constant observation, and partly also to furnish her with an entertaining pet. The following are my sister's notes *in extenso* and without alteration :—

Brown Capuchin (Cebus fatuellus—Linn.), *Brazil.*
DIARY, 1880.

December 18th. Arrived in box with keeper. Seemed rather frightened and screamed a good deal on being transferred from small box to a larger one.

19th. Took him out of the box he had been in all night and fastened chain on to collar. Was meek and subdued, hiding his face in my lap.

20th. Has become much more lively and somewhat aggressive, especially towards the servants. He has taken a fancy to my mother, and (she holding his chain) he plays with her in a gentle and affectionate manner in her bed, but flies angrily at any of the servants who come near the bed. I observed to-day that he breaks walnuts (which are too hard for him to crack with his teeth) by striking them with the flat bottom of a dish he has for drinking out of. He is ceaselessly active all day, and at night covers himself very neatly with warm shawls, and sleeps soundly till about eight o'clock.

21st. I notice that the love of mischief is very strong in him. To-day he got hold of a wine-glass and an egg-cup. The glass he dashed on the floor with all his might, and of course broke it. Finding, however, that the egg-cup would not break for being thrown down, he looked round for some hard substance against which to dash it. The post of the brass bedstead appearing to be suitable for the purpose, he raised the egg-cup high over his head and gave it several hard blows. When it was completely smashed he was quite satisfied. He breaks a stick by passing it down between a heavy object and the wall, and then hanging on to the end, thus breaking it across the heavy object. He frequently destroys an article of dress by carefully pulling out the threads (thus unripping it) before he begins to tear it with his teeth in a more violent manner. If he gets hold of anything that he sees we do not care about, he soon leaves it again; but if it is an article of value (even if it be only a scrap of paper) which he sees we are anxious about, nothing will induce him to give it up. No food, however inviting, will distract his attention: scolding only makes him more angry, and he keeps the article until it is quite destroyed. To-day I gave him a hammer to break his walnuts with, and he uses it in a proper manner for that purpose.

22nd. To-day a strange person (a dressmaker) came into the room where he is tied up, and I gave him a walnut that she might see him break it with his hammer. The nut was a bad one, and the woman laughed at his disappointed face. He then became very angry, and threw at her everything he could lay hands on; first the nut, then the hammer, then a coffee-pot which he seized out of the grate, and, lastly, all his own shawls. He throws things with great force and precision by holding them in both hands, and extending his long arms well back over his head before projecting the missile, standing erect the while.

23rd. There is continual war between him and Sharp [a small terrier], but they both seem to have a certain mutual respect for each other. The dog makes snatches at nuts, &c., and runs away with them beyond the reach of his chain, and the monkey catches at the dog, but seems afraid to hold him or hurt him. He however pelts him with nuts or bits of carrot, and chatters at him. At other times he holds out his hand as if to make friends, but the dog is too suspicious to go near him. His hostility towards the servants (one especially) increases, so that he will not even take a nut from her without catching fiercely at her hand; he also frequently throws things at her. On the other hand, he allows my mother to do anything with him.

24th. He bit me in several places to-day when I was taking him away from my mother's bed after his morning's game there. I took no notice, but he seemed ashamed of himself afterwards, hiding his face in his arms and sitting quiet for a time.[1] In accordance with his desire for mischief, he is of course very fond of upsetting things, but he always takes great care they do not fall on himself. Thus he will pull a chair towards him till it is almost over-balanced, then he intently fixes his eyes on the top bar of the back, and when he sees it coming over his way, darts from underneath and watches the fall with great delight; and similarly with heavier things. There is a washhand-stand, for example, with a heavy marble top, which he has with great labour upset several times, and always without hurting himself.[2]

25th. I observed to-day that if a nut or any object he wishes to get hold of is beyond the reach of his chain, he puts out a stick to draw it towards him, or, if that does not succeed, he stands upright and throws a shawl back over his head, holding it by the two corners so that it falls down his back; he then throws it forward with all his strength, still holding on by the corners; thus it goes out far in front of him and covers the nut, which he then draws towards him by pulling in the shawl. When his chain becomes twisted round the bars of a 'clothes-horse' (which is given him to run about upon), and thus too short for his comfort, he looks at it intently and pulls it with his fingers this way and that, and when he sees how the turns are taken, he deliberately goes round and round the bars, first this way, then that, until the chain is quite disentangled. He often carries his chain grasped in his tail and held high over his back to keep it from getting into the way of his feet. He is always rather excited in the morning when I loosen his chain preparatory to taking him to my mother's bed; jumps about and tugs at the chain. Sometimes, however, if the chain is entangled, and I am rather long in getting it unfastened, he sits quietly down beside me, and begins picking at the chain with

[1] On subsequent observation (January 14, 1881), I find this quietness was not due to shame at having bitten me, for whether he succeeds in biting any person or not he always sits quiet and dull-looking after a fit of passion, being, I think, fatigued. He has bitten me often since December 24, and seems to enjoy the fun on the whole.

[2] These heavy objects he overturns with exceeding caution, balancing them several times carefully, and studying them before finally throwing or pulling them over.

his fingers as if to help me to untie it. I cannot say, however, that he succeeds in helping me at all.

26th. He seems very fond of spinning things round. If he gets a whole apple or orange he generally sits spinning it on one end, before beginning to eat it. He eats an orange by biting off a tiny piece of the peel, and putting his long, thin finger deep into the fruit; he then lays the whole orange under a piece of wire netting he has near him, and, putting his mouth to the hole he has made, presses the wire netting down upon the fruit, thus squeezing the juice up into his mouth. When a good deal of juice begins to run out, he holds the orange up over his head and lets the juice run into his mouth.

27th. To-day he obtained possession of a rather valuable document, and, as usual, nothing I could do would persuade him to give it up. He neglected any kind of food I offered him, and only chattered when I coaxed him. When at last I tried threatening him with a cane, he only became savage and flew at me, chattering. My mother now came and sat down in a chair beside him. He immediately jumped into her lap, and remained quite still while she took the paper out of his hands. When, however, she handed it to me and I laughed at her success, he showed his teeth and screamed and chattered at me angrily. I find laughing generally irritates him. Thus, when he is playing with my mother in the bed in the best of humour, as long as I sit quietly on the bed all is well, but if I laugh, for example at any of his affectionate glances, he makes a dart at me to send me off, and then returns with renewed demonstrations of affection to my mother, tumbling head over heels and lying on his back, grinning in a most comical manner, and making a sound very like slight laughter.

28th. His chain is fastened to the marble slab of a washhand-stand, placed on the floor against the wall. It is too heavy for him to pull along by his chain without hurting himself, so when he desires to do any mischief which is beyond the reach of his chain, he deliberately goes to the marble and pushes an arm down between an upright part of it and the wall, until he has moved the whole slab sufficiently far from the wall to admit of his slipping down behind the upright part himself. He then places his back against the wall and his four hands against the upright part of the marble, and pushes the slab as far as he can stretch his long legs. He only does this, however, when he is bent on mischief, as the fact of food being beyond the reach of his chain does not furnish a strong enough inducement to lead him to take so much exertion. Thus to-day he began to pull the glazed leather cover off a trunk which was near him. I pulled the trunk away, and when he found it was out of his reach he ran and pushed the marble towards the trunk in the manner I have described, and when he knew his chain was then sufficiently long to reach the trunk, he ran to the latter and hastily resumed his destructive process.

29th. I notice that nothing the person does who has hold of his chain offends him. I mean, although he is furiously angry at having anything taken away from him, he is not at all angry if he is pulled away by his chain. If he is trying to bite a person, and another person takes hold of his chain behind him and so prevents his spring forward, he does not turn to bite the person who has taken hold of his chain, as a dog would do under similar circumstances, but quietly submits to be thus held. He seems to look upon his confinement and management by a chain as a natural law against which it is useless to struggle. On the other hand, he seems to be quite aware of the place where his chain is fastened, and to know that if he were clever enough to undo it he would be free. After we found he could move about the marble slab of the washband-stand in the way described, we had a ring sunk in the floor to tie him to. The moment the chain was fastened to that [1] he began to investigate its new connection, and continued to do so for hours, passing the chain rapidly backward and forwards through the ring. When he found this did not loosen it, he began to hammer it and the ring also with all his strength, and this he continued to do for the rest of the day.

30th. He still continues to work at the chain where it is fastened to the ring. He passed the whole of the chain through the ring so many times with his fingers that it became quite blocked up in the ring, which made it very short, and it took me a quarter of an hour to disentangle it. He was very much interested in this process, sitting quietly beside me and watching my fingers intently, sometimes gently pulling my fingers on one side in order to see better, and sometimes casting a quick intelligent glance into my face as if asking how I did it. After I had disentangled and lengthened the chain he worked at it again for hours, but took care not to twist it into the ring a second time.

31st. To day he hurt himself by getting one of his toes caught in a hinge of the clothes-horse. He did not make any

[1] January 14, 1881. The marble slab was left with him after the chain had been fastened to the ring; but since that time he has never attempted to move the marble.

fuss, although the accident must have been somewhat a painful one, nor did he try to pull the toe out, which would have been useless and only hurt him more; but he sat almost motionless, making slight complaining noises until I discovered that there was something wrong with him. When I began to extricate his foot, he remained perfectly passive—although I dare say I hurt him a good deal—and only looked at me gratefully.

January 1, 1881. He has now quite given up trying to loosen his chain himself; having tried every way and failed, he has evidently become hopeless about it. He now resents being tied up. When I loosen him he is quite pleased, and when I tie him he waits until he is quite sure he is being tied, and not loosened, and then he flies at me and bites me.

10th. As he is always tied up in the same place he has no new opportunities given him of showing his intelligence. His attachment to my mother has increased. When she goes out he immediately gives up all play and mischief, and does nothing but run round and round in a restless manner, making a peculiar sweet calling noise, such as he never makes when she is in the room, listening intently between times. As long as she remains away he takes no rest or amusement, nor does he ever, or hardly ever, become angry; but the moment she returns he begins all his old ways again, usually becoming more savage at other people than before.

My mother frequently takes things away from him, and he never resents it to her as he would do to any other person. He generally, however, chatters angrily at some one else when my mother removes anything he wishes to keep. At first I thought he was deceived in the matter—that he could not believe it possible that his best friend could deprive him of what he valued, and so thought someone else must have done it. But the same thing has now happened so frequently that I can hardly think he is not really aware of who takes the things away. He seems rather to think it politic to keep on good terms with one person, and that although he does see her remove the things, and feels angry in consequence, he thinks it more prudent to vent his anger upon someone with whom he has already quarrelled. He always shows more irritation when my mother gives anything to me after having taken it away from him, than when she keeps it herself (as mentioned on December 26), and this may be the reason partly why he resents these matters to me; he thinks when I obtain possession of anything he wants that it is a sort of triumph to me. In the same way my mother may laugh as much as she likes whether he is with her or not, but if I laugh at all at anything it generally results in something being thrown at me. If my mother calls out to the servants—if, for instance, a servant has left the room and my mother calls her back—he becomes very angry at the servant, and salutes her on her return with a shower of missiles. Sometimes my mother pretends to scold or beat the servants, and then he joins with great energy, by way of supporting his friend. If I scold or beat the servants he does not mind so much. When my mother comes back after being out he does not show any great demonstrations of joy. He screams out with pleasure when he hears her voice approaching on the stairs, but does not make much ado when she enters the room. While my mother is out I can do anything I like with him, just as she can when she is at home. Perhaps being in low spirits he does not feel angry, or perhaps he thinks it prudent to be amiable when his best friend is away. When my mother comes back, all his ill-temper returns at once and even in an increased degree towards other people, and he immediately resumes playing with all his toys.

11th. When he throws things at people now he first runs up the bars of the clothes-horse; he seems to have found out that people do not much care for having things thrown at their feet, and he is not strong enough to throw such heavy objects as a poker or a hammer at people's heads; he therefore mounts to a level with his enemy's head, and thus succeeds in sending his missile to a greater height and also to a greater distance.

14th. To-day he obtained possession of a hearth-brush, one of the kind which has the handle screwed into the brush. He soon found the way to unscrew the handle, and having done that he immediately began to try to find out the way to screw it in again. This he in time accomplished. At first he put the wrong end of the handle into the hole, but turned it round and round the *right way for screwing*. Finding it did not hold, he turned the other end of the handle and carefully stuck it into the hole, and began again to turn it the right way. It was of course a very difficult feat for him to perform, for he required both his hands to hold the handle in the proper position and to turn it between his hands in order to screw it in, and the long bristles of the brush prevented it from remaining steady or with the right side up. He held the brush with his hind hand, but even so it was very difficult for him to get the first turn of the screw to fit into the thread; he worked at it, however, with the most unwearying perseverance until he got the first turn of the screw to catch, and he then quickly turned it round and round until it was screwed up to the end. The most remarkable thing was that, however often he was disappointed in the beginning, he never was induced to try turning the handle the wrong way ; he always screwed it from right to left. As soon as he had accomplished his wish, he unscrewed it again, and then screwed it in again the second time rather more easily than the first, and so on many times. When he had become by practice tolerably perfect in screwing and unscrewing, he gave it up and took to some other amusement. One remarkable thing is that he should take so much trouble to do that which is no material benefit to him. The desire to accomplish a chosen task seems a sufficient inducement to lead him to take any amount of trouble. This seems a very human feeling, such as is not shown, I believe, by any other animal. It is not the desire of praise, as he never notices people looking on ; it is simply the desire to achieve an object for the sake of achieving an object, and he never rests nor allows his attention to be distracted until it is done.

16th. When he is angry, and has at hand only those things which he wishes to keep, he makes a great show of throwing them at people, but always retains a hold. Thus if he has had a plaything a long time and is tired of it, he throws it right at a person without the least hesitation ; but if he has a new thing which he values, he goes through all the appropriate motions for throwing, but only brings the object down with a noise upon the ground, taking care not to let go his hold. He beats people with a long cane he has, and when he cannot reach people he strikes it with all his strength upon the ground to show what he would do if he had the chance. There is no more comical sight than to see him hurriedly climbing his screen in fierce anger, taking (not without great difficulty) his long and awkward stick up with him in order to be high enough to give a good blow to a person. The dog is quite afraid of the stick in the monkey's hands, although he is too petted to be afraid of it in a person's. The monkey is jealous of the dog lying in the armchair in which he sometimes seats himself with my mother, so he pokes the stick at the dog (as the chair is beyond the reach of his chain) and makes him get off.

18th. He was very angry to-day at a servant girl sweeping out his place with a long brush, and he seized the brush every time the servant attempted to sweep. My mother then took it, and he at once became not only quite good-tempered, but assisted her in sweeping, by gathering the rubbish in the corners of his place into little heaps with his hands, and putting the heaps into the way of the brush.

20th. To-day he broke his chain, and flew at a servant

savagely, but seeing my mother he immediately jumped into her lap. While another chain was being prepared he got to the trunk where his nuts are kept. I have long noticed that he looks upon that trunk as in some special sense his own property. There are other things kept in the trunk as well as the nuts, and if any person goes to the trunk for anything he becomes furiously angry. Indeed nothing makes him so angry as people opening the trunk, and this is not because he wants nuts out of it, for he always has more than he can eat beside him, and generally refuses to take any that are offered to him. Well, to-day, as soon as the breaking of his chain enabled him to get to the trunk, he began picking at the lock with his fingers. I then gave him the key, and he tried for two full hours without ceasing to unlock the trunk with this key. It was a very difficult lock to open, being slightly out of order, and requires the lid of the trunk to be pressed down before it would work, so I believe it was absolutely impossible for him to open it, but he found in time the right way to put the key in, and to turn it backwards and forwards, and after every attempt he pulled the lid upwards to see if it were unlocked. That this was the result of observing people is obvious, from the fact that after every time he put the key into the lock and failed to open the trunk, he passed the key round and round the outside of the lock several times. The explanation of this is that, my mother's sight being bad, she often misses the lock when putting in the key, and then feels round and round the lock with the key; the monkey therefore evidently seems to think that this feeling round and round the lock with the key is in some way necessary to the success of unlocking the lock, so that, although he could see perfectly well how to put the key in straight himself, he went through this useless operation first.

21st. To-day I gave him a wooden box with the lid nailed on, and an iron spoon, to see if he would use the latter as a lever wherewith to raise the lid. The experiment was somewhat spoiled by my mother putting the handle of the spoon into the crack between the lid and the box to show him how to do it. Therefore I cannot tell whether or not he would have taken this first step himself, if he had had time to do so. However, when the handle of the spoon was in he certainly used it in the proper manner, pulling it down with all his strength at the extreme end, thus drawing the nails out of the box and raising the lid.

22nd. He was sitting on my mother's knee, and she washing his hands with a little sponge, a process of which he is very fond; she tried to wash his face, and that he disliked very much. Every time she began, the expression of his face became more angry; at last he suddenly jumped off her knee, and made a violent attack on one of the servants who is usually his favourite, although she was doing nothing at all to anger him. This is a good instance of his habit of venting his anger at my mother on other people. He always eats vigorously when he is angry, or after a fit of passion. After a prolonged fit of passion he always lies down on his side as if dead, probably from exhaustion.

30th. He quite understands the meaning of shaking hands. He always holds out his own hand when he wishes to be friendly, especially when a friend is entering or leaving the room. To-day he had been a long time playing with his toys, taking no notice of any one. Suddenly my mother remembered that to-day was my birthday, and (for the first time since he came to the house) shook hands with me in congratulation. He immediately became very angry with me, screamed and chattered and threw things at me, being evidently jealous of the attention my mother was paying me.

February 1st. He has now been moved down to the diningroom, where he is chained between the fireplace and the window. He seems quite miserable on account of the change, as he does not see so much of my mother.

4th. His low spirits continue, and threaten to make him ill. He will not play with anything, but sits moping and shivering in a corner. To-day I found him very cold and unhappy, and warmed his hands for him. He is very meek and gentle, and seems to be getting fond of me.

8th. He has quite recovered his spirits since he took a fancy to me. He likes me now apparently as well as he used to do my mother; that is to say, he allows me to nurse him, and walk about in his place, and even take things away from him. When, however, my mother comes to see him, he does not care for me, although he shows none of his old hostility. To the servants, however, he continues to do so when my mother is present.

10th. We gave him a bundle of sticks this morning, and he amused himself all day by poking them into the fire and pulling them out again to smell the smoking end. He likewise pulls out hot cinders from the grate and passes them over his head and chest, evidently enjoying the warmth, but never burning himself. He also puts hot ashes on his head. I gave him some paper, and, as he cannot, from the length of his chain, quite reach the fire, he rolled the paper up into the form of a stick, and then put it into the fire, pulling it out as soon as it caught light, and watching the blaze in the fender with great satisfaction. I gave him a whole newspaper, and he tore it in pieces, rolled up each piece as I have described, to make it long enough to reach the fire, and so burnt it all piece by piece. He never once burnt his own fingers during the operation.

13th. He can open and shut the folding shutters with ease, and this seems to be an amusement to him. He also unscrewed all the knobs that belong to the fender. The bell-handle beside the mantelpiece he likewise took to bits, which involves the unscrewing of three screws.

15th. He is so amiable to me now that he constantly gives me bits of things that he himself is eating, evidently expecting me to share his repast with him. Sometimes this attention on his part is not altogether agreeable. For instance, to-day he thrust into my hand, when I was not looking, a quantity of sopped bread and milk out of his pan, no doubt thinking himself very kind-hearted thus to supply me with food.

17th. He offered the dog a bit of toast which he himself was eating, and the dog took a part of it. I think, however, that he had at the same time a sly design of catching the dog with the other hand, but he did not do so—perhaps because I was looking on, and he knows the dog is a friend of mine; but he had a wicked look in his eye while feeding the dog, which he has not when he extends his bounty to me.

19th. When I was brushing him to-day he took the brush away from me. Playthings are especially valuable to him now, as he is not allowed to have any lest he should break the windows with them. For this reason I was afraid to leave the brush with him, but found he was not at all disposed to give it up. I threw other things within his reach, but he carried the brush in his hind hand while going after the other things. At last I sat down and called him gently, when he mildly came up to my lap and put the brush into my two hands, evidently resolving that he would not now quarrel with his only friend.

22nd. His manner of showing his humours is interesting, as illustrating the principle of antithesis. Thus when he is angry he springs forward on all four hands with tail very erect and hair raised, so making himself look much bigger. When affectionate he advances slowly *backwards* with his body in the form of a hoop, so that the crown of his head rests on the ground, face inwards. He walks on three hands (hair very smooth), and puts the fourth fore-hand out at his back in advance of his body. He expects this hand to be taken kindly, and he then assumes his natural attitude. In that manner of advancing it is obviously impossible that he could bite, as his mouth is towards his own chest, so it is the best way of showing how far he is from thinking of hostility.

February 28, 1881.

The above account may be taken as fully trustworthy. Most of the observations recorded I have myself subsequently verified numberless times. From the account, however, several observations which I happened to make myself in the first instance are designedly omitted, and these I shall therefore now supply.

I bought at a toy-shop a very good imitation of a monkey, and brought it into the room with the real monkey, stroking and speaking to it as if it were alive. The monkey evidently mistook the figure for a real animal, manifesting intense curiosity, mixed with much alarm if I made the figure approach him. Even when I placed the figure upon a table, and left it standing motionless, the monkey was afraid to approach it. From this it would appear that the animal trusted much more to his sense of sight than to that of smell in recognising one of his own kind.

I placed a mirror upon the floor, and the monkey at once mistook his reflection in it for a real animal. At first he was a little afraid of it; but in a short time he gained courage enough to approach and try to touch it. Finding he could not do so, he went round behind the mirror and then again before it a great number of times; but he did not become angry, as the monkey of which Prof. Brown Robertson wrote me. Strange to say, he appeared to mistake the sex of the image, and began in the most indescribably ludicrous manner to pay to it the addresses of courtship. First placing his lips against the glass he rose to his full height on his hind legs, retired slowly, and while doing so turned his back to the mirror, looking over his shoulder at the image, and, with a preposterous amount of 'pinch' in his back, strutted up and down before the glass with all the appearance of the most laughable foppery. This display was always gone through when at any subsequent time the mirror was placed upon the floor.

From the first time that he saw me, this monkey took as violently passionate an attachment to me as that which he took to my mother. His mode of greeting, however, was different. When she entered the room after an absence, his welcome was of a quiet and contented character; but when I came in, his demonstrations were positively painful to witness. Standing erect on his hind legs at the full length of his tether, and extending both hands as far as he could reach, he screamed with all his strength, in a tone and with an intensity which he never adopted on any other occasion. So loud, indeed, were his rapidly and continuously reiterated screams, that it was impossible for any one to hold even a shouting conversation till I took the animal in my arms, when he became placid, with many signs of intense affection. Even the sound of my voice down two flights of stairs used to set him screaming in this manner, so that whenever I called at my mother's house I had to keep silent while on the staircase, unless I intended first of all to pay a visit to the monkey.

It has frequently been noticed that monkeys are very capricious in forming their attachments and aversions; but I never knew before that this peculiarity could be so strongly marked as it was in this case. His demonstrations of affection to my mother and myself were piteous; while towards every one else, male or female, he was either passively indifferent or actively hostile. Yet no shadow of a reason could be assigned for the difference. My sister, to whom animals are usually much more attached than they are to me, used always to be forbearingly kind to this one—taking all his bites, &c., with the utmost good humour. Moreover, she supplied him with all his food, and most of his playthings, so that she was really in every way his best friend. Yet his antipathy to her was only less remarkable than his passionate fondness of my mother and myself.

Another trait in the psychology of this animal which is worth observing was his quietness of manner towards my mother. With me, and indeed with every one else, his movements were unrestrained, and generally monkey-like; but with her he was always as gentle as a kitten: he appeared to know that her age and infirmities rendered boisterousness on his part unacceptable.

I returned the monkey to the Zoological Gardens at the end of February, and up to the time of his death in October 1881, he remembered me as well as the first day that he was sent back. I visited the monkey-house about once a month, and whenever I approached his cage he saw me with astonishing quickness—indeed, generally before I saw him—and ran to the bars, through which he thrust both hands with every expression of joy. He did not, however, scream aloud; his mind seemed too much occupied by the cares of monkey-society to admit of a vacancy large enough for such very intense emotion as he used to experience in the calmer life that he lived before. Being much struck with the extreme rapidity of his discernment whenever I approached the cage, however many other persons might be standing round, I purposely visited the monkey-house on Easter Monday, in order to see whether he would pick me out of the solid mass of people who fill the place on that day. Although I could only obtain a place three or four rows back from the cage, and although I made no sound wherewith to attract his attention, he saw me almost immediately, and with a sudden intelligent look of recognition ran across the cage to greet me. When I went away he followed me, as he always did, to the extreme end of his cage, and stood there watching my departure as long as I remained in sight.

In conclusion, I should say that much the most striking feature in the psychology of this animal, and the one which is least like anything met with in other animals, was the tireless spirit of investigation. The hours and hours of patient industry which this poor monkey has spent in ascertaining all that his monkey-intelligence could of the sundry unfamiliar objects that fell into his hands, might well read a lesson in carefulness to many a hasty observer. And the keen satisfaction which he displayed when he had succeeded in making any little discovery, such as that of the mechanical principle of the screw, repeating the results of his newly earned knowledge over and over again, till one could not but marvel at the intent abstraction of the 'dumb brute'—this was so different from anything to be met with in any other animal, that I confess I should not have believed what I saw unless I had repeatedly seen it with my own eyes. As my sister once observed, while we were watching him conducting some of his researches, in oblivion to his food and all his other surroundings—'when a monkey behaves like this, it is no wonder that man is a scientific animal!' And in my next work I shall hope to show how, from so high a starting-point, the psychology of the monkey has passed into that of the man.

INDEX.

ACC

ACCOUCHEUR, fish, 246; toad, 254
Acerina cernua, 246
Acinia prehensa, 233
Actinia, 233, 234
Actinophrys, apparent intelligence of, 20
Adamsia, 234
Adaptive movement, as evidence of mind, 2, 3
Addison, his definition of instinct, 11
Addison, Mrs. K., on gesticulating signs made by a jackdaw, 316
Ælian, on division of labour in harvesting ants, 98
Æsthetic emotions of birds, 279-82
Affection, sexual, parental, and social, of snails, 27; of ants, 45-9 and 58, 59; of bees, 155, 156, and 162; of earwig, 229; of fish, 242-6; of reptiles, 256, 258, 259; of birds, 270-6; of kangaroo, 326, 327; of whale, 327; of horse, 329; of deer, 334; of bat, 341; of seal, 341-6; of hare, 338-40; of rats, 340; of mice, 341; of beaver, 367; of elephant, 387-92; of cat, 411, 412; of dog, 437, 440, 441; of monkeys, 471-5 and 484-98
Agassiz, Professor A., on instinct of hermit-crab, 232; nest of fish, 242-3; on beaver-dams, 384, 385
Agassiz, Professor L., on intelligence of snails, 26
Alison, Professor, on curious instinct of polecat, 347

ANT

Allen, J. A., on breeding habits of pinniped seals, 341-6
Alligators, 256-8 and 263
Alopecias vulpes, 252
Amœba, apparent intelligence of, 21
Anemones, sea, 233, 234
Anger, of ants and bees, *see* under; of fish, 246, 247; of monkeys, 478, 479 and 484-96
Angler-fish, 247, 248
Annelida, apparent intelligence of, 24
Antennæ, effects of removal in ants, 142; in bees, 197
Antithesis, principle of, in expression of emotions by monkeys, 494, 495
Ant-lion, 234, 235
Ants, powers of special sense, 31-37, of sight, 31-33; of hearing, 33; of smell, 33-37; sense of direction, 37, 38; memory, 39-45; recognition of companions and nestmates, 41-45; emotions, 45-49; affection, 45-48; sympathy, 48, 49; communication, 49-57; habits general in sundry species, 57-93; swarming, 57, 58; nursing, 58, 59; education, 59, 60; keeping aphides, 60-64; making slaves, 64-68; wars, 68-83; keeping domestic pets, 83, 84; sleep and cleanliness, 84-7; play and leisure, 87-89; funeral habits, 89-93; habits peculiar to certain species, 93-122; leaf-cutting, 93-96; harvesting, 96-110; African, 110, 111; tree, 110, 111; honey making, 111-114 and 142; ecitons, or mili-

APE

tary, 114-122; general intelligence, 122-142; Sir John Lubbock's experiments on intelligence, 123-128; intelligence displayed in architecture, 128-130; in using burrows made by elater larvæ, 130; in artificial hives, 130; in removing nest from shadow of tree, 131; in cutting leaves off overshadowing tree, 131, 132; in bending blades of grass while cutting them, 132, 133; in cooperating to glue leaves together, 133, 134; in getting at food in difficult places, 134, 135; in making bridges, &c., 135-139; in tunnelling under rails, 140; anatomy and physiology of nerve-centres and sense organs, 140-2
Apes, see Monkeys
Arachnidæ, 204-225, see Spiders and Scorpions
Arago, his observation regarding sense of justice in dog, 443
Arderon, on taming a dace, 246
Argyroneta aquatica, 212
Arn, Capt., on sword- and thresher-fish, 252, 253
Articulata, see under divisions of
Ass, general intelligence of, 328 and 333
Association of ideas, see under various animals
Atenchus pilularius, 226
Athealium, apparent intelligence of, 19-20
Atkinson, the Rev. J. C., on reasoning power of a dog, 458, 459
Audubon, on ants making beasts of burden of bugs, 68; plundering instincts of white-headed eagle, 284; variations in instinct of incubation, 299, 300
Auk, nidification of, 292
Automatism, hypothesis of animal, 6

BABOON, sympathy shown by Arabian, 474; rage of, 478; revenge of, 478
Badcock, on dog making peace-offerings, 452

BEC

Baer, Van, on organisation of bee, 241
Bailey, Professor W. W., on dog stopping a runaway horse, 459
Baines, A. H., on dog communicating wants by signs, 446, 447
Baker, on sticklebacks, 245
Baldamus, Dr., on cuckoo laying eggs coloured in imitation of those of the birds in whose nests they lay them, 307
Ball, Dr. Robert, on commensalism of crab and anemone, 234
Banks, Sir Joseph, on intelligence of tree-ants, 133; fish coming to sound of bell, 250
Bannister, Dr., on cat trying to catch image behind mirror, 415, 416; on intelligence of the Eskimo dogs, 461, 462
Barrett, W. F., on instincts of young alligator, 256
Barton, Dr., on alleged fascination by snakes, 264
Bastian, on termites, 198
Bates, on ants' habit of keeping pets, 84; cleaning one another, 87; play and leisure, 88, 89; leaf-cutting, 93-95; tunnelling, 99; ecitons, 114-21; on sand-wasp taking bearings to remember precise locality, 150; mygale eating humming-birds, 208; on nidification of small crustacean, 232, 233; habits of turtles, and alligators, 257, 258; intelligence of vultures, 314; bats sucking blood, 341
Batrachians, 254, 255
Bats, 341
Baya-bird, nidification of, 294
Bears, 350-352
Beattie, Dr., on dog communicating desires by signs, 447
Beaver, 367-85; breeding habits, 367, 368; lodges, 368-73; dams, 373-79; canals, 379-83; general remarks upon, 368, 377, 379, 383; age of their buildings, 384; effects of their buildings on the configuration of landscapes, 384, 385
Bechstein, on birds dreaming, 312

INDEX.

BEE

Bee, mason, 178, 179; tapestry, 179; carpenter, 179; rose, 179; carding, 179, 180
Bees, sense of sight, 143, 144; of smell and hearing, 144; of direction, 144-51; remembering exact locality of absent hive, 148-49; following floating hives, 149; memory, 151-55; sympathy, 155, 156; distances over which they forage, 150; powers of communication, 156-60; economy of hive, 160-8; food and rearing, 160-163; swarming and battles of queens, 163, 164; drone-killing, 164-68; plunder and wars, 168-170; architecture, 170-8; way-finding, 181, 182; instinct of neuters, 181; recognising companions, 183, 184; barricading doors against moths, 184, 185; strengthening combs in danger of falling, 185, 186; mode of dealing with surfaces of glass, 186; with strange hives, 186, 187; evacuating fallen hive, 187; ceasing to store honey in Barbadoes and California, 187, 188; recognising persons, 188, 189; biting holes in corollas, 189; ventilating hives, 191, 192; covering slugs, &c., with propolis, 190, 191; effects of removing antennæ, 197
Beetles, *see Coleoptera*
Belshaw, on cat knocking knockers, 422
Belt, on ants, duration of memory in, 39, 40; sympathy, 48; division of labour, 99; ecitons, 114-19 and 138; tunnelling under rails, 140; on sand-wasp taking precise bearings to remember locality, 150, 151; struggle between wasps and ants for secretion of frog-hoppers, 194, 195; intelligence of spiders in protecting themselves from ecitons, 219, 220; beetles undermining stick supporting a dead toad, 228; intelligence of monkeys, 480
Benedictson. on navigating habits of Iceland mice, 364, 365

BLA

Bennet, on birds dreaming, 312
Bennett, on conjugal fidelity of duck, 270, 271
Berkeley, G., on beetle storing its food, 228, 229
Bettziech-Beta, on termites, 199
Bidie, on suicide of scorpion, 222, 223; on reasoning power of cat, 415
Bingley, on intelligence of ants, 133; carpenter-bees, 179; account of alleged training of bees, 189; co-operation of beetles, 226, 227; ant-lion, 230, 235; domestication of toad, 255; fascination by snakes, 264; sympathy in birds, 272; eccentricity of nest building instinct, 295; education of birds, 312; pigs pointing game, 339, 340; intelligence of otter, 346; memory of elephant. 387; vindictiveness of elephant, 387, 389; elephants enduring surgical operations, 399, 400
Bird, Miss, on combined action of crows in obtaining food from dogs, 320
Birds, 266-325; memory of, 266-70; emotions, 270-82; special habits of procuring food, 283-6; of incubation and taking care of offspring, 287-310; general intelligence, 310-25; dreaming and imagination, 311-12; learning to avoid telegraph wires, 313; recognising painting of birds, 311; submitting to surgical operation, 313-14; honey-guide, 315-16; appreciation of mechanical appliances, 315-16; concerted action, 318-322
Birgus latro, 233
Bison, 334-5
Blackbirds, breaking shells against stones, 283; removing eggs, 289; mobbing cat, 291
Blackburn, Professor H., on distances over which bees forage, 150
Blackhouse, R. O., on dog being alarmed at a statue, 453
Blackman, on cats learning to beg for food, 414-15

BLA

Blackwall, on early display of instincts by spiders, 216
Blanchard. on mason-bee, 178
Blood, on reasoning power of a dog, 464
Boa-constrictor, really a Python, which *see*
Bodley, W. H., on dogs crossing a river to fight undisturbed, 451-2
Bold, on canary singing against own image in mirror, 276
Bombyx moth, larva of, 238-40
Bonnet, on spider following her eggs into pit of ant-lion, 205; his experiments on instincts of caterpillars, 236; observations on ditto, 238
Boobies, plundered by frigate pelicans, 284
Bosc, on migrating fish, 248
Bower-bird, instincts of, 279-81, 325
Bowman, Parker, his cat opening swivel of window, 425
Boys, C. V., his experiments with a tuning-fork on spiders, 206, 207
Brehm, on wasps recognising persons, 188; intelligence of lapwing, 315, 316; curiosity of monkeys, 477
Broderip, on vindictiveness of elephant, 389
Brodie, Sir B., his definition of instinct, 15; on bees strengthening their combs, 185, 186
Brofft, Herr L., on powers of communication in bees, 160
Brougham, Lord, on hexagonal form of bees' cells, 172; on intelligence of a dog, 450
Brown, Capt., on vindictiveness of a stork, 277-8
Brown, W., on a cat extinguishing fire by water, 425
Browne, Dr. Crichton, on cat ringing bell, 423
Browne, Murray, on fox allowing itself to be extricated from trap, 431
Browning, A. H., on intelligence of a dog, 450
Brydon, Dr., on collective instinct of jackals, 434

BUL

Buchanan, Dr., on climbing perch, 249; on nidification of baya-bird, 294
Büchner, Professor, on ants: nursing habits, 59; stocking trees with aphides, 63; warfare, 71-9; play, 87-88; leaf cutting, 95-96; intelligence in making a bridge of aphides over tar, 136; of themselves over a space, 136-37; and of a straw over water, 137; ecitons, 139; anatomy and physiology of brain, 141-42. On bees and wasps: powers of communication, 158-60; swarming habits, 168; wars and plunder, 169; cell-building, 177-78; evacuating dangerous hive, 187; keeping hives clean, 190; carrying dead from hive and burying them, 191; ventilating hives, 191-92; hornet and wasp dismembering heavy prey, and carrying it to an eminence in order to fly away with it, 196; on termites, 198-202. On spiders: web-building, 211-12; wolf spider, 213; trap-door spiders, 217-18; intelligence of a spider habitually fed by Dr. Moschkau, 218-19; spiders weighting their webs, 221. On beetles: co-operation of, 227-28
Buck, E. C., on intelligence of crocodiles, 263; on collective instinct of wolves, 433; on combined action of pelicans, 319
Buckland, F., on pigeon remembering voice of mistress, 266; crows breaking shells by dropping them on stones, 283; birds avoiding telegraph wires, 313
Buckley, on harvesting ants, 103
Buckton, G. B., on caterpillars, 236
Buffalo, 335-37
Buffon, on hexagonal form of bees' cells, 171-72; association of ideas in parrot, 269; sympathy in ditto, 275; goat sucker removing eggs, 289
Bufo obstetricans, 254
Bull, intelligence of, 338

INDEX. 503

BUR

Burmeister, on powers of communication in ants, 49
Byron, Lord, lines on alleged tendency to scorpion to commit suicide, 222

CADDIS-WORMS, 240
Cairns, Mr. W., on reasoning power of a dog, 461
Campbell, Mrs. G. M. F., on intelligence of goose, 316
Canary, jealousy of, 276; modification of incubating instinct in cage, 287; flying against mirror, 311; trained, 312
Canning, J., his dog knowing value of different coins, 452-3
Carassius auratus, 246
Carbonnier, M., on telescope-fish, 246
Carlisle, Bishop of, on congregation or court held by jackdaws, 324
Carpenter, Dr., on intelligence of rats, 361
Carreri, Gemelli, on monkeys thrusting stones between oyster-shells to keep them from closing, 481
Carter, H. J., on apparent intelligence of *athealium*, 19; of *actinophrys* and *amœba*, 20-1
Carus, Professor, on spiders weighting their webs, 221
Cat, the, 411-25; general remarks upon, 411-14; emotions of, 412-13; general intelligence of, 413-42; showing zoological discrimination, 414; punishing kittens for misbehaviour, 414; begging for food, 414-15; feeding kittens on bread when milk fails, 415; carrying kittens to be protected by master, 415; trying to catch image behind mirror, 416; communicating by signs, 419; devices for catching prey, 417-20; appreciation of mechanical appliances, 420-25; extinguishing fire by water, 425
Caterpillars, instinct of assisted by intelligence, 236-8; migrating, 238-40

CON

Catesby, on co-operation of beetles, 226, 227; on frigate-pelican plundering boobies, 284
Cattle, fear exhibited by in slaughterhouses, 334; pride of, 334
Cebus fatuellus, observations on intelligence of, 484-98
Cecil, H., on tactics displayed by hunting wasps, 194
Cephalopoda, intelligence of, 29-30
Cetacea, 327-28
Challenge, mode of, in gulls, 291
Charming of snakes, 264
Cheiroptera, 341
Chelmon rostratus, 248
Chimpanzee, play of, 476-77
Chinese swallow, nidification of, 292
Chironectes, 243
Choice, as evidence of mind, 2
Clark, G., on intelligence of a bat, 341
Clark, Rev. H., on harvesting ants, 99; on dog recognising portrait, 454-5
Clarville, on co-operation of beetles, 228
Clavigero, on sympathy of pelicans for wounded companions, 275
Claypole, on intelligence of horse, 331-2
Cnethocampii pityocampa, 244
Cobra, sexual affection of, 256; charming, 265; intelligence of, 262
Cock, domestic, killing hen upon hatching out eggs of other birds, 278
Cœlenterata, movements of, and question concerning their intelligence, 22
Coleoptera, 226-9; co-operation of, 226-8; other instances of intelligence, 228-9
Colquhoun, on reasoning power of a dog, 463-4
Commensalism, between crab and anemone, and between mollusk and anemone, 233
Communication, *see* Co-operation
Concerted action, *see* Co-operation

CON

Cones, Captain Elliot, on intelligence of wolverine, 348–50
Conilurus constructor, 326
Conklin, W. A., on elephants thatching their backs, 409
Consciousness, as evidence of mind, 2; gradual dawn of, 13
Conte, John Le, on reasoning power of a dog, 460–1
Cook, Capt., on tree ants, 111; intelligence of tree-ants, 133
Cook, George, on dog dragging mat about to lie upon, 466
Co-operation, of ants, 48–49, 51- 59, 64 *et seq.* (in making slaves and waging war), 85–96; (in sundry occupations), 96–100; (in harvesting), 108–10, 111–14; (of apparently different species), 114–122; (of military ants) 127-30, 132–4, 136–40; of bees, 159–74; (in general work, wars, and architecture), 177, 178, 184–6, 190–2; of termites, 198–203; of beetles, 226–8; of birds, 318–22; of horses and asses, 333; of bison and buffalo, 335; of pigs, 339; of rats, 361, 362; of mice, 364; of beavers, 367–83; of elephants, 401; of foxes, 433; of wolves, 433 and 436; of jackals, 432–5; of baboons, 483
Corse, on memory of elephant, 386, 387; emotions of elephant, 393
Corvus cornice, punishing offenders, 323, 324
Couch, on maternal instinct of hen, 272; mode in which guillemots catch fish, 285; mode of escape practised by swan, 290; birds removing dung from neighbourhood of their nests, 290; blackbirds mobbing cat, 291; nidification of swan, 296–8; crows punishing offenders, 323–4; intelligence of hare, 359; cat unlocking door, 424; fox avoiding trap, 428; catching crabs with tail, 432; mode by which a dog killed crabs, 459
Cowper, on intelligence of hare, 359, 360

DAR

Cox, C., play-houses of bower-birds presented by him to Sydney Museum, 280
Crabs, 231–4
Craven, on intelligence of a sow, 340
Crehore, on foxes avoiding traps, 428, 429; on dog recognising portrait, 453
Cripps, his elephant dying under emotional disturbance, 396
Criterion of mind, 4–8
Crocodiles, 263
Crow, Capt. Hugh, on sympathy shown by monkeys for sick companion, 473, 474
Crows, memory of, 266; breaking shells by dropping them on the stones, 283; punishing offenders, 323–5
Cruelty, of cat, 413
Crustacea, 231–34
Cuckoo, parasitic instincts of, 301–7; eggs of coloured like those of the bird in whose nest they are laid, 307–9; American, 305, 306
Curiosity, of fish, 247; of birds, 278, 279; of ruminants and swine, 335; of monkeys, 477
Curlew, nidification of, 292
Cuvier, his orang drawing chair to stand upon to reach a latch, 481; on birds dreaming, 312

DACE, tamed, 246
Dampier, on frigate-pelicans plundering boobies, 284; on monkeys hammering oyster shells with stones, 481
Daphnia pulex, seeking light, especially yellow ray, 23
Darwin, Charles, on apparent intelligence of worms, 24; of oyster, 25; of snail, 27; Mr. Hague's letter to, on powers of communication in ants, 54–7; observations on ants keeping aphides, 60, 61; on ants making slaves, 64, 66, 67; communications of Lincecum to, on harvesting ants, 103, 107; on proportional size of ants' brain, 140; communication of Müller on

INDEX.

DAR

powers of communication in bees, 157; origin and development of cell-making instinct, 173-7; instincts of neuters, 181; quotation in MS. from Sir B. Brodie on bees supporting their combs, 185-6; his 'law of battle' in relation to spiders, 205; intelligence of crab, 233; his theory of sexual selection, 279-82; sense of smell in vultures, 286; on Wallace's theory of correlation between colour of sitting birds and form of their nests, 299; instincts of cuckoo, 304-6; birds dreaming, 312; Gauchos taming wild horses, 329; memory of horse, 330; intelligence of bear, 352; of elephant, 398, 402; collective instinct of wolves, 436; duration of memory in dogs, 438; intelligence of Eskimo dogs, 462; reasoning of retriever, 463-4; maternal care and grief of monkey, 472; sense of ludicrous in monkeys, 476; curiosity and imitativeness of monkeys, 477; imitativeness of man, 477-8; intelligent observation displayed by monkeys, 479, 480

Darwin, Erasmus, on bees ceasing to store honey in Barbadoes, 187; wasp dismembering fly to facilitate carriage, 195; unmoulted crab guarding moulted, 233; crows breaking shells by dropping them on stones, 283; bird shaking seed out of poppy, 286; elephant acting nurse to young child, 408

Darwin, F., on bees biting holes through corollas, 189

Davis, on instincts of larvæ of bombyx moth, 239

Davy, Dr., on instincts of alligators, 256, 257; taming cobra, 265; performing operation on elephants, 400

Davy, Sir H., on eagles teaching young to fly, 290

Day, F., on intelligence of fish, 244-52

Deceitfulness, of elephant 410; of

DZI

dog, 443, 444, 450-52, 457, 458; of monkey, 494

Deer, intelligence of, 336, 338, 339

De Fravière, on powers of communication in bees, 158; their scouts, 168

Descartes, his hypothesis of animal automatism, 6

Dicquemase, on intelligence of oyster, 25

Dipterous insects, intelligence in finding way out of a bell-jar, 153, 154; gad-fly, 230; house-fly, 230, 231

Division of labour, *see* Co-operation

Dog, ringing bell, 423; knocking knocker, 423; collective instinct of, 435, 436; general remarks on psychology of, as influenced by domestication, 437, 438; memory of, 438; emotions of, 438-45; pride and sensitiveness, 439-42; intolerance of pain, 441; emulation and jealousy, 442, 443; sense of justice, 443; deceitfulness, 443, 444; sense of ludicrous and dislike of ridicule, 444, 445; general intelligence of, 445-70; communicating ideas, 445-7; instances of reason, 447-69

Doldorff, on climbing perch, 248, 249

Dolomedes fimbriata, 213

Doras, 248

D'Osbonville, on monkeys administering corporal chastisement to their young, 482, 483

Dreaming, of birds, 269, 312; of ferrets, 347

Duchemin, M., on toads killing carp, 254

Duck, conjugal fidelity of, 270, 271; conveying young on back, 289

Dugardin, on communication among ants, 49; in bees, 156

Duncan, on cunning of a dog, 451

Dzierzon, on cause determining sex of bees' eggs, 162; bees repairing injuries to their cells, 186

EAG

EAGLE, plundering instinct of white-headed, 284; teaching young to fly, 290; variations in nest-building, 299; submitting to surgical operations, 313, 314
Earwig, 229, 230
Ebrard, on co-operation of ants, 132
Echinodermata, movements of, 23
Edmonson, Dr., on crows punishing offenders, 323, 324
Edward, on intelligence of frogs, 255; sympathy of terns for wounded companion, 274, 275; crows breaking shells by dropping them on stones, 283; co operation of turnstones, 321
Edward, H., on honey-making ants, 111-14
Eimer, Dr., on voluntary and involuntary movements of *Medusæ*, 22, 23
Elephant, general remarks upon, 386; memory of, 386, 387; emotions of, 387-96; vindictiveness, 387-9; sympathy, 389-90; rogue, 393, 394; dying under effects of emotion, 395, 396; general intelligence of, 396-410; enduring surgical operations, 399-400; vigilance, 401; formation of abstract ideas, 401, 402; intelligence of tame decoys, 402-6; of tame workers, 306-8; thatching their backs, 308, 309; removing leeches, and fanning away flies, 309, 310; concealing theft, 410
Ellendorf, Dr. F., on leaf-cutting ants, 95, 96; on ants making a bridge, 137
Elliot, on collective instinct of wolves, 433
Emery, J., on powers of communication in bees, 157
Emulation, of birds, 277; of dogs, 442
Encyclopædia Britannica, on bees following floating hives, 149; battles of queen-bees, 163, 164; parasitic instincts in birds, 306
Endurance, of pain by wild dogs, 441; of surgical operations by eagle, 313, 314; by elephants, 399, 400; by monkey, 482

FLE

Engelmann, on *Daphnia pulex* seeking yellow light, 23
Epeira aurelia, Mr. F. Pollock on perfection of web built by young, 217
Erb, G. S., on intelligence of deer, 338, 339
Esox lucius, 246
Espinas, on co-operation of ants, 130

FABRE, on instincts of sphexwasp, 180, 181
Faister, Mdlle de, her tame weasel, 346
Falcon, variations in nest-building, 299
Faraday, J., on intelligence of skate, 251
Fascination, alleged, by snakes, 263, 264
Fayrer, Sir J., on fascination by and charming of snakes, 264
Fear, in horses, 329; in ruminants, 334; in rabbits, 355; in rats, 360 excited in dogs by portraits, 455-7; in monkey by snakes, 477, and by imitation monkey, 495
Ferret, 347
Fire-flies, stuck on nests by bayabirds, 294
Fish, 241-53; comparison of brain with that of invertebrata, 241; emotions, 242-7; ' nidification, courtship, and care of young, 242-6; pugnacity, and social feelings, 242; anger, 246, 247; play, jealousy, curiosity, 247; angler, 247, 248; jaculator, 248; travelling over land, 248; climbing trees, 248, 249; migrations, 249, 250; general intelligence, 250-53
Fisher, J. F., on hen removing eggs with her neck, 288
Fleeson, Captain B., on honey-making ants, 111-14
Fleming, W. J., on intelligence of horse, 330
Fleury, Cardinal, on intelligence of ants in making bridges, 135

FOR

Forbes, on nidification of tailor-bird, 293
Forbes, James, on monkey begging for dead body of companion, 472
Forel, on ants; recognising slaves, 43; and fellow-citizens, 44; swarming habits, 58; experiment in rearing together hostile species, 59, 60; tunnelling to obtain aphides, 61; warfare, 68-77; play, 88; intelligence shown in architecture, 129
Forsteal, on termites, 198
Forster, W., on intelligence of a bull, 338
Fothergill, Percival, on reasoning power of a dog, 466
Fouillouse, J. de, on intelligence of hares, 357, 358
Fox, 426-33; lying in wait for hares, 426, 427; avoiding traps, 427-30; allowing itself to be extricated from trap, 431; catching crabs with tail, 432; collective instinct in hunting, 433
Fox, C., on intelligence of porpoises, 328
Frankland, Mrs., on cock bullfinch recognising portrait of hen, 311
Franklin, on powers of communication in ants, 49
Franklin, Dr., on sympathy in parrots, 276
Frogs, 254, 255
Frost, Dr., on cat sprinkling crumbs to attract birds, 418, 419
Furniss, J. J., on elephants thatching their backs, 408, 409

GAD-FLY, instinct of, 230
Gander, *see* Goose
Gaphaus, H. A., on cat opening thumb-latch, 421
Gardener, on intelligence of crab, 233
Garraway, Dr., on beetle concealing its store of food, 229
Gasteropoda, intelligence of, 26-29
Gasterosteus pungitius, 243; *G. spinachia*, 243
Geer, M., on earwig incubating young, 229
Gelasinnus, 233

GRO

Gentles, W. Laurie, on intelligence of a sheep-dog, 448, 449
Geoffrey, on pilot fish, 252
Gibbons, their sympathy for suffering companions, 472, 473
Gleditsch, on beetles undermining stick supporting a dead toad, 228; on spiders weighting their webs, 221
Glutton, 347-50
Goat, intelligence of, 337, 338
Goat-sucker, removing eggs, 289; nidification of, 292
Goldfinch, trained, 312
Goldsmith on habits of rooks, 322, 323
Goldsmith, Dr., on intelligence of otter, 346
Gollitz, Herr, on co-operation of beetles, 227
Goodbehere, S., on intelligence of a pony and ass, 332, 333; on cunning of sheep-killing dogs, 450; on dog knowing value of different coins, 452, 453
Goose, affection and sympathy of, 272, 273; removing eggs from rats, 288; noting time, 314; opening latch of gate, 316
Gosse, on commensalism of crab and anemone, 234
Gould, on bower-bird, 279-81; on humming-birds, 281; on talegallus, 294, 295
Graber, Titus, on proportional size of ant's brain, 141
Grapsus strigosus, 231
Gray, Sir George, on nidification of talegallus, 295
Gredler, Vincent, on division of labour among leaf-cutting ants, 99, 100
Green, on intelligence of pigs, 339
Green, Seth, on tactics displayed by hunting wasps, 193
Griffiths, on intelligence of elephant, 388, 389
Grosbeak, nidification of, 295, 296
Grouse, learning to avoid telegraph wires, 312, 313
Groves, J. B., on cat trying to catch image behind mirror, 416

Guana, *see* Reptiles
Guerinzius, on wasps recognising persons, 188
Guillemots, plundering of by gulls, 283, 284; mode of catching fish, 285
Gulls, plundering guillemots, 283, 284; mode of challenge, 291; nidification, 292
Guring, Thomas, on intelligence of geese, 314, 315

HAGEN, on termites, 202
Hague, on powers of communication in ants, 54-7
Hamilton, R., on fear exhibited by cattle in slaughterhouses, 334
Hancock, Dr., on fish quitting water, 248; crows breaking shells by dropping them on stones, 283
Harding, S., on intelligence of a pig, 340
Hare, 357-60
Hartmann, Von, his definition of instinct, 15; on fondness of spiders for music, 206
Harvesting-ants, 96-110; mice, 365, 366
Hawkshaw, J. Clarke, on limpet remembering locality, 28-9
Hayden, on monkey keeping door open with blanket, 481
Hayes, Dr., on intelligence of Eskimo dogs, 462
Heber, Bishop, on sympathy of elephant, 289
Helix pomatia, intelligence of, 26, 27
Hemerobius chrysops, 240
Hen, maternal instinct of, 272; removing eggs with neck, 288; and young chicken on back, 288, 289
Henderson, on navigating habits of Iceland mice, 364, 365
Heron, variations in nest-building, 290
Hogg, on intelligence of his sheep-dog, 448
Holden, on starlings learning to avoid telegraph wires, 312, 313
Hollmann, on intelligence of *octopus*, 30

Homarus marinus, 233
Hooker, Sir Joseph, on navigating habits of Iceland mice, 364
Hooper, W. F., on intelligence of a dog, 463
Horn, Mrs., on reasoning powers of a dog, 462
Hornet, carrying heavy prey up an elevation in order to fly away with it, 196
Horse, emotions of, 328-30; memory, 330; general intelligence, 328, 330-3
Horse-fly, tamed, 230, 231
Horsfall, on dog finding his way about by train, 467, 468
Hoste, Sir W., on wounded monkey showing its blood to the sportsman, 476
Houzeau, on hen transporting young chicken on her back, 288, 289; parrots not being deceived by mirrors, 310, 311; birds dreaming, 312; mules counting their journeys, 332; monkeys destroying poison-fangs of snakes, 483
Hubbard, Mrs., on intelligence of a cat, 414
Huber, F. and P., on instinct, 16. On ants: sense of smell in, 33; recognising companions, 44; powers of communication, 49, 50; observations on slave-making instinct, 65; on warfare, 76; play, 87, 88; harvesting, 97; carrying one another, 109; intelligence shown in architecture, 128, 129. On bees: sense of hearing in, 144; duration of memory, 155; powers of communication, 156, 159; manipulation and uses of propolis, 161; battles of queen-bees, 164, 165; form of cells, 173; building cells, 177, 178; barricading doors against moths, 184; strengthening combs, 185; biting holes in corollas, 189; ventilating hives, 191, 192; effects of removing antennæ of bees, 197
Hudson, on habits of *Melothrus*, 309, 310
Hugen, on termites, 198

HUM

Humboldt, on instincts of young turtles, 257
Humming-birds, æsthetic instincts of, 281
Hutchings, J., on intelligence of a cat, 417
Hutchinson, on alleged tendency of scorpion to commit suicide, 225
Hutchinson, Col., on reasoning power of a dog, 463, 464
Hutchinson, Dr. H. F., on wolf-spider stalking own image in mirror, 213
Hutchinson, S. J., on intelligence of polar bear, 351, 352
Hutton, Mrs., on ants burying their dead, 91, 92
Hydrargyra, 248
Hymenoptera, *see* Ants and Bees

IBEX, does assisting wounded buck to escape, 334
Idealism, cannot be refuted by argument, 6
Ideas, *see* Association
Imitation, shown by talking birds, monkeys, and idiots, 477, 478
Instinct, defined and distinguished from reason and reflex action, 10-17; of medusæ, 23; of worms, 24; of mollusca, 25; of ants with reference to colour, 32, 33; to smell, 33-7; to sense of direction, 37-9; to recognising friends, 41-5; to swarming, 57, 58; to nursing, 58; to education, 59, 60; to keeping aphides, 60-4; to making slaves, 64-8; to wars, 68-83; to keeping pets, 83, 84; to sleep and cleanliness, 84-7; to play and leisure, 87-9; to treatment of dead, 89-95; of leaf-cutting species, 93-6; of harvesting species, 97-110; of tree-inhabiting species, 110, 111; of honey-making species, 111-14; of ecitons, 114-22; of driver and marching species, 121-2; of bees and wasps, with reference to colour, 143-4; to sense of direction, 144-51; to

JEN

food-collecting and wax-making, 160-2; to propagation, 162-8; of queens, 162-5; of killing drones, 165-8; with reference to wars, 169, 170; to architecture, 170-80; of sphex-wasp, 180, 181: of termites, 198-203; of spiders, 204-18; of scorpion, 222-5; of beetles, 226-9; of earwig, 229, 230; of flies, 230, 231; of crustacea, 231, 232; of larvæ, 234-40; of fish, 242-53; of batrachians, 254; of reptiles, 256-9; of birds, with reference to procuring food, 283-7; to incubation, 287-91; to nidification, 291-301; of cuckoo, 301-10; of marsupials, 320; of whale, 327; of ruminants, 335; of swine, 339; of bats, 341; of seals, 341-8; of wolverine, 348-50; of rodents, 353, 354; of rabbit, 354-7; of hare, 354-9; of rats, 360; of mice, 364-5; of rat-hare, 365, 366; of beaver, mixed with intelligence, 367; with reference to propagation and lodges, 367-71; to procuring food, 371-3; to dams, 373-80; to canals, 380-4; of cat, 411-12; of dog, 437, 438; of monkey, 471

JACKAL, 426; collective instinct in hunting, 432-35
Jackdaw, gesticulating signs made by, 316; congregation for court held by, 324
Jacob, Sir G. Le Grand, on crows punishing offender, 324-5; ibexes assisting wounded mate to escape, 334
Japp, on dog spontaneously learning use of coin, 452
Jealousy, of fish, 242; of birds, 276-7; of horse, 329, 330; of dogs, 442, 443; of monkey, 493
Jenkins, H. L., on formation of abstract ideas by elephants, 401, 402
Jenner, on instinct of young cuckoo, 301-4

Jerdon, Dr., on harvesting-ants, 97; on birds dreaming, 312

Jervoise, Sir J. C., on bee biting hole in a corolla, 189; on combined action of rooks in obtaining food from pheasants, 321

Jesse, on intelligence of bees in adapting their combs to smooth surface, 186; spider protecting eggs from cold, 219; tame housefly, 230, 231; affection of male for female pike, 246; attachment between alligator and cat, 258, 259; conjugal fidelity of swan, and pigeon, 271; sympathy of rooks, 273, 274; lapwing stamping on ground to make worms rise, 285; goose removing eggs from rats, 288; birds removing dung from neighbourhood of their nests, 290; swallows killing and imprisoning hostile sparrows, 318, 319; kangaroo throwing young from pouch when pursued, 326, 327; stag shaking berries from trees, and manifesting intelligence in escaping from dogs, 336; intelligence of buffalo, 336, 337; intelligence of rats, 360–2; of elephants, 398; collective instinct of foxes, 433; wounded monkey showing its blood to the sportsman, 476

Jilson, Professor, on habits of the 'prairie-dog,' 366

John, St., on intelligence of fox, 426, 427; idea of caste in dog, 442

Johnson, on termites, 198; on orang-outangs removing their dead companions, 472

Johnson, Capt., on wounded monkey showing its blood to the sportsman, 475

Johnson, Dr., his definition of reason, 14

KANGAROO, throwing young from pouch when pursued, 326, 327

Kaup, on fish, 246

Kemp, Dr. L., on battles of queen-bees, 164; robber bees, 170; on intelligence of decoy elephants, 402

Kent, Saville, on intelligence of porpoises, 327, 328

Kesteven, Dr. W. H., on cat knocking knocker, 424

Kingfisher, nidification of, 292

Kirby, on water-spider, 212; shore crabs, 232; migration of salmon, 249, 250; intelligence of carp, 250

Kirby and Spence, on powers of communication in ants, 49; sense of direction in bees, 148; hexagonal form of bees' cells, 172; ceasing to store honey in tropics, 188; co-operation of beetles, 226; caterpillars, 236, and 238, 239

Klein, Dr., on intelligence of a cat, 418, 419

Kleine, Herr, on behaviour of bees when finding empty combs substituted for full ones, 186, 187

Klingelhöffer, Herr, on co-operation of beetles, 227–8

König, on termites, 198

Kreplin, Herr H., on ecitons, 130

LABRUS, 247

Lacepède, on fish coming to sound of bell, &c., 250

Lacerta iguana, 255

Lagomys, provident habits of, 365

Landois, on powers of communication in bees, 158

Langshaft, on bees recognising hive companions, 183; on robber bees, 183–4

Lapwing, stamping on ground to make worms rise, 285; intelligence of, 315, 316

Larvæ, of insects, intelligence of, 234–40

Latreille, on an's, sympathy of, 47

Lauriston, Baron, on sympathy of elephant, 390 .

Layard, Consul, on intelligence of cobra, 262; on nidification of baya-bird, 294; on cat pulling bell-wire, 424

Lee, Mrs., on intelligence of robin, 314; of goats, 337; of rats, 361; on vindictiveness of elephant, 389
Leeches, apparent intelligence of, 24
Lefroy, Lieut-Gen, Sir John, on terrier communicating wants by signs, 446
Lehr, Herr H., on bees draining their hive, 190
Leroy, C. G., on nidification of birds, 300; on migration, 301; on collective instinct of wolves, 436
Lespes, on ants: slave-making instinct, 65, 66; warfare, 68, 69; division of labour, 98, 99: on termites, 198
Leuckart, Prof., on intelligence of ants in surmounting obstacles, 135
Lever, Sir Ashton, his experiment on eccentricity of nest-building instinct, 295
Limpet, remembering locality, 28, 29
Lincecum, Dr., on harvesting ants, 97 and 103-7; carrying one another, 109
Lindsay, Dr. L., on birds dreaming, 312
Linnæus, on swallows imprisoning sparrows, 318
Linnet, intelligence of in not flying against mirror, 311; trained, 312
Liparis chrysorrhaea, 238
Livingstone, Dr., on certain ants of Africa, 110; honey-guide, 315; intelligence of buffalo, 335, 336; reasoning power of dog, 457
Lobster, 233
Lockman, J., on fondness of pigeon for a particular air of music, 282
Lonsdale, on intelligence of snails, 27
Lophius piscator, 247-8
Lophobranchiate fish, incubating eggs in mouth, 245-6
Loudoun's 'Magazine of Natural History,' quotations from, 357
Love-bird, conjugal affection of, 270
Löwenfels, Herr H., on a wasp dis-membering a fly to facilitate carriage, 196.
Lubbock, Sir John, on ants: sense of sight in, 32; of hearing, 33; of smell, 33-7; of direction, 37-8; recognising companions and nest-mates, 41-3 and 44-5; deficiency of affection and sympathy. 45-7; powers of communication, 50-3; collecting hatching eggs of aphides, 61-2; keeping pets, 84; general intelligence, 123-8. On bees and wasps: sense of sight in, 143; of smell and hearing, 144; of direction, 144-8; memory, 151-4; taming wasps, 153; experiment on comparative intelligence of wasp and fly in finding way out of a bell-jar, 153-4; experiments to test sympathy, 155-6; way-finding, 181-3; recognising one another, 183-4. On co-operation of beetles, 226.
Ludicrous, sense of, in dogs, 444-5; in monkeys, 476, 485, 487, and 490
Lukis, F. C., on limpet remembering locality, 29

MACLACHLAN, on caddis-worms, 244
MacLaurin, on mathematical principles observed by bees in constructing their cells, 171
Macropodus, 244
Malcolm, Sir James, on sympathy shown by monkey, 474-5
Malle, Dureau de la, on dog knocking knocker, 423-4; collective instinct of dogs, 435-6
Mammals, 326-498
Mann, Mr. and Mrs., their tame snakes, 256, 260-2
Mansfield, nest of fish, 242-43
Marsupials, 326-7
Martin, nidification of house, 292; of land, 292
Martin, John, on reasoning power of cat, 415
M'Cook, the Rev. Dr., on ants: recognising fellow-citizens, 44; feed-

512 INDEX.

M'CR

ing comrades with aphides-secretion, 63-4; keeping cocci and caterpillars, 64; warfare, 78, 81-3; sleep and cleanliness, 84-87; play, 88; funeral habits, 89-91; agricultural, 97, 103-10; modes of mining, 108; swarming habits of agriculturals, 108-9; carrying one another, 109-10; removing nest from shade of tree, 131; cutting leaves from shading tree, 131-2; co-operation in cutting grass, 132

M'Cready, on larva of *Medusæ* sucking nutriment from parent, 34

Meek, his cat trying to catch image behind mirror, 415-16

Meenan, on a wasp carrying heavy prey up an elevation in order to fly away with it, 197

Melanerpes formicivorus, 285

Melia tessellata, 233-4

Melipona domestica, form of its cells, 173-6

Melothrus, 309-10

Memory, of mol'usca, 25-9; of ants, 39-45; of bees, 151-5; of beetles and earwig, 226-30; of batrachians, 255; of reptiles, 259 *et seq.*; of birds, 266-70; of horse, 330; of elephant, 386-7; of dog, 438; of monkey, 497

Menault, on eagle submitting to surgical operation, 313-14; on mason bee, 178-9

Merian, Madame, on ants of visitation, 130; mygale spider eating humming-birds, 208

Merrell, Dr., on instinct of American cuckoo, 305-6

Mice, 360-4

Migration, of caterpillars, 238; of crabs, 232; of fish, 248-50; of reptiles, 257-8; of birds, 266; of mammals, 341-50, and 368

Mildmay, Sir Henry, on pigs learning to point game, 339-40

Mill, John S., on instinct of cruelty in man, 413

Miller, Prof., calculations regarding form of bee's cell, 173

Mind, subjective and objective

MON

analysis of, 1; evidence of, 2; criterion of, 4-8

Mischievousness, fondness of, shown by monkeys, 485 *et seq.*

Mitchell, on fish removing eggs from disturbed nest, 251

Mitchell, Major, on habits of *Conilurus constructor*, 326

Mivart, on instincts of sphex-wasps, 181

Mobbing instinct in birds, 291

Möbius, Prof., on commensalism between crab and anemone, 233

Moggridge, on ants: sympathy of, 48; suggestion to Mr. Hague, 56; warfare of, 79-81; keeping pets, 83; harvesting, 97-8 and 100-2; division of labour, 98; harvesters using burrows made by elater, 130; intelligent adaptation to artificial conditions, 130; co-operation in cutting grass, &c., 133. On trap-door spiders covering trap-doors with moss, &c., 214-15; making trap-door at exposed end of accidentally inverted tube, 215-216; perfection of dwellings built by young spiders, 216-17; manner in which instinct of making trap-doors probably arose, 217-18

Mollusca, intelligence of, 25-30

Monboddo, Lord, on snake finding way home, 262

Monkeys, 471-98; general remarks on psychology of, 471 and 497-98; emotions of, 471-8; affection and sympathy, 471-5; reproach, 475-6; ludicrous, 476, 485, 487, 490; play, 476-77; curiosity, 477; imitation, 477; rage, jealousy, and revenge, 478; memory of, 497; general intelligence of, 478; behaviour with mirror, 478-9 and 495-6; picking shells off eggs, and taking care not to be stung by wasps in paper, 479; intelligence of Mr. Belt's, 480; disentangling chains, 480 and 486 8; raking in objects with sticks or cloths, 480 and 486; drawing chair to stand upon,

481; using levers, 481 and 492; using hammers, 481 and 485; divining principle of screw, 490-91; keeping door open with blanket, 481-2; allowing tooth to be drawn, 482; punishing young, 482-3; destroying snake's fangs, 483; concerted action, 483; love of mischief, 485 *et seq.*; throwing things in rage, 485 *et seq.*; pushing slab to which tied, 484-7; capricious attachments and dislikes, 484 *et seq.*; trying to unlock a box, 492; playing with fire, 493-4; expression of emotions, 494-5; dread of imitation monkey, 495

Morgan, L. A., on spider conveying insect to larder, 220

Morgan, L. H., on the beaver, 367-83

Moschkau, Dr., on intelligence shown by a spider which he habitually fed, 218-19

Moseley, Lewin, performing operation on a monkey, 482

Moseley, Prof., on intelligence of crabs, 231-2

Mossman, Rev. J. W., on wasps coming out of small aperture backwards, 192-3

Mule, alleged counting by, 332; intelligence of, 333-4

Müller, Adolph, on instinct of cuckoo, 306-7

Müller, F., on powers of communication in bees, 157; on termites, 198 and 201

Murray, S., intelligence of his dog, 450

Music, fondness of spiders for, 205-7; of parrots and pigeon, 282

Mygale spider eating humming-birds, 208

Myriophyllum spicatum, 243

Myrmeleon formicarium, 234-5

NADAULT, Madame, the association of ideas shown by her parrot, 269

Napier, Commander, on pigeon making a horse shake oats from nose-bag, 317

Napier, Lady, recollection in parrot, 269, 270; emulation in parrot, 276, 277

Nest, *see* Nidification

Newall, R. S., on wasp dividing caterpillar to facilitate carriage, 195, 196

Newbury, on absence of beaver dams in California, 370, 371

Newton, Professor A., on instincts of cuckoo, 306-9

Nichols, W. W., on intelligence of pigeons, 317

Nicols, A., on reasoning power of a retriever, 464, 465

Nicrophorus, 228

Nidification, of crustacean, 232, 233; of fish, 242-5; of birds, 291-301; petrels and puffins, 291, 292; auks, curlew, goatsucker, ostrich, gulls, sandpipers, plovers, kingfisher, Chinese swallow, house-martin, 292; tomtit, woodpecker, starling, weaver, 293; baya, tale-gallus, 294; grosbeak, 295, 296; swan, 296-8; Wallace's theories concerning, 298, 299; variability of 299-301; of harvesting mice, 365

Nightingales, removing nest, 289

Niphon, Professor, on intelligence of a mule, 333, 334

Noctua Eningii, 238

Noctura verbasci, 236

North, the Rev. W., on intelligence of mice, 361, 362

Nottebohm, Herr, on ants stocking trees with aphides, 63

OBSTETRIC-FISH, 246; toad, 254 Octopus, intelligence of, 29, 30

Œcypoda ippeus, 231

Oldham, A., on jealousy in dog, 442, 443

Orang-outang, removing dead companions, 472; sense of humour in, 478; drawing chair to stand upon to reach high places, 481

Orthotomus, 293

OST

Ostrich, conjugal affection of, 270; nidification, 292
Otter, 346
Oyster, intelligence of, 25

PALLAS, on provident habits of Lagomys, 365
Parrot, memory of, 267-9; recollection, 269, 270; talking, &c., 267-70; sympathy, 275, 276; exultation on baffling imitative powers of master, 277; vindictiveness, 277; fondness of music, 282; difficult to deceive by mirrors, 310, 311
Parry, Captain, on instincts of wild swan, 297
Partridge, removing eggs, 289
Peach, C. W., on dog recognising portrait, 453, 454
Peal, G. E., on elephants removing leeches and fanning away flies, 409, 410
Pearson, Colonel, the reasoning power of his dog, 466, 467
Peeweet, *see* Lapwing.
Pelicans, sympathy of for wounded companions, 275; frigate, 284; combined action of in fishing, 319
Penky, the Rev. Mr., on reasoning power of a dog, 466, 467
Pennant, on navigating habits of Iceland mice, 364
Pennent, on domestication of toad, 255; on fascination by rattlesnake, 263
Perca scandens, 248, 249
Perception, 9
Perch, climbing, 248, 249
Percival, Dr., on cock killing hen when she hatched out eggs of partridge, 278
Petrels, nidification of, 291, 292
Phillips, J., his portrait-painting recognised by a dog, 454
Picton, Mrs. E., on sensitiveness of a terrier, 440, 441
Pieris rapæ, 236
Pigeon, memory of, 266; conjugal affection and fidelity, 270,

PRI

271; fondness for a particular air of music, 282; intelligence in avoiding turtles, 317; in making horse shake oats from nose-bag 317
Pigs, 339-41
Pike, affection of male for female, 246
Pilot-fish, 251, 252
Pinnipeds, breeding habits of, 342, 346
Pipe-fish, 246
Piracy, instinct of, in birds, 283, 284, 301-7
Pisces, *see* Fish
Play, of ants, 87, 89; of fish, 242; of birds, 279; of porpoise, 327, 328; of dogs, 445; of monkeys, 476, 477
Pliny, on ants burying their dead, 91; sexual affection of snakes, 256; on intelligence of elephant, 386; on memory of elephant, 387
Ploceus textor, 293
Plover, *see* Lapwing; nidification of, 292
Plutarch, on intelligence of elephant, 386
Podocerus capillata, 332
Polar bear, 352, 353
Polecat, curious instinct of, 347
Polistes carnifex, taking precise bearings to remember locality, 150, 151
Polistes Gallica, tamed by Sir John Lubbock, 153; robber, 169
Pollock, F., on perfection of webs built by young spiders, 217
Pollock, W., on association of ideas in parrot, 269
Polydectes cupulifer, 233
Pope, on instinct and reason, 15
Porpoise, intelligence of, 327, 328
Portraits, recognised by birds, 311; by dogs, 453-7
Pouchet, on improvement in nidification of swallows, 300, 301
Powelsen, on navigating habits of Iceland mice, 364
Prairie dog, 366
Pride, of birds, 279; of horse, 330; of ruminants, 334; of dog, 439-42

PRI

Prinia, 293
Protozoa, movements of, 18; apparent intelligence of, 19-21
Provident instincts, of ants, 97-110; of bees, 160-162; of a bird, 285; of rodents, 353, 354, and 365, 366; of beaver, 368-70
Puffins, nidification of, 291, 292
Pugnacity, of ants, 45; of bees, 165-70; of spiders, 204-5; of fish, 242; of seal, 341-6; of rabbits, 355; of rat-hare, 365, 366; of canine animals, 426
Python, tame, affection of, &c., 256 and 260-2

QUARTERLY REVIEW, on intelligence of rats, 360, 361
Quatrefages, on termites, 198

RABBIT, 354-7
Rabigot, on fondness of spiders for music, 206
Rae, Dr. John, on intelligence of horse, 331; of wolverine, 348; of wolves and foxes, 429, 430; of dog, 465, 466
Rae, on dog ringing bell, 423
Ransom, Dr., on sticklebacks, 245
Rarey, his method of taming horses, 328, 329
Rats, 360-3
Rattlesnake, alleged fascination by, 263
Ravens, breaking shells by dropping them on stones, 283
Razor-fish, intelligence of, 25
Reason, definition of, and distinguished from instinct, 13-17; exhibitions of, by various animals, *see* under sections headed 'general intelligence'
Réaumur, on intelligence of ants, 128; sympathy of bees, 156; carpenter-bee, 179; encasing snail with propolis, 190; conveying carrion out of hive, 191; experiments on instincts of caterpillars, 237; on larvæ chasing aphides, 240

ROG

Reclain, Professor C., on spider descending to violin-player, 205, 206
Recognition of persons, by bees, 188; by snakes and tortoises, 269-61: of places, by mollusca, 27-9; by ants, 33 *et seq.*; by bees, 144 *et seq.*: of offspring, by earwig, 229: of portraits, *see* Birds and Dogs: of other members of a hive by ants and bees, *see* Ants and Bees
Reeks, H., on collective instinct of wolves, 436
Reflex action, 2-4
Reid, Dr., on mathematical principles observed by bees in constructing their cells, 171
Rengger, on maternal care and grief of a cebus, 472; monkeys displaying intelligent observation, 479; using levers, 481
Reproach, shown by gestures of monkeys 475-478
Reptiles, 255-265; emotions of, 255, 256, and 260-2; incubating eggs, sexual and parental affection of, 256; general intelligence of, 256-263; fascination by, 263, 264; charming of, 264, 265
Reyne, his observations on snake-charming, 264, 265
Rhizopoda, apparent intelligence of, 19-21
Richards, Captain, on pilot-fish, 252
Richardson, Mrs. A. S. H., on elephant concealing theft, 410; on dog finding its way home by train, 468, 469
Ridicule, dislike of, by dogs and monkeys, *see* Ludicrous
Risso, M., on habits of pipe-fish, 246
Robertson, Professor G. Croom, on behaviour of an ape with a mirror, 478, 479
Robin, intelligence of, 314
Rodents, 353
Rodwell, on intelligence of rats, 360-2
Rogue-elephants, *see* Elephant

Romanes, Miss C., on dog recognising portrait, 455, 456; on intelligence of cebus, 484–95
Romanes, G. J., on movements of rotifer, 18, 19; of medusæ, 22; of echinodermata, 23; emotions of stickleback, 246, 247; piracy of terns and gulls, 283–4; mode of challenge practised by gulls, 291; birds deceived by mirrors, 311; grouse learning to avoid telegraph wires, 313; intelligence of horse, 330; intelligence of ferrets, 347; instincts of rabbits, 354; intelligence of rabbits, 354, 355; rabbits fighting rats, 355; drawing dead companions out of holes, 356, 357; intelligence of hare, 357; hares and rabbits allowing themselves to be caught by weasels, 359; rats using their tails for feeding purposes, 363; cat opening thumb-latch, 420, 421; collective instinct of jackals, 434, 435; of dogs, 435; duration of memory in dog, 438; pride and sensitiveness in dog, 439, 440; intolerance of dog towards pain, 441; emulation and jealousy in dog, 442; deceitfulness and dislike of ridicule in dog, 444; sense of ludicrous in dog, 444, 445; dogs communicating ideas, 445, 446; dogs slipping into their collars to conceal their sheep-killing, &c., 435 and 450, 451; dog recognising portrait, 456, 457; reasoning of dog, 457, 458; caution of a dog in killing snakes, 460; sympathy of an Arabian baboon, 474; sense of ludicrous and dislike of ridicule in monkey, 476; intelligence of *Cebus fatuellus*, 484–98
Rooks, sympathy of, for wounded companions, 273, 274; concerted action of, in obtaining food from dogs, 319, 320; from pheasants, 321; nesting habits and punishment of culprits, 322–5.
Rotifera, movements of, 18
Ruminants, 334
Russell, Lord Arthur, witnessing tameness of snakes, 261

SAGARTIA parasitica. 234
Salmon, migration of, 249, 250
Salticus scenicus, 213
Sandpipers, nidification of, 292
Sarsia, seeking light, 23
Saunders, S. S., on trap-door spiders, 215
Savage, on play of chimpanzees, 476, 477
Schiller, on pride of bell-wether steers, 334
Schipp, Lieut., on combined action of baboons, 483
Schlosser, on jaculator-fish, 248
Schlüter, Herr A., on a hornet carrying heavy prey up an elevation in order to fly away with it, 196
Schneider, on intelligence of *octopus*, 29, 30; on fish guarding eggs, 242; jealousy of fish, 247
Sclater, Dr., on instincts of cuckoo, 325; lending a cebus for observation, 483
Scoresby, on maternal affection of whale, 327; on intelligence of polar bear, 351
Scorpion, alleged suicide of, when surrounded by fire or heat, 222–25
Sea-anemones, 233, 234
Seals, intelligence of, and breeding-habits of pinnipeds, 341–6
Seebohm, on instincts of cuckoo, 325
Semnopithecus entellus, destroying poison fangs of snakes, 483
Sensation, 8
Severn, H. A., on nidification of baya-bird, 294
Severn, W., on snakes, 260, 261
Sheep, pride of leaders, 334
Shelley, lines on curiosity of fish, 247
Shipp, Capt., on vindictiveness of elephant, 387, 388; on intelligence of elephant, 397, 398
Siebold, on robber-wasps, 169
Sieur, Roman, his trained birds, 312

INDEX.   517

Signs, made by ants, 49 *et seq.*; by bees, 157 *et seq.*; by termites, 200; by birds, 315, 316; by elephants, 391 and 401; by cat, 416; by dog, 445-7; by monkey, 472, 475, 476
Simiadæ, *see* Monkeys
Simonius, on fondness of spiders for music, 206
Sinclair, W., on intelligence of horse, 33
Skate, supposed intelligence of, 251
Skinner, Major, on intelligent vigilance of elephants, 400, 401; on training of cobra, 265
Slingsby, his experiment in training a house-fly, 230, 231
Smeathman, on termites, 198-203
Smeaton, Th. D., on dog making peace-offerings, 452
Smiles, Dr. S., on observation of Stephenson, 247; on observations of Edward, 255, 275, 283, 321
Smith, A. P., on intelligence of a cat, 414
Smith, Colonel, on pilot-fish, 252
Smith, Colonel Hamilton, on intelligence of cattle-dogs, 449
Smith, Sir Andrew, on revenge of a baboon, 478
Snails, intelligence of, 26-28
Snakes, incubating eggs, sexual and parental affection of, 256; tamed, 256, 260-3, 265; finding way home, 262; intelligence of, 262-3; fascination by, 263-4; charming of, 264-5
Social feelings, *see* Sympathy and Affection; habits common to Hymenoptera and termites, 202
Sow, pointing game, 339, 340
Sparman, on termites, 198
Spencer, Herbert, on migration of salmon, 249; on play as allied to artistic feeling, 279
Sphex, *see* under Wasp
Spiders, emotions of, 204-7; courtship, 204, 205; strength of maternal instinct, 205; fondness of music, 205-7; web-building, 207-12; geometric, 209; water, 212; wolf or vagrant, 213; trap-door, 213-18; admit of being tamed and distinguish persons, 218-19; protecting eggs from cold, 219; protecting themselves from ecitons, 219; conveying prey to larder, 220; suspending weights to steady web, 220-2; wide geographical range of trap-door spiders, 216
Stag, intelligence of, 336
Starlings, nidification of, 293; learning to avoid telegraph-wires, 312-13
Stephenson, on curiosity of fish, 247
Stevens, J. G., on intelligence of a cat, 417-18
Sticklebacks, 243-5, 246-7
Stickney, on bees remembering in successive years the position of a disused hive, 154
Stodmann, on wasps recognising persons, 188
Stone, on reasoning power of a dog, 460
Stork, vindictiveness of, 277-8
Strachan, on elephants dying under emotional disturbance, 395-6
Strange, F., on habits of bower-bird, 281
Strauss, on co-operation of beetles
Street, J., on blackbirds removing their young, 289
Strickland, on intelligence of a mare, 332
Swainson, on vindictiveness of elephant, 389
Swallows, memory of, 266; improvement in their nidification and adopting new modes of, 300; migration, 301; making tunnels, 318; killing imprisoned hostile sparrows, 318-19
Swan, conjugal fidelity of, 271; mode of escaping with young, 290; nidification, 496-8
Swine, 339-41
Sword-fish, 252-3
Sykes, Colonel, on harvesting ants, 97; on tree ants, 110-11; intelligence of ants in getting at food in difficult situations, 134, 135; on nidification of tailor-bird, 293

SYL

Sylvia, 293
Sympathy, of ants, 46-9; of bees, 155-6; of fish, 242; of birds, 270-6; of horse, 331-2; of ruminants, 334; of elephants, 387-92, and 397, 398; of cat, 416; of monkeys, 471-5

TAIT, LAWSON, on cat signing to have bell pulled, 423
Talegallus, nidification of, 294
Taylor, the Rev. Mr., cunning of his dog, 451
Tegetmeier, on amount of sugar required by bees to make honey, 176
Telescope-fish, 246
Tennent, Sir E., on apparent intelligence of land-leeches, 24; intelligence of tree-ants, 134; mygale eating humming birds, 208; climbing-perch, 249; sexual affection of cobra, 246; snake-charming, 264, 265; taming of cobra, 265; nidification of baya-bird, 294; combined action of crows, 319, 320; of buffaloes, 335; use of tame buffalo, 335; on emotions and intelligence of elephant, 389, 390, 393-6, 400-8; collective instinct of jackals, 432, 433
Tepper, Mr. Otto, on intelligence of a cat, 424
Termites, 198-203; architecture, 198, 199, and 201, 202; workers and soldiers, 200, 201; swarming, breeding, &c., 202; remarkable similarity of instincts to those of Hymenoptera, 202; instincts detrimental to individual but beneficial to species, 202, 203
Terns, sympathy of, for wounded companions, 274, 275; robber, 284; mobbing robber-terns, 291
Theda isocrates, 238
Theuerkauf, Herr G., on intelligence of ants in making a bridge of aphides over tar, 136
Thompson, E. P., on bees remembering exact position of absent hive, 149; on garden-spider's

VIL

mode of web-building, 210, 211; ant-lion, 234, 235; emotions of guana, 255, 256; fascination by snakes, 264; nidification of sociable grosbeak, 295, 296; birds dreaming, 312; maternal affection of whale, 327; bisons defending themselves from wolves, 334, 335; pigs defending themselves from wolves, 339; cleanliness of pig, 340, 341; intelligence of weasel, 346; of mouse, 361; harvesting-mice, 365, 366
Thomson, Dr. Allen, on scorpions committing suicide, 223-5
Thornton, Colonel, his sow trained to point game, 340
Thresher-fish, 252, 253
Thrushes, breaking shells against stones, 283
Timea, 237
Toads, 254, 255
Tomtit, nidification of, 293
Topham, Dr. J., on spiders weighting their webs, 222
Topham, Mr. J., on bees remembering exact position of absent hive, 149
Tortoises, knowing persons, 259
Townsend, the Rev. W., on elephant concealing theft, 410; on dog finding its way about by train, 468-9
Truro, Lord, on intelligence of a dog, 450
Turner, George, on bees remembering exact position of absent hive, 149
Turnstones, intelligence of, 321
Turtles, 257, 258, and 262

VAILLANT, Le, on fascination by tree-snake, 263, 264
Valiant, L., on nidification of sociable grosbeak, 296
Venn, on association of ideas in parrot, 267, 268
Vigot, Dr., on snake finding way home, 262
Villiers, De, on instincts of larvæ of bombyx moth, 240

INDEX. 519

VIN

Vindictiveness, of birds, 277, 278, and 318-25; of horse, 330, 331; of elephant, 387-9; of monkeys, 478, and 484-96

Virchow, on difficulty of distinguishing between instinct and reason, 12

Vogt, Karl, on duration of memory in ants, 41; bridge-making, 136

Vultures, finding carrion by sight and not by smell, 286, 287; intelligence, 314

WAFER, on monkeys hammering oyster-shells with stones, 481

Wakefield, P., on intelligence of goats, 337, 338

Wallace, A. R., on philosophy of birds'-nests, 298-300

Warden, on frogs going straight to nearest water, 254

Wasp-mason, 180; butcher, 180, 181; sphex, 181; hunting, 193, 194; common, tamed by Sir John Lubbock, 153

Wasps, sense of direction in, 147; teaching themselves, 154; killing larvæ, 167, 168; making cells, 180; instincts of neuters, 181; recognising persons, 188; coming out of small aperture backwards, 192, 193; struggles with ants for secretion of frog-hoppers, 194, 195; dismembering heavy prey for convenience of carriage, and mounting eminences for same purpose, 195, 197

Wasser, on nidification of puffins, 291

Waterhouse, on hexagonal form of bee's cell, 173

Water-rail, its mode of escape, 289

Waterton, on nidification of swan, 295, 296

Watson, on spiders weighting their webs, 221; cock killing hen on her hatching out eggs of other birds, 278; intelligence of rats, 360-62; vindictiveness of elephant, 389; elephant enduring surgical operation, 399; intelligence of

WIN

sheep-dogs, 448; of cattle-dogs, 449

Weasel, 346, 347

Weaver, nidification of, 293

Web, see Spider

Web-building, see Spiders

Webb, Dr., performing operation on elephant, 399

Weber, Professor E. H., on spiders weighting their webs, 221

Wedgewood, the Rev. R. H., on memory of horse, 330

Westlecombe, on reasoning power of a dog, 462, 463

Westropp, on intelligence of bear, 352

Westwood, on instinct of caterpillars, 288

Weygandt, on robber-bees, 170

Whale, maternal affection of, 327; attacks on, by sword- and thresher-fish, 252, 253

Whately, Archbishop, on cat ringing bell, 423

White ants, see Termites

White, the Rev. Gilbert, on nests of harvesting-mice, 365; on nidification of house-martin, 292, 293

White, W., on intelligence of snails, 26

White, the Rev. W. W. F., on sympathy of ants, 49; keeping pets, 84; burying dead, 92, 93

White-headed eagle, see Eagle

Wildman, his alleged training of bees, 189

Wilks, Dr. S., observations on talking of parrot, 267, 268; on dog recognising a portrait, 455

Williams, on intelligence of sheep-dogs, 448

Williams, B., on cunning of sheep-killing dogs, 450, 451

Wilson, on memory of crow, 266

Wilson, Dr. Andrew, on reasoning power of a dog, 460

Wilson, Charles, on intelligence of swallows, 318

Wilson, Dr. D., on elephant enduring surgical operation, 399

Winkell, Dietrich aus dem, on intelligence of fox, 428

WOL

Wolf, 426-36; avoiding gun-traps, 431; drawing up fish-lines to take fish, 431; collective instinct in hunting, 433, 436
Wolverine, 347-50
Wood, Rev. G. J., on spiders weighting their webs, 221
Woodcock, conveying young on back, 289
Woodpecker, ant-eating, its instinct of storing food, 285; nidification, 293
Words, understanding of, by bees, 189; by talking birds, 267-9
Worms, apparent intelligence of, 24

YUL

Wright, his portrait-painting recognised by a dog, 454-5

YARRELL, on fish, 246; on intelligence of hare, 358-9
Youatt, on pigs learning to point game, 340
Young, the Rev. Charles, on emotions and intelligence of elephant, 390-92
Young, Miss E., on dog finding his way about by train, 468
Yule, Captain, on elephants dying under emotional disturbance, 395

THE END.

Animal Intelligence.

By GEORGE J. ROMANES, F. R. S.,
Zoölogical Secretary of the Linnæan Society, etc.

12MO.   CLOTH, $1.75.

"My object in the work as a whole is twofold: First, I have thought it desirable that there should be something resembling a text-book of the facts of Comparative Psychology, to which men of science, and also metaphysicians, may turn whenever they have occasion to acquaint themselves with the particular level of intelligence to which this or that species of animal attains. My second and much more important object is that of considering the facts of animal intelligence in their relation to the theory of descent."—*From the Preface*.

"Unless we are greatly mistaken, Mr. Romanes's work will take its place as one of the most attractive volumes of the INTERNATIONAL SCIENTIFIC SERIES. Some persons may, indeed, be disposed to say that it is too attractive, that it feeds the popular taste for the curious and marvelous without supplying any commensurate discipline in exact scientific reflection; but the author has, we think, fully justified himself in his modest preface. The result is the appearance of a collection of facts which will be a real boon to the student of Comparative Psychology, for this is the first attempt to present systematically well-assured observations on the mental life of animals."—*Saturday Review*.

"The author believes himself, not without ample cause, to have completely bridged the supposed gap between instinct and reason by the authentic proofs here marshaled of remarkable intelligence in some of the higher animals. It is the seemingly conclusive evidence of reasoning powers furnished by the adaptation of means to ends in cases which can not be explained on the theory of inherited aptitude or habit."—*New York Sun*.

"The high standing of the author as an original investigator is a sufficient guarantee that his task has been conscientiously carried out. His subject is one of absorbing interest. He has collected and classified an enormous amount of information concerning the mental attributes of the animal world. The result is astonishing. We find marvelous intelligence exhibited not only by animals which are known to be clever, but by others seemingly without a glimmer of light, like the snail, for instance. Some animals display imagination, others affection, and so on. The psychological portion of the discussion is deeply interesting."—*New York Herald*.

"The chapter on monkeys closes this excellent work, and perhaps the most instructive portion of it is that devoted to the life-history of a monkey."—*New York Times*.

"Mr. Romanes brings to his work a wide information and the best of scientific methods. He has carefully culled and selected an immense mass of data, choosing with admirable skill those facts which are really significant, and rejecting those which lacked sustaining evidence or relevancy. The contents of the volume are arranged with reference to the principles which they seem to him to establish. The volume is rich and suggestive, and a model in its way."—*Boston Courier*.

"It presents the facts of animal intelligence in relation to the theory of descent, supplementing Darwin and Spencer in tracing the principles which are concerned in the genesis of mind."—*Boston Commonwealth*.

"One of the most interesting volumes of the series."—*New York Christian at Work*.

"Few subjects have a greater fascination for the general reader than that with which this book is occupied."—*Good Literature, New York*.

For sale by all booksellers; or sent by mail, post-paid, on receipt of price.

New York: D. APPLETON & CO., 1, 3, and 5 Bond Street.

ANTS, BEES, AND WASPS.

A Record of Observations on the Habits of the Social Hymenoptera.

By Sir JOHN LUBBOCK, Bart., M. P., F. R. S., etc.,

Author of "Origin of Civilization, and the Primitive Condition of Man," etc., etc.

With Colored Plates. 12mo. Cloth, $2.00.

"This volume contains the record of various experiments made with ants, bees, and wasps during the last ten years, with a view to test their mental condition and powers of sense. The principal point in which Sir John's mode of experiment differs from those of Huber, Forel, McCook, and others, is that he has carefully watched and marked particular insects, and has had their nests under observation for long periods —one of his ants' nests having been under constant inspection ever since 1874. His observations are made principally upon ants because they show more power and flexibility of mind; and the value of his studies is that they belong to the department of original research."

"We have no hesitation in saying that the author has presented us with the most valuable series of observations on a special subject that has ever been produced, charmingly written, full of logical deductions, and, when we consider his multitudinous engagements, a remarkable illustration of economy of time. As a contribution to insect psychology, it will be long before this book finds a parallel."—*London Athenæum.*

"These studies, when handled by such a master as Sir John Lubbock, rise far above the ordinary dry treatment of such topics. The work is an effort made to discover what are the general, not the special, laws which govern communities of insects composed of inhabitants as numerous as the human beings living in London and Peking, and who labor together in the utmost harmony for the common good. That there are remarkable analogies between societies of ants and human beings no one can doubt. If, according to Mr. Grote, 'positive morality under some form or other has existed in every society of which the world has ever had experience,' the present volume is an effort to show whether this passage be correct or not."—*New York Times.*

"In this work the reader will find the record of a series of experiments and observations more thorough and ingenious than those instituted by any of the accomplished author's predecessors. . . . Sir John has been a close observer of the habits of ants for many years, generally having from thirty to forty communities under his notice, and not only watching each of these in its carefully isolated glass house, but, by the use of paint-marks, following the fortunes of individuals. . . . One notable result of this system has been the correcting of previous theories as to the age to which ants attain: instead of living merely a year, as the popular belief has been, some of Sir John's queens and workers are thriving after being under observation since 1874 and 1875."—*New York World.*

"Sir John Lubbock's book on 'Ants, Bees, and Wasps' is mainly devoted to the crawlers, and not the fliers, though he has some observations upon honey-bees and more interesting ones upon the unpopular wasp, which he fondly deems to be capable of gratitude. Darwin made a strong case for the monkeys, but Lubbock may yet make us out to be, as Irishmen say, 'The sons of our ants.' For he begins his entertaining book thus: 'The anthropoid apes no doubt approach nearer to man in bodily structure than do any other animals, but, when we consider the habits of ants, their large communities and elaborate habitations, their roadways, their possession of domestic animals, and, even in some cases, of slaves, it must be admitted that they have a fair claim to rank next to man in the scale of intelligence.'"—*Springfield Republican.*